Bibliographic information published by the Deutsche Nationalbibliothek

The Deutsche Nationalbibliothek lists this publication in the Deutsche Nationalbibliografie; detailed bibliographic data are available on the Internet at http://dnb.d-nb.de .

ISBN 978-3-8325-4094-4

Logos Verlag Berlin GmbH
Comeniushof, Gubener Str. 47,
10243 Berlin
Tel.: +49 (0)30 42 85 10 90
Fax: +49 (0)30 42 85 10 92
INTERNET: http://www.logos-verlag.de

Self-propelled colloidal particles: from isotropic to anisotropic microswimmers

Inaugural-Dissertation

zur Erlangung des Doktorgrades
der Mathematisch-Naturwissenschaftlichen Fakultät
der Heinrich-Heine-Universität Düsseldorf

vorgelegt von

Borge Sebastian ten Hagen

aus Steinfurt

Düsseldorf, Februar 2015

aus dem Institut für Theoretische Physik II: Weiche Materie
der Heinrich-Heine-Universität Düsseldorf

Gedruckt mit der Genehmigung der
Mathematisch-Naturwissenschaftlichen Fakultät der
Heinrich-Heine-Universität Düsseldorf

Referent: Prof. Dr. Hartmut Löwen
1. Korreferent: Prof. Dr. Stefan U. Egelhaaf
2. Korreferent: Prof. Dr. Jan Kierfeld

Tag der mündlichen Prüfung: 26. März 2015

Preface

This thesis is based on the work I have performed at the *Institut für Theoretische Physik II: Weiche Materie* of the *Heinrich-Heine-Universität Düsseldorf* since 2012. The results have been published (or are accepted for publication) in nine different articles in peer-reviewed journals, namely *Nature Communications*, *Physical Review Letters*, *Physical Review E*, *The Journal of Chemical Physics*, *Journal of Physics: Condensed Matter*, and *Journal of Statistical Mechanics: Theory and Experiment*. Moreover, I have already been able to present the results at several international conferences. Parts of the work have been done in collaboration with experimental groups from Stuttgart, Beijing, New York, and Chicago. At the end of the individual chapters, statements about my own contribution to the corresponding publications are given. In the following, the detailed bibliographical data of the articles this thesis is based on are provided.

1. **Gravitaxis of asymmetric self-propelled colloidal particles**
   Borge ten Hagen, Felix Kümmel, Raphael Wittkowski, Daisuke Takagi, Hartmut Löwen, and Clemens Bechinger,
   Nature Communications **5**, 4829 (2014).

2. **Can the self-propulsion of anisotropic microswimmers be described by using forces and torques?**
   Borge ten Hagen, Raphael Wittkowski, Daisuke Takagi, Felix Kümmel, Clemens Bechinger, and Hartmut Löwen,
   Journal of Physics: Condensed Matter (accepted for publication), preprint arXiv:1410.6707 (2014).

3. **Circular motion of asymmetric self-propelling particles**
   Felix Kümmel, Borge ten Hagen, Raphael Wittkowski, Ivo Buttinoni, Ralf Eichhorn, Giovanni Volpe, Hartmut Löwen, and Clemens Bechinger,
   Physical Review Letters **110**, 198302 (2013).

4. **Non-Gaussian statistics for the motion of self-propelled Janus particles: experiment versus theory**
   Xu Zheng, Borge ten Hagen, Andreas Kaiser, Meiling Wu, Haihang Cui, Zhanhua Silber-Li, and Hartmut Löwen,
   Physical Review E **88**, 032304 (2013).

5. **Swimming path statistics of an active Brownian particle with time-dependent self-propulsion**
   Sonja Babel, Borge ten Hagen, and Hartmut Löwen,
   Journal of Statistical Mechanics: Theory and Experiment (2014) P02011.

6. **Reply to Comment on "Circular motion of asymmetric self-propelling particles"**
   Felix Kümmel, Borge ten Hagen, Raphael Wittkowski, Daisuke Takagi, Ivo Buttinoni, Ralf Eichhorn, Giovanni Volpe, Hartmut Löwen, and Clemens Bechinger,
   Physical Review Letters **113**, 029802 (2014).

7. **Brownian motion and the hydrodynamic friction tensor for colloidal particles of complex shape**
   Daniela J. Kraft, Raphael Wittkowski, Borge ten Hagen, Kazem V. Edmond, David J. Pine, and Hartmut Löwen,
   Physical Review E **88**, 050301(R) (2013).

8. **Dynamics of a deformable active particle under shear flow**
   Mitsusuke Tarama, Andreas M. Menzel, Borge ten Hagen, Raphael Wittkowski, Takao Ohta, and Hartmut Löwen,
   Journal of Chemical Physics **139**, 104906 (2013).

9. **Transport powered by bacterial turbulence**
   Andreas Kaiser, Anton Peshkov, Andrey Sokolov, Borge ten Hagen, Hartmut Löwen, and Igor S. Aranson,
   Physical Review Letters **112**, 158101 (2014).

Moreover, I have also been involved in the following publications which are not part of this thesis.

10. **Two-dimensional colloidal mixtures in magnetic and gravitational fields**
    Hartmut Löwen, Tobias Horn, Tim Neuhaus, and Borge ten Hagen,
    European Physical Journal Special Topics **222**, 2961–2972 (2013).

11. **How does a flexible chain of active particles swell?**
    Andreas Kaiser, Sonja Babel, Borge ten Hagen, Christian von Ferber, and Hartmut Löwen,
    submitted.

12. **Depolarized light scattering from anisotropic particles: the influence of the particle shape on the field autocorrelation function**
    Christopher Passow, Borge ten Hagen, Hartmut Löwen, and Joachim Wagner,
    in preparation.

Based on the results of this thesis, several additional projects are currently under way. These predominantly focus on self-propelled particles with nonconstant swimming velocities. On the one hand, the theory for time-dependent self-propulsion protocols presented in this thesis is applied to study an experimental system of light-activated Janus particles. On the other hand, the motion of microswimmers with spatially varying swimming velocity is investigated. Furthermore, the gravitactic behavior of bottom-heavy particles in three dimensions is studied.

Abstract

The ability to swim is found in very different life forms on our planet. Locomotion in a liquid medium is also essential for many technological applications over all orders of magnitude. The present thesis deals with self-propelled colloidal particles, also called microswimmers, for which a directed swimming motion is particularly difficult due to the lack of inertia and as a consequence of Brownian noise. The main focus is directed at the influence of the body shape of the swimmers. Different systems ranging from spherical particles, via deformable and rodlike microswimmers, to completely asymmetric objects are investigated. The theoretical models are mostly developed in the context of specific experimental situations, and the computer simulations are designed accordingly. Thus, the theoretical predictions can directly be compared to related measurements.

The main part of this thesis consists of eight chapters. First, spherical self-propelled Janus particles are considered. These represent one of the most simple but simultaneously most prominent realizations of artificial microswimmmers. They allow for a quantitative investigation of the dynamical behavior, which is governed by the interplay between active particle displacements and diffusive motion. In chapter 2, a theoretical model based on the Langevin equations for self-propelled Brownian particles is used together with experimental data analysis in order to identify different regimes of motion. For this purpose, the mean square displacement is studied as well as higher moments of the displacement probability distribution function. The presented theoretical model provides the basics for the description of more complicated microswimmers in the subsequent chapters of this thesis. In chapter 3, the theory is first of all generalized towards active particles with time-dependent self-propulsion. This enables a more accurate modeling of many biological systems, where the propulsion is governed by a certain periodic process or is influenced by external conditions, such as the food supply in the vicinity of a microorganism.

Before studying the dynamical behavior of completely asymmetric microswimmers, in chapters 4–6 several systems are investigated which already show clear deviations from the case of spherical particles and provide some important preparations for the later chapters of this thesis. In chapter 4, a theoretical model for deformable particles is presented. It is relevant to describe active droplets, for example. Building upon previous related work, in particular deformable particles under shear flow are considered. The corresponding coupled nonlinear equations of motion yield a manifold of dynamical states, which are systematically analyzed and characterized. Chapter 5 focuses in detail on the grand resistance matrix and, correspondingly, the diffusion tensor of rigid particles. By using passive colloids, it is demonstrated how the various diffusion and coupling coefficients can be obtained from experimental measurements. The method is applicable to particles with arbitrary shapes, which is very important in the context of asymmetric microswimmers. In addition to the experimental determination method, a way of predicting the diffusion and coupling coefficients theoretically is discussed. This procedure is also used in the following chapter 6, where a system consisting of a mesoscopic wedgelike carrier in a bacterial bath is investigated. It is shown that the turbulentlike motion of the bacteria can be exploited for the directed transport of passive objects through the suspension.

Chapter 7 is eventually devoted to the core issue of this thesis, i.e., the theoretical description of real asymmetric self-propelled particles. As opposed to spherical or rodlike particles, here the propulsion does in general not only induce active translational displacements but also a corresponding rotational motion. Thus, due to the coupling, circular trajectories evolve. The theoretical modeling is facilitated by using effective forces and torques which enter the Langevin equations. The theory is explicitly used in order to obtain detailed predictions concerning the motional behavior of L-shaped microswimmers. As confirmed by the analysis of corresponding experimental data, a consistent and comprehensive theoretical framework for the dynamics of asymmetric self-propelled particles is provided. In chapter 8, the validity and applicability of the presented model is discussed in detail.

Finally, in chapter 9, the theory is used in order to contribute to the understanding of the gravitactic behavior which is observed in many microorganisms. Although several explanatory approaches exist, the detailed origin of gravitaxis is still an unsolved question in microbiology. Here, by studying artificial microswimmers, a possible explanation based on an asymmetric body shape is given. The theoretical and experimental results demonstrate that a physiological gravity sensor is not required for biological microorganisms to systematically counteract gravitation. Furthermore, a detailed state diagram of the various types of motion occurring for L-shaped microswimmers is provided and theoretically explained.

By explicitly considering the particle shape, in many situations the results of this thesis help to improve the theoretical description of the dynamics of biological and artificial microswimmers. Thus, they provide an important contribution to the understanding of biophysical phenomena such as gravitaxis. Regarding artificial self-propelled particles, the motional behavior can be systematically tuned by appropriately designing the particle shape. In this context, a particularly promising field is the development of novel micromachines. Such devices have great potential for medical purposes as well as for many other technological applications.

Zusammenfassung

Schwimmen ist eine Fähigkeit, die bei den verschiedensten Formen des Lebens auf unserem Planeten weit verbreitet ist. Auch für viele technologische Anwendungen in ganz unterschiedlichen Größenordnungen ist aktive Fortbewegung in einem flüssigen Medium von essentieller Bedeutung. Die vorliegende Dissertation beschäftigt sich mit selbstangetriebenen kolloidalen Teilchen beziehungsweise Mikroschwimmern, für die eine gerichtete Fortbewegung aufgrund der fehlenden Trägheit und wegen der Auswirkungen der Brownschen Bewegung besonders schwierig ist. Der Hauptfokus ist dabei auf den Einfluss der Körperform der Schwimmer gerichtet. Es werden verschiedene Systeme von kugelförmigen Teilchen über verformbare und stäbchenförmige Mikroschwimmer bis hin zu komplett asymmetrischen Objekten untersucht. Die Entwicklung der theoretischen Modelle und Computersimulationen erfolgt größtenteils in Anlehnung an konkrete experimentelle Systeme, sodass die theoretischen Vorhersagen mit entsprechenden Messdaten verglichen werden können.

Der Hauptteil der vorliegenden Dissertation umfasst insgesamt acht Kapitel. Zunächst werden kugelförmige selbstangetriebene Janus-Teilchen betrachtet. Diese stellen eine der einfachsten aber zugleich bedeutendsten Realisierungen von künstlichen Mikroschwimmern dar und ermöglichen eine quantitative Untersuchung des dynamischen Verhaltens, das durch das Wechselspiel zwischen aktiven Bewegungskomponenten und Brownschen Einflüssen geprägt ist. In Kapitel 2 werden anhand eines theoretischen Modells, das auf den Langevin-Gleichungen für selbstangetriebene Brownsche Teilchen basiert, und durch Auswertung von experimentellen Daten verschiedene Bewegungstypen identifiziert. Dazu werden neben dem mittleren Verschiebungsquadrat auch höhere Momente der Wahrscheinlichkeitsverteilung für die Teilchenposition analysiert. Das vorgestellte theoretische Modell bildet den Grundstein für die Beschreibung von komplizierteren Mikroschwimmern in den folgenden Kapiteln dieser Dissertation. In Kapitel 3 wird die Theorie zunächst auf aktive Teilchen mit zeitabhängigem Antrieb verallgemeinert. Dies ermöglicht eine präzisere Modellierung vieler biologischer Systeme, bei denen der Antrieb einem bestimmten periodischen Ablauf folgt oder durch externe Faktoren wie etwa das Nahrungsangebot in der Umgebung eines Mikroorganismus beeinflusst wird.

Auf dem Weg zu einer Beschreibung des dynamischen Verhaltens von komplett asymmetrischen Mikroschwimmern werden in den Kapiteln 4–6 verschiedene Systeme betrachtet, die bereits klare Abweichungen von der zuvor behandelten Kugelsymmetrie der Teilchen aufweisen und zum Teil wichtige Vorbereitungen für die späteren Kapitel dieser Dissertation liefern.

In Kapitel 4 wird ein theoretisches Modell für verformbare Teilchen präsentiert, das zum Beispiel für die Beschreibung von aktiven Tröpfchen relevant ist. Aufbauend auf früheren diesbezüglichen Arbeiten werden speziell verformbare Teilchen in einem Scherfluss untersucht. Die entsprechenden gekoppelten nichtlinearen Bewegungsgleichungen führen zu einer Vielzahl dynamischer Zustände, die systematisch analysiert und charakterisiert werden. Kapitel 5 befasst sich detailliert mit dem Reibungs- beziehungsweise Diffusionstensor von formstabilen Teilchen. Am Beispiel von passiven Kolloiden wird gezeigt, wie sich die einzelnen Diffusions- und Kopplungskoeffizienten auch für komplizierte Teilchenformen aus experimentellen Daten ermitteln lassen. Dies ist essentiell für die spätere Beschreibung von asymmetrischen aktiven Teilchen. Neben der experimentellen Bestimmungsmethode wird in Kapitel 5 zusätzlich eine Möglichkeit zur theoretischen Vorhersage der Koeffizienten diskutiert. Davon wird auch im anschließenden Kapitel 6 Gebrauch gemacht, in dem ein System bestehend aus einem mesoskopischen V-förmigen Objekt in einem Bakterienbad untersucht wird. Es wird gezeigt, dass die ungerichtete Bewegung der stäbchenförmigen Bakterien ausgenutzt werden kann, um passive Objekte gezielt durch die Suspension zu transportieren.

Kapitel 7 widmet sich schließlich dem zentralen Anliegen dieser Arbeit, nämlich der theoretischen Beschreibung von realen asymmetrischen selbstangetriebenen Teilchen. Im Unterschied zu kugel- oder stäbchenförmigen Teilchen resultiert hier aus dem Antrieb im Allgemeinen nicht nur eine aktive Translation, sondern auch eine entsprechende Rotationsbewegung, sodass durch die Kopplung kreisförmige Trajektorien entstehen. Dies lässt sich anschaulich mithilfe von effektiven Kräften und Drehmomenten beschreiben, die in den Langevin-Gleichungen berücksichtigt werden. Konkret wird das theoretische Modell genutzt, um detaillierte Vorhersagen über das Bewegungsverhalten von L-förmigen Mikroschwimmern zu machen. Bestätigt durch die Auswertung entsprechender experimenteller Daten wird somit eine umfassende, konsistente Theorie für die Dynamik asymmetrischer selbstangetriebener Teilchen geliefert. In Kapitel 8 wird die Gültigkeit und Anwendbarkeit des vorgestellten Modells detailliert diskutiert.

In Kapitel 9, dem letzten Hauptkapitel der vorliegenden Arbeit, wird die entwickelte Theorie angewendet, um zum Verständnis des gravitaktischen Verhaltens, das bei vielen Mikroorganismen beobachtet wird, beizutragen. Obwohl es verschiedene Erklärungsansätze gibt, ist die genaue Ursache der Gravitaxis bis heute eine ungeklärte Frage aus dem Bereich der Mikrobiologie. Anhand von künstlichen Mikroschwimmern wird hier eine mögliche Erklärung basierend auf der asymmetrischen Körperform der Teilchen geliefert. Die präsentierten theoretischen und experimentellen Ergebnisse zeigen, dass ein physiologischer Gravitationssensor nicht benötigt wird, damit Mikroorganismen sich gezielt entgegengesetzt zur Gravitationskraft bewegen können. Für die untersuchten L-förmigen Mikroschwimmer wird außerdem ein detailliertes Zustandsdiagramm der verschiedenen auftretenden Bewegungsformen erstellt und theoretisch erklärt.

Die Resultate dieser Dissertation erlauben durch die explizite Berücksichtigung der Teilchenform in vielen Fällen eine verbesserte Beschreibung der Dynamik von biologischen und künstlichen Mikroschwimmern. Somit leisten sie einen wichtigen Beitrag zur Erklärung biophysikalischer Phänomene wie zum Beispiel der Gravitaxis. Bei künstlichen Mikroschwimmern lässt sich durch ein geeignetes Design der Teilchenform das Bewegungsverhalten gezielt steuern. Dadurch ergeben sich vielfältige Anwendungsmöglichkeiten bei der Entwicklung neuartiger Mikromaschinen, die sowohl in der Medizin als auch in vielen anderen technologischen Bereichen großes Potential aufweisen.

Danksagung

An dieser Stelle danke ich allen, die mich während meiner Promotion fachlich oder außerfachlich unterstützt haben.

Mein ganz besonderer Dank gilt Herrn Prof. Dr. Hartmut Löwen, der mir die Möglichkeit gegeben hat, an seinem Lehrstuhl über ein spannendes und hochaktuelles Thema zu promovieren. Vor allem danke ich ihm für die hervorragende Betreuung. Einerseits konnte ich mich jederzeit bei auftretenden Problemen auf hilfreiche Ratschläge verlassen und hatte andererseits immer auch Gelegenheit zur Realisierung eigener Ideen.

Wesentlich geprägt war meine Promotionsphase durch die Kollaboration mit mehreren experimentellen und teils auch theoretischen Gruppen. Zuallererst ist hier die äußerst ergiebige Zusammenarbeit mit Felix Kümmel und Clemens Bechinger von der Universität Stuttgart zu nennen. Dadurch wurde es möglich, die theoretischen Überlegungen in enger Anlehnung an experimentelle Beobachtungen und Messergebnisse durchzuführen. Bereits von Anfang an hatte ich damit Anwendungsmöglichkeiten der entwickelten Theorie vor Augen.

Ungeachtet der herausragenden Bedeutung der Kollaboration mit den Experimentatoren aus Stuttgart möchte ich mich auch bei allen anderen Kooperationspartnern bedanken. Auf experimenteller Seite sind dies insbesondere Xu Zheng und Zhanhua Silber-Li vom State Key Laboratory of Nonlinear Mechanics in Beijing. Über die Arbeit an der gemeinsamen Veröffentlichung hinaus danke ich ihnen für die herzliche Gastfreundschaft während meines Besuches im Oktober 2013.

Auf theoretischer Seite gilt mein Dank an erster Stelle Raphael Wittkowski und Andreas Kaiser für die sehr effektive Zusammenarbeit und zahlreiche hilfreiche Diskussionen. Mitsusuke Tarama danke ich außerdem für die Gastfreundschaft und Hilfsbereitschaft während meiner Teilnahme am Workshop "East Asia Joint Seminars on Statistical Physics" 2013 in Kyoto. Des Weiteren bedanke ich mich bei Sonja Babel, deren Bachelorarbeit ich mitbetreuen durfte, sowie bei Andreas Menzel von der Heinrich-Heine-Universität Düsseldorf und bei Daisuke Takagi von der University of Hawaii at Manoa, Honolulu für die Kollaborationen. Mein Dank gilt natürlich auch allen anderen, die an den während meiner Promotionszeit entstandenen Veröffentlichungen mitgewirkt haben, sei es als Koautor oder durch hilfreiche Ratschläge und Diskussionen.

Ebenfalls danke ich meinem Zweitbetreuer Prof. Dr. Stefan U. Egelhaaf, meinem Mentor Dr. Michael Schmiedeberg sowie allen Mitarbeitern des Instituts für Theoretische Physik II der Heinrich-Heine-Universität Düsseldorf, die meine Zeit hier so angenehm und unvergesslich gemacht haben. Für die Unterstützung in allen bürokratischen Angelegenheiten bedanke ich mich bei Karin Wildhagen und Brigitte Schumann-Kemp. Unserem Systemadministrator Joachim Wenk danke ich für die Hilfe bei technischen Fragen jeglicher Art.

Schließlich möchte ich mich ganz herzlich bei meiner Familie und meinen Freunden bedanken, mit denen ich die Erfolge und Herausforderungen der letzten Jahre teilen durfte. Ihre anhaltende Unterstützung in allen Phasen der Promotion hat diese Zeit für mich nicht nur wissenschaftlich, sondern auch persönlich zu etwas ganz Besonderem gemacht.

Contents

CHAPTER 1 Introduction

In this introductory chapter, an overview of different systems of active or self-propelled particles is given. To illustrate their great abundance, various macroscopic examples are mentioned first. After that, the distinctive features of micron-sized self-propelled particles, which are investigated in this thesis, are discussed. As ubiquitous realizations in nature, a classification of the most important biological microswimmers is provided. In the subsequent section, several systems of artificial self-propelled colloidal particles are described. The focus is on the different propulsion mechanisms that have been developed in recent years. Next, a short review on tactic behavior of microorganisms is provided, which is particularly relevant to the last main chapter of this thesis. Following this general introduction, the motivation for the various studies presented in the individual chapters is outlined. They are placed in context of related previous work, and the relevance of the considered issues is illustrated.

1.1 Active particles in general

In a very general sense, active particles are objects that are able to perform an autonomous directed motion. According to this definition, a significant part of all living beings is included. While the investigation of the detailed processes enabling the locomotion is usually a biological question, the resulting motional patterns are a fascinating topic from a physical point of view. This holds for the single-particle motion as well as for collective phenomena.
Detailed studies have been conducted on various animal species of all orders of magnitude [1,2]. On the macroscopic scale, frequently considered examples are fish [3,4], birds [5–7], and insects [8–11]. All these animals are well suited to study their collective motion as they often occur in large shoals, flocks, or swarms. To elucidate the dynamics of schooling fish, recently the details of the interactions between golden shiners have been analyzed based on small shoals of two or three individuals [4]. With regard to birds, field studies have been performed on starlings [5], sea ducks [6], and pigeons [7], for example. For the flocking behavior of starlings it could be shown that each bird typically interacts with about six to seven neighbors [5]. Some of the largest animal swarms occurring in nature are formed by migrating desert locusts [9,10].

These swarms can consist of up to several billion individuals and may thus cause severe plagues. Further examples of insects which exhibit characteristic collective behavior are ants [8] and midges [11].
Following the definition given above, even humans are yet another realization of self-propelled particles. Similar methods as used for the description of large groups of animals can also be applied to study the behavior of crowds of humans. Thereby, new insights have been obtained about panic and jamming mechanisms [12]. Moreover, lane formation in pedestrian streams has been investigated [13], and different ordered and disordered states of motion have been identified in crowds at heavy metal concerts [14]. Surprising effects, such as "phantom traffic jams" [15], also occur in the context of vehicle traffic. In addition to the collective behavior of people in various situations, the motion of single human walkers has been analyzed. Here, characteristic ring wandering has been observed for blindfolded people although they intended to walk on straight lines [16]. As discussed later in this thesis, such circular trajectories are actually the generic type of motion of most unguided self-propelled particles. Interestingly, for many active systems certain universal features have been identified although the situations seem to be completely different [1, 17].

Besides the macroscopic examples mentioned so far, a ubiquitous group of natural self-propelled particles are microorganisms such as bacteria and algae swimming in a liquid.[1] However, there is one crucial difference between the swimming motion of mammals and other macroscopic living beings on the one hand, and microscopic biological organisms on the other hand. While the former can accelerate themselves by specific swimming strokes and then glide through the water for a certain time, this is not possible for microorganisms. The main problem is that the latter are so small that they have no inertia when moving through a viscous medium, i.e., their motion is overdamped. Therefore, they will immediately stop when they do not propel themselves anymore. For this reason, microswimmers have to continuously convert energy into directed propulsion.
These illustrative considerations are typically quantified by means of the Reynolds number Re. It is given by

$$\mathrm{Re} = \frac{\rho v d}{\eta} \tag{1.1}$$

as a function of the swimming velocity v, a characteristic length scale d of the moving object, and the density ρ and the dynamic viscosity η of the fluid. The Reynolds number determines the ratio between inertial and viscous forces and thus characterizes the fluid flow for a given setup. A small Reynolds number $\mathrm{Re} \ll 1$ corresponds to laminar flow, while a high Reynolds number $\mathrm{Re} \gg 1$ indicates turbulent behavior. The calculations and considerations in this thesis focus on the low-Reynolds-number case, which is characteristic of the motion of microswimmers.
The basic challenge of mechanical self-propulsion under noninertial conditions is described by

[1] Realizations of self-propelled particles can even be found on a subcellular scale. As an example, the motional behavior of microtubules propelled by surface-bound dyneins has been studied [18].

the scallop theorem [19]. It states that a simple reciprocal movement as realized by the opening and closing of a scallop cannot generate locomotion in the low-Reynolds-number regime. This would only make a swimmer move back and forth, but in the end it arrives at the initial position again. Therefore, microorganisms had to develop more sophisticated propulsion mechanisms in order to solve the problem of locomotion at low Reynolds number (see the following section 1.2).

1.2 Biological microswimmers

Biological microswimmers include some of the oldest living beings on earth. Certain *prokaryotes* already existed on our planet more than 3.5 billion years ago. These microorganisms, which comprise all bacteria and cyanobacteria, are characterized by the fact that they do not have a cell nucleus [20]. Many prokaryotes are able to actively move through a surrounding liquid by means of flagella and can thus be considered as a typical example of biological microswimmers. Bacteria cells are usually a few micrometers in size and have their flagella either polarly or laterally attached. While some bacteria have only a single flagellum responsible for the locomotion, others have many flagella, which are often organized in bundles. In most cases, the flagella push a bacterium forward by performing an oscillating or rotating motion [21, 22]. Therefore, such bacteria are called “pushers.” The swimming strokes are executed in a nonreciprocal way so that locomotion is also possible at extremely low Reynolds number. Typically, a flagellum performs 40–60 revolutions per second and can thus propel the cell with a velocity of up to $200\,\mu\mathrm{m}\,\mathrm{s}^{-1}$ [20]. A single flagellum usually has a diameter of about 10–20 nm and can have a maximal length of approximately $20\,\mu\mathrm{m}$. Flagellar propulsion is characteristic for most swimming bacteria, including the extensively studied examples *Escherichia coli* and *Bacillus subtilis*. However, some prokaryotes, namely cyanobacteria, rely on a different driving mechanism. Although they lack flagella, cyanobacteria of the *Synechococcus* group have been observed to swim at speeds of about $25\,\mu\mathrm{m}\,\mathrm{s}^{-1}$ [23]. It has been suggested that their propulsion is based on traveling surface waves [24, 25].

As opposed to prokaryotes, the cells of *eukaryotes* have a membrane-bound nucleus. In addition to a great variety of unicellular microorganisms [26], eukaryotes include animals, plants, and fungi. The first unicellular eukaryotes appeared about two billion years ago. Many eukaryotic microorganisms also have flagella, but these are usually more complicated than in prokaryotes. A lot of different kinds of eukaryotic flagellates and ciliates exist [26].

Flagellates usually have only a rather small number of flagella which generate the self-propulsion. The main flagella are often larger than the cell itself. Typical sizes of unicellular eukaryotes are several tens of micrometers. Thus, they are in general about one order of magnitude larger than prokaryotes. Some eukaryotes such as *Paramecium* even have a size of more than $100\,\mu\mathrm{m}$.

Several flagellates with two flagella exist. A well-known example is the green alga *Chlamydomonas*, where both flagella are equal. They are attached to the front part of the cell. Such microswimmers with their propulsion apparatus in the front are often referred to as "pullers." The distinction between pushers and pullers is important in the context of hydrodynamic theoretical modeling [27–29]. As opposed to flagellates with two identical flagella such as *Chlamydomonas*, dinoflagellates possess only a single usual longitudinal flagellum and a second one which is realized as a transverse flagellum [30]. Furthermore, there are various kinds of flagellates with flagella bundles or undulating membranes, which are sinuous expansions of the cytoplasmic membrane.

Ciliates such as *Paramecium* do not have normal flagella but instead of that a large number (several thousand) of cilia, which are shorter and often enclose the whole cell body. Their movement is coordinated by the cell so that they can collectively generate a swimming motion [31]. Sometimes, they are combined to specific organelles, which enable the microorganism to perform a crawling motion along surfaces or on the ground.

In most cases, flagellates and ciliates have an asymmetric cell shape, which is particularly relevant in the context of the present thesis. This asymmetry is often induced by the position of the oral groove [32]. As a consequence, *Paramecium*, for example, does typically not move on straight lines but on circling trajectories. Normally, the sense of rotation is counterclockwise, but for some species of *Paramecium* clockwise rotation has been observed as well [33].

The primeval mechanism of locomotion by using flagella is also essential for the reproduction of all higher animals, including humans. In order to reach the female ovum, male gametes, the spermatozoa, use this type of self-propulsion when swimming through liquids.

The motional behavior of the previously mentioned microswimmers has been investigated both in experiments and in theory. With respect to bacteria, *Escherichia coli* is probably most popular as a laboratory model organism [34]. This rod-shaped bacterium commonly occurs in the intestine of animals and humans [35]. It has been used for countless experimental studies, including many on its motile behavior. Besides three-dimensional tracking of single bacteria [36] and detailed investigations of the run-and-tumble motion[2] [37–39], a lot of interest has been directed at the collective behavior in active suspensions [40–42]. Moreover, the interaction of *Escherichia coli* bacteria with geometrical structures has been addressed [43]. Another widely studied prokaryote is the rod-shaped aerobic bacterium *Bacillus subtilis*, which is most commonly found in soil. Being about 2 μm long, it has a similar size as *Escherichia coli*. An important distinction is the fact that *Bacillus subtilis* is Gram-positive [44], while *Escherichia coli* is Gram-negative [45], i.e., their cell walls have different structures which lead to contrasting responses when treated with Gram stain [46]. Many collective phenomena

[2]Run-and-tumble motion is characterized by a successive alternation between active translational displacements and reorientation phases. This behavior is determined by changes in the flagellar activity [37].

in bacterial suspensions have been investigated based on *Bacillus subtilis* bacteria as model microorganisms [47–51].

Although for the investigation of bacterial dynamics *Escherichia coli* and *Bacillus subtilis* are most frequently used, experiments have also been performed on other species such as *Caulobacter crescentus* [52], for example, which is abundant in fresh water. In contrast to the previously mentioned bacterial species, which are peritrichously flagellated (i.e., they have multiple flagella), *Caulobacter crescentus* has only a single flagellum. Such monotrichous bacteria cannot perform a tumbling motion in order to adapt their swimming direction. Therefore, rotational Brownian motion is vital for reorientation [52].

While *Escherichia coli* can be considered as the most important prokaryotic model microorganism, the biflagellated green alga *Chlamydomonas* has a similar role among unicellular eukaryotic microswimmers [53]. *Chlamydomonas* usually lives in fresh water or in moist soil. Typically, it has an ellipsoidal shape and a body size of approximately $10\,\mu\text{m}$. Thus, it is larger than most bacterial species and can therefore be more easily tracked in experiments, even without specialized tracking microscopes [54]. Moreover, Brownian motion can often be neglected, which is advantageous for many purposes [54]. Detailed three-dimensional measurements including high-speed imaging have shown that *Chlamydomonas* can switch between different kinds of flagellar activity [55]. While synchronous flagellar action leads to nearly straight motion, certain asynchronous beating patterns induce significant reorientations [55]. Thus, the swimming behavior is similar to the run-and-tumble motion which is characteristic for many bacteria [38]. Further insights have been obtained for the flow field in the solvent around a swimming *Chlamydomonas* cell [56] and concerning the scattering at surfaces [57].

Spermatozoa are no autonomous microorganisms such as bacteria and algae. Nevertheless, they are able to perform an active swimming motion by using their single flagellum. A lot of research has been devoted to the question of how exactly spermatozoa find the way to the female egg cell [58–60]. The motile behavior has been studied in many laboratory experiments based on various kinds of mammalian sperm cells, including human [61], bull [57, 62, 63], and mouse spermatozoa [64]. Furthermore, experimental studies have been performed on sea urchin sperm cells [59, 65]. These have the advantage that they can easily be obtained in sufficiently high numbers, which is often necessary for research purposes. The wide availability is due to the fact that male sea urchins emit large amounts of sperm into the sea in order to fertilize the female's eggs, which usually float freely in the water.

As illustrated by the given examples, there are multitudinous realizations of biological microswimmers. New techniques and devices that have become available over the last few years make it possible to unravel previously unresolved aspects of certain biological and physical phenomena. Vice versa, microorganisms also serve as valuable model systems to study collective phenomena or to address fluid dynamical problems.

1.3 Artificial self-propelled colloidal particles

In particular in the last decade, a lot of research has been dedicated to the development of artificial microswimmers. Various types of such devices have already been fabricated, and their dynamical behavior has been investigated in detail. The high interest in this field is due to several reasons. First, man-made micron-sized swimming objects mimic the motion of certain microorganisms and can therefore be used as idealized model systems. A main advantage as compared to experiments with living microswimmers is the fact that certain features such as the size and shape of the particles can easily be modified. Thus, the influence of these parameters on the motional behavior can be systematically studied. Secondly, in colloid physics[3] [66] it is an important issue of fundamental research to generalize previous findings for systems of passive particles to active systems [67, 68]. Far-reaching insights have been obtained into the behavior of colloidal dispersions in external fields [69, 70], for example. While the focus has been on passive particles here, even more versatile dynamics may be expected for similar setups with active particles [71]. To study these in detail, well-defined artificial microswimmers represent an ideal model system. Thirdly, self-propelled particles offer a manifold of fascinating application possibilities. This holds for specifically designed individual micromachines as well as for many-particle systems. Very promising examples include transport of colloidal cargoes [72, 73], targeted drug delivery [74, 75], and other medical tasks [76].

The bulk of the realizations of artificial self-propelled particles available so far are based on phoretic driving mechanisms [77]. Phoretic motion is in general induced by fields that interact with the surface of a particle or, more precisely, with the interfacial region between a particle and the solvent [78]. The term *phoresis* is used when colloidal particles move as a consequence of such effects, while the term *osmosis* refers to the situation when solvent flows occur at fixed surfaces due to similar interfacial processes [79, 80].
The most common examples of phoretic transport are electrophoresis, diffusiophoresis, and thermophoresis. In all cases, the motion is caused by gradients in the solvent surrounding a particle. These gradients can either be generated by externally applied fields [78], or they are created by the particle itself [81, 82]. The latter situation is usually referred to as *self-phoretic motion*.

In *electrophoresis*, the motion is induced by gradients of the electrostatic potential. Although it is often not possible to trace back the propulsion of artificial microswimmers to one single mechanism as will be discussed in more detail further below, self-electrophoresis is very likely to be mainly responsible for the motion of bimetallic swimmers [83–85]. Theoretical studies including numerical simulations and scaling analyses have been performed in order to understand the underlying physics [86–88].

[3]Colloidal particles are typically on the order of several nanometers to a few micrometers. Hence, the upper end of this range corresponds to the size of many biological microswimmers.

Self-electrophoresis has also been suggested as a propulsion mechanism for certain biological microorganisms [89, 90], in particular for cyanobacteria which can swim although they do not have flagella. However, later this mechanism has been ruled out as the source of the swimming motility in cyanobacteria [91].

Diffusiophoretic movement is generated by spatial differences in the concentration of certain chemical species. Usually, these concentration gradients are caused by a chemical reaction which leads to an asymmetric distribution of the reaction products [81]. Detailed theoretical studies are available both for freely moving particles [81, 82] and for diffusiophoretic motion under spatial confinement [92]. The first experimental realization of the concept of self-diffusiophoretic propulsion as proposed in reference [81] was presented in reference [93], where polystyrene spheres with one platinum-coated side have been used.

The third well-known example of phoretic motion is *thermophoresis*, which is caused by temperature gradients in the solvent. Detailed experiments including measurements of the temperature distribution in the vicinity of a colloidal particle have shown that thermophoretic mobility is not significantly affected by the particle size [94]. The characteristics of self-thermophoresis as a propulsion mechanism for artificial microswimmers have been investigated in reference [95]. In the described setup, irradiated laser light is absorbed by the metal side of a half-coated colloidal particle. Thus, local temperature gradients responsible for the active motion are generated. As has been shown recently, thermophoretic motion can also be induced by applying a magnetic field [96]. With regard to the collective behavior of thermophoretically driven particles, a theoretical model based on a stochastic formulation has been provided [97].

Various designs of artificial microswimmers moving due to the described self-phoretic effects have been developed. A substantial part of them is based on the catalytic decomposition of hydrogen peroxide (H_2O_2) to oxygen and water [83, 85, 98–100]

$$2H_2O_2 \rightarrow O_2 + 2H_2O\,. \tag{1.2}$$

In most setups, the chemical reaction is catalyzed by platinum, but also other materials such as nickel [100] can function as catalyst. Before utilizing this catalytic reaction for the self-propulsion of artificial microswimmers, it was first studied in the context of autonomously moving mesoscopic particles [98]. The considered platelike particles were several millimeters large and had a small area of platinum attached to one surface. The propulsion has been attributed to the recoil of bubbles which are generated by the catalytic reaction [98]. While for these mesoscopic particles the direction of motion points away from the side with the platinum catalyst, the opposite is the case for bimetallic nanorods[4] [83, 85]. This suggests that bubble recoil does not play a significant role in the propulsion of these micron-sized particles. While Pt-Au rods were the first realization of bimetallic mircoswimmers [83], the

[4]These bimetallic rods typically have a length of approximately 2 µm [83, 85]. Nevertheless, we use the term "nanorod" as is commonly done in literature.

motional behavior is very similar for other combinations of metals [85]. The relevance of various propulsion mechanisms [101] including interfacial tension gradients [83], viscosity gradients [102], self-diffusiophoresis [99], and self-electrophoresis [85] has been investigated in detail. Most likely, self-electrophoresis is the dominant mechanism mainly responsible for the self-propulsion [84, 85].
While the speeds of the first catalytic nanorods were already on the order of $10\,\mu\mathrm{m\,s^{-1}}$ [83], in subsequent years the particle design could be optimized toward even higher propulsion speeds [103, 104]. A substantial enhancement of the swimming velocity of Pt-Au nanorods has been achieved by incorporating carbon nanotubes into the platinum segment [103]. Even faster motion could be generated by using an Ag/Au alloy instead of the Au component [104]. Thus, the obtained swimming velocities correspond to those of biological microorganisms such as bacteria, which typically move at up to $200\,\mu\mathrm{m\,s^{-1}}$ [20].

Besides bimetallic nanorods, catalytic self-propulsion has been realized in various other systems. One example are sphere dimers which consist of one catalytic and one inactive sphere. Such a setup was first discussed theoretically based on computer simulations and analytical estimates in reference [105]. Later, corresponding experiments were implemented as well [106]. An even simpler setup consists of just one spherical particle which has two hemispheres with different chemical properties. Such particles are usually referred to as *Janus particles*. The probably most popular realization are polystyrene spheres with one platinum-coated side, which catalyzes the chemical reaction in a hydrogen peroxide solution. After the first characterization in reference [93], this system has been investigated in detail in several subsequent studies [107–109]. A very similar setup is also used in chapter 2 of this thesis in order to verify theoretical predictions regarding the non-Gaussian behavior of self-propelled Janus particles. Moreover, active platinum-coated polystyrene spheres provide the building blocks for more complicated autonomously moving objects. By combining active Janus particles to doublets or larger agglomerates, microswimmers with specific well-defined dynamical properties can be designed [72, 110].

It had long been assumed that the propulsion mechanism of platinum-polystyrene Janus particles in hydrogen peroxide solutions is purely due to self-diffusiophoresis [81]. However, recent observations, such as a reduction of the swimming velocity by adding salt, cannot be explained by diffusiophoresis of *neutral* solute molecules [111, 112]. Moreover, in reference [111] it is argued that *ionic* self-diffusiophoresis alone cannot induce propulsion speeds as high as observed in the experiments. One possible explanation of the new findings is based on the assumption that electrophoretic processes significantly contribute to the self-propulsion [111, 112]. Thus, the situation would be similar to bimetallic nanorods [84, 85]. Due to these uncertainties about the detailed processes responsible for the active motion of artificial Janus particles, it is important to have reliable theoretical models that do not depend on the sophisticated propulsion mechanism but describe the dynamics on a more coarse-grained level. Such a model is put forward in this thesis. It will be shown that it can be applied to many different types of self-propelled particles and predicts important results, such as the gravitactic behavior of asymmetric microswimmers (see chapter 9).

One disadvantage of self-propelled particles moving due to the catalytic decomposition of hydrogen peroxide is the fact that the propulsion velocity is set by the H_2O_2 concentration and can in general not easily be modified during a single experiment. An exception is the setup of Palacci and coworkers [113]. In that case, the catalytic reaction, and thus the activity of the particles, can be switched on and off by illumination with blue light. The usage of different photocatalytic materials also allows for a wavelength-dependent activation of certain species in a mixture of self-propelled particles [114].

The system of Bechinger and coworkers, which was first described in reference [115] and characterized in more detail in reference [116], even enables a continuous tuning of the self-propulsion velocity by varying the intensity of a highly defocused illuminating laser beam. The silica spheres under study have a gold coating at one side, which is heated due to the absorption of light. As the particles are suspended in a slightly subcritical water-lutidine mixture, this heating leads to a local demixing of the solvent. As a consequence, concentration gradients arise and propulsion based on self-diffusiophoresis occurs. By increasing the illumination intensity, stronger spatial concentration differences are created so that the particles move faster. This propulsion mechanism is also used for the experiments discussed in chapters 7 and 9 of this thesis. It is explained in more detail in section 1.5 on page 23.

In addition to the previously described examples of artificial microswimmers based on self-phoretic propulsion mechanisms, several other realizations with different driving mechanisms have been proposed. A useful tool in this context are external magnetic fields. On the one hand, they have been applied to control the motion of catalytically driven Janus particles [117, 118]. This is possible by changing the orientation of the magnetic field, but also by just modifying its strength [118]. On the other hand, under certain conditions the propulsion itself can be magnetically actuated [119–122]. In the setup studied in reference [119], doublets consisting of two differently sized paramagnetic colloidal particles linked by DNA bridges have been used. As the motion occurs near a solid surface, a precessing external magnetic field induces a translational net motion. Measurements have also been performed on more complicated assemblies of colloidal particles [120], and a comprehensive corresponding theoretical framework has been provided [123].
Moreover, oscillating magnetic fields have been used to mimic the flagellar beating pattern which is found in many biological microorganisms. A magnetically actuated artificial flagellum has been realized by attaching a DNA-linked linear chain of colloidal magnetic particles to a red blood cell [124]. By modulating the external magnetic field, the beating motion of the artificial flagellum is tuned. Thus, both the magnitude and the direction of the swimming velocity can be controlled [124, 125]. However, strictly speaking, these particles are not *self*-propelled

because their motion is fully governed by the external magnetic field. Only very recently, a synthetic swimmer moving *autonomously* by means of a biohybrid artificial flagellum based on contractile cells has been reported [126].

Further propulsion strategies that have been realized in artificial systems are based on shear-thinning or shear-thickening properties of the surrounding fluid [127], bioelectrochemical processes [128], cooperative motion in self-assembled complexes [129], or ultrasonic standing waves [130, 131]. For a non-Newtonian solvent, which shows shear-thinning or shear-thickening behavior, it has been demonstrated that locomotion at low Reynolds number is also possible by reciprocal movements [127]. This illustrates that the scallop theorem is not valid for non-Newtonian fluids.
An important advantage of particles propelled by ultrasound as compared to catalytically driven systems is the biocompatibility. While hydrogen peroxide, which is frequently used in catalytic self-propulsion, is toxic to cells, this problem does not occur for acoustically actuated systems [130, 131]. Although there are still some challenges to be overcome before artificial self-propelled particles can actually be systematically used for medical and other purposes, they are likely to play an important role for future applications.

1.4 Tactic behavior of microorganisms

Artificial self-propelled particles often represent useful model systems in order to study the dynamical behavior of biological microswimmers. This also holds if the motion is affected by external fields. In nature, microorganisms are usually exposed to a large number of different stimuli in their environment. Many of them are able to respond to various factors such as light, chemical gradients, solvent flows, gravity, magnetic fields, and temperature differences, for example. They often adapt their motion by either swimming in the direction of the stimulus or away from the source. The former case is called *positive taxis*, while in the latter case the behavior is referred to as *negative taxis*.

In general, one can differentiate between active and passive taxis mechanisms. An active tactic behavior means that a microorganism is able to *sense* the stimulus. This triggers an active response, such as a reorientation of the microorganism. On the other hand, passive taxis mechanisms are purely physical phenomena. They are determined by external forces and do not require active sensing. One can often decide whether a tactic mechanism is active or passive by comparing the behavior of live and dead organisms. As opposed to active taxis, passive mechanisms usually also affect the motion of dead cells.

A widely spread active tactic behavior is *chemotaxis*. It is one of the most studied tactic phenomena and has been observed for many biological microswimmers. Chemotaxis is most common in bacteria [38, 132, 133] and spermatozoa [58, 59, 134, 135], but it has also been

observed for algae [136, 137] and amoebae [138–140], for example. The detailed processes enabling microorganisms to respond to chemical gradients in their environment have been subject of intense research in biology. With regard to bacteria, the signaling network has been studied in detail [141], and various possible response strategies have been compared [142]. In some microorganisms the chemotactic response is so fast that they can also exploit transient nutrient patches in the ocean before they are dispersed by the flow. This has recently been shown for the marine bacterium *Pseudoalteromonas haloplanktis* [143].
While the chemotactic behavior which is characteristic for many biological microorganisms is explained by active sensing, in artificial non-biological systems a similar behavior caused by purely physical effects has been observed [144]. In that case, the mechanism is based on gradients of the diffusion coefficient.
To describe chemotactic motion, various theoretical models have been developed [145, 146]. These include a predator-prey model describing the dynamics of a predator which senses a chemical emitted by the prey [147]. Vice versa, the prey also senses a diffusing secretion emitted by the predator. Moreover, *autochemotaxis* has been addressed, where a particle interacts with the distribution of its own previously emitted chemoattractant or chemorepellent [148–150].

A special form of chemotaxis is *aerotaxis* [151], which refers to the response of microorganisms to variations in oxygen concentration. Therefore, it is also called *oxytaxis* [152–154]. In many cases, aerobic bacteria swim up oxygen gradients [152]. Thus, by performing a positive oxytactic motion, they are able to accumulate in areas with proper oxygen concentration.

In *thermotaxis*, microswimmers respond to temperature gradients in their environment. *Escherichia coli*, for example, modifies its swimming behavior when it encounters temperature changes [155]. Moreover, some observations indicate that thermotaxis is an important mechanism for the guidance of mammalian sperm cells [156]. It has been suggested that thermotaxis is relevant to the long-range guiding, while chemotaxis plays a major role as a short-range mechanism [156, 157].

Another active tactic mechanism that is essential for some microorganisms is *phototaxis*. Certain ciliates and flagellates, such as *Euglena gracilis* [158], have the ability to move toward or away from a light source. While some microorganisms always show positive or negative phototaxis, others can respond differently when exposed to illumination by light with high or low intensities [159]. For that purpose, they often have an eyespot apparatus, which is a photoreceptive organelle that enables the microorganisms to sense differences in light intensity. Thus, they can move to areas with ideal light conditions by changing the beating pattern or the activity of their flagella. This is especially important for phototrophic microorganisms in order to find a region with an optimal light exposure for photosynthesis. The processes responsible for the ability of *Chlamydomonas reinhardtii* to switch between positive and negative phototaxis have been investigated in reference [160]. Moreover, the detailed role of the two flagella and the changes of the beating pattern during phototactic movement have been analyzed [161, 162]. Phototaxis is also relevant to marine zooplankton [163] and certain prokaryotes [164].

In addition to the previously mentioned active types of tactic behavior, there are several examples how microswimmers respond to their environmental conditions based on passive physical effects. When their motion is affected by magnetic fields, the term *magnetotaxis* is used. Magnetotactic behavior has been observed for certain bacterial species [165–167] and also for some algae [168]. In reference [165], it has first been reported that the migration of *Spirochaeta plicatilis* bacteria is influenced by the earth's magnetic field. Subsequently, the evolutionary origin of bacterial magnetotaxis has been investigated [169], and the resulting swimming behavior has been analyzed in detail. Typically, magnetotactic bacteria in the northern hemisphere swim toward the geomagnetic north, while southern hemisphere bacteria swim in the direction of the geomagnetic south [166]. Thus, anaerobic bacteria are able to move downward to areas with lower oxygen concentration and sufficient food supply. However, more recently, some northern hemisphere populations of bacteria have been identified which swim toward the geomagnetic south in response to high oxygen concentrations [167]. The reason for this opposite behavior is still unclear.
Sometimes, magnetotaxis is combined with other mechanisms, such as aerotaxis, in order to more efficiently migrate to a position with the preferred oxygen concentration [170]. Magnetotaxis can also be artificially induced in biological organisms by using ferromagnetic nanoparticles [171]. Thus, the motion is finely controllable as illustrated for the eukaryotic ciliate *Tetrahymena pyriformis* [171].

Rheotaxis is the directed movement in response to a flow in the surrounding fluid. While rheotaxis is an active mechanism in most macroscopic animals like fish [172] or humans swimming against the current in a river or in the sea, the response of sperm cells [173–175] or bacteria [176, 177] to shear flow is often purely passive. Experiments in Poiseuille flow between two horizontal plates have been conducted both with live and dead spermatozoa [173, 174]. While swimming against the flow only occurs for live sperm cells, of course, similar orientational effects have been observed for dead cells as well [173]. These early experiments already suggest the passive nature of sperm rheotaxis. Detailed measurements have shown that swimming spermatozoa under shear perform a characteristic spiraling upstream motion [175]. This results from the flow in combination with steric surface interactions and the flagellar beating pattern [175]. Furthermore, the effects of the wall shape on the motion of sperm cells swimming through microchannels have been analyzed [178]. The observations confirm that rheotaxis is likely to be an important factor for the guidance of sperm cells in the female reproductive tract of mammals [178, 179].
With regard to bacteria, characteristic upstream swimming in shear flow has been observed for *Escherichia coli* in the vicinity of solid surfaces [176]. Moreover, detailed experiments on *Bacillus subtilis* swimming freely in a bulk fluid have shown that under these conditions a

rheotactic motion is caused by the interplay of fluid velocity gradients and the specific helical shape of the flagella [177].

In microalgae, rheotaxis is often accompanied by gravitational effects originating from an inhomogeneous mass distribution of the involved particles [180–182]. The directed motion resulting from the interplay between gravitational and viscous torques is referred to as *gyrotaxis* [180]. This passive mechanism plays an important role in the formation of phytoplankton layers in the sea [183]. Here, gyrotaxis often induces self-focusing [184] and can thus generate macroscopic phenomena [183]. Similarly, dense aggregations of gyrotactic particles have been predicted to occur in a steady vortical flow [185]. Moreover, it has been investigated how the gyrotactic behavior of phytoplankton is affected by turbulent flows [186]. In bioconvection, often different kinds of taxes act simultaneously. In addition to gyrotaxis, phototaxis is also likely to significantly contribute to the occurring patterns of motion [187, 188].

As far as gyrotaxis of phytoplankton is concerned, the effect of gravity is usually attributed to an inhomogeneous mass distribution within the individual organisms. However, for most biological microswimmers the origin of *gravitaxis*,[5] i.e., the response to gravity, is highly debated [189, 190]. In many cases, it is still unclear whether the underlying mechanisms are active as in chemotaxis and phototaxis, or if the phenomenon can be completely explained by passive physical effects similar to magnetotaxis and rheotaxis. The bottom heaviness of phytoplankton involved in gyrotaxis is one realization of such a passive mechanism. This behavior has been reproduced in experiments with platinum-coated Janus particles [191]. In that case, the inhomogeneous mass distribution results from the coating material, which is heavier than the colloidal sphere itself. As a consequence, the particles preferably move upward against gravity, i.e., they show negative gravitaxis. In the theoretical description, this effect is taken into account by an explicit gravitational torque [180, 192, 193]. In combination with intrinsic and viscous torques, characteristic helical trajectories are obtained [192, 194]. This is in agreement with observations of many phytoplankton species [195–197].

However, gravitaxis has also been observed for microorganisms with a homogeneous mass distribution. It is often argued that in these cases the gravitactic response is triggered by a physiological gravity sensor [198–200]. Various methods and experimental setups have been developed in order to evaluate the importance of such active sensing mechanisms as compared to purely passive effects [201, 202]. One microorganism where the existence of a gravity receptor is indeed likely is *Euglena gracilis* [159, 203, 204]. Measurements conducted during sounding rocket experiments suggest that physical effects cannot exclusively be responsible for gravitaxis in this case [205]. To understand the processes enabling *Euglena gracilis* to actively sense the direction of gravity, further detailed studies have been performed [159, 204].

An alternative explanation for gravitaxis in microswimmers with homogeneous mass density is based on an asymmetric body shape [206]. Corresponding sedimentation experiments have

[5] In older literature, gravitaxis is often called geotaxis.

shown that passive particles with a pronounced fore-rear asymmetry assume a specific orientation when they move downward in the gravitational field [207]. This suggests that the particle shape might also play an important role in the gravitactic behavior of microswimmers. Although it had previously been supposed that gravitaxis in *Paramecium* is caused by an inhomogeneous mass distribution [208, 209], more recent experiments have shown that this is very unlikely [210]. In experiments with immobilized *Paramecium* cells it has been observed that the orientation depends on the sedimentation direction in a hypo- or hyperdensity medium. While an upward orientation occurred during downward sedimentation in a hypodensity medium, the orientation was opposite for the upward motion in a hyperdensity medium [210]. This behavior cannot be explained by a bottom heaviness, which would lead to the same orientation in both situations. Instead, the findings strongly suggest an alignment mechanism originating from the fore-rear asymmetry [190, 206, 207]. The same mechanism is also likely to contribute to the gravitactic behavior of *Chlamydomonas reinhardtii* [211, 212], although here the effect of the flagella is probably more important than the body shape [207].
To further elucidate the question to what extent an asymmetric body shape can be responsible for gravitaxis in microswimmers, in chapter 9 of this thesis, the motion of artificial self-propelled particles with a well-defined shape and mass distribution is investigated.

1.5 Motivation of this work

The main purpose of this work is to enlighten how the motion of self-propelled particles is influenced by their shape. Therefore, the various chapters are organized according to increasing complexity of the studied particles. In chapters 2 [213] and 3 [214], spherical Janus particles are considered. First, a theoretical model based on the Langevin equations for an active particle is tested against experiments with Pt-silica Janus particles. Subsequently, a generalization toward time-dependent self-propulsion is provided. Chapter 4 [215] also takes particle deformations into account and in particular focuses on deformable active particles in shear flow. For the study of asymmetric self-propelled particles it is necessary to predict the grand resistance matrix and thus the diffusion and coupling coefficients for the specific considered particle shapes. How these quantities can be obtained either from theory or from experimental data for passive Brownian particles with complex shapes is illustrated in chapter 5 [216]. Before considering biaxial particles without any symmetry axes in chapters 7–9, in chapter 6 [217] experimental and numerical results for the motion of a mesoscopic wedgelike carrier submersed in a suspension of rod-shaped swimming bacteria are presented.

The tools and methods introduced in chapters 2–6 enable the investigation of fully asymmetric active particles: as the first of the two centerpieces of this thesis, in chapter 7 [218] a theory for the Brownian dynamics of asymmetric self-propelled particles is presented and verified by experimental observations of L-shaped artificial microswimmers. Chapter 8 [219] provides a detailed theoretical justification of this concept and focuses on the applicability of effective forces and torques in the context of asymmetric microswimmers. Finally, in chapter 9 [220], which is the second centerpiece of this thesis, the intriguing question of the origin of gravitaxis in microswimmers is addressed. An explanation based on the particle shape is put forward and illustrated by means of experiments with asymmetric self-propelled colloidal particles. The most important results of this thesis are summarized in chapter 10, and the consequences for future research are outlined.

Spherical Janus particles can be considered as one of the by now most popular artificial realizations of self-propelled particles. Due to their simple shape and the relatively easy implementation of the experimental setup, in **CHAPTER 2** they serve as the model system to introduce some of the basic theoretical concepts used in this thesis. Although the detailed self-propulsion mechanism is not essential in the context of this thesis, we briefly comment on the most important underlying processes. The main precondition for the generation of a directed motion are the different properties of the two hemispheres of a Janus particle. These can be made of different materials or have different physical or chemical functionality [221]. If the combination of the particle composition and the solvent medium is appropriately chosen, a chemical reaction is induced. It is usually catalyzed at one surface of the particle. The resulting concentration gradients finally lead to a self-propelled motion of the Janus particle. The swimming behavior has been studied intensely in recent years. Insights have been obtained, e.g., with regard to the direction of motion [107, 108], the swimming efficiency [88], and the dependence of the propulsion velocity on the particle size [109]. What is still unclear is whether the self-propulsion is solely due to diffusiophoresis [78, 81, 93] or if additional mechanisms are largely responsible for the initiation of the deterministic translation [111, 112] (see also section 1.3). However, it is uncontroversial that the resulting motion evolves from a superposition of a finite self-propulsion velocity and the Brownian translational and rotational contributions. Thus, the orientation of a Janus particle fluctuates, and in the mean square displacement a transition from a ballistic regime, where the particle on average is self-propelled along its orientation, to a long-time diffusive behavior is observed.

Quantitatively, the dynamics of a single Janus particle is described by appropriate Langevin equations [222, 223]. In its simplest form, for the two-dimensional motion of a spherical Janus particle with rotational diffusion in the x-y plane, they are given by

$$\begin{aligned} \dot{\mathbf{r}} &= \beta F D_{\mathrm{T}} \hat{\mathbf{u}} + \boldsymbol{\zeta}_{\mathbf{r}} \,, \\ \dot{\phi} &= \zeta_{\phi} \,, \end{aligned} \tag{1.3}$$

where $\dot{\mathbf{r}}$ and $\dot{\phi}$ denote the time derivatives of the center-of-mass position $\mathbf{r}(t) = (x(t), y(t))$ of the particle and its orientation angle $\phi(t)$, respectively. This angle defines the instantaneous particle orientation $\hat{\mathbf{u}}(t) = (\cos\phi(t), \sin\phi(t))$, which coincides with the direction of the effective self-propulsion force[6] $\mathbf{F} = F\hat{\mathbf{u}}$. Further parameters entering into equations (1.3) are the translational diffusion coefficient D_{T} and the inverse effective thermal energy $\beta = 1/(k_B T)$ of the system. Finally, $\boldsymbol{\zeta}_{\mathrm{r}}(t)$ and $\zeta_\phi(t)$ are Gaussian noise terms characterized by zero mean and variances $\langle \boldsymbol{\zeta}_{\mathrm{r}}(t_1) \otimes \boldsymbol{\zeta}_{\mathrm{r}}(t_2) \rangle = 2D_{\mathrm{T}}\mathbb{1}\delta(t_1 - t_2)$ and $\langle \zeta_\phi(t_1)\zeta_\phi(t_2) \rangle = 2D_{\mathrm{R}}\delta(t_1 - t_2)$, where $\mathbb{1}$ is the 2×2-dimensional unit tensor and D_{R} is the rotational diffusion constant. As far as spherical particles are concerned, $\boldsymbol{\zeta}_{\mathrm{r}}$ and ζ_ϕ are statistically independent.

In chapter 2 of this thesis, the basic equations (1.3) are adapted to an experimental setup of spherical Pt-silica Janus particles moving in solutions with different concentrations of hydrogen peroxide. Different aspects of the motion are highlighted and studied in detail.
First, we address the mean square displacement, which can be considered as the standard quantity for the characterization of different dynamical regimes in the motion of self-propelled particles. Basically, three stages are predicted by our theoretical model and verified by the experimental data. While simple Brownian motion dominates at short times, superdiffusion is found at intermediate times. Finally, the system returns to diffusive behavior again at long times. The existence of the superdiffusive regime is characteristic for self-propelled particles and plays an important role in the context of many biological phenomena in active systems, such as for the rectification of swimming bacteria [224]. While we focus on the bulk motion of Janus particles in chapter 2, in particular in systems with spatial heterogeneity or when obstacles are present in the environment of the particles, also subdiffusion of self-propelled particles occurs [225].
As the second quantity of interest, the excess kurtosis indicating non-Gaussian behavior is studied. Non-Gaussianity is an important feature in many different areas of physics. To give some examples, we refer to the cosmic inflation in the early universe [226–228], magnetohydrodynamic turbulence as observed in solar wind [229], glass-forming materials [230–233], and rare events such as earthquakes [234] and stock crashes [235, 236]. Here, the concept is used to get further insights into the dynamics of self-propelled particles. While higher displacement moments of active particles have already been addressed in pure theoretical studies [223, 237, 238], chapter 2 [213] is the first analysis based on experimental data. We consider displacement moments up to fourth order and are thus able to predict and measure skewness and kurtosis. Subsequent to the publication of our work, the method of calculating the displacement moments analytically has been applied to obtain even higher moments [239].
Lastly, we also analyze the full probability distributions for the translational displacements as well as for the orientation of the particles. With regard to the displacement probability distribution, a surprising difference between the experimental data for Janus particles in solutions with low and high H_2O_2 concentrations is observed. While in the former case the originally Gaussian curve exhibits a significantly broadened peak after some time, in the latter case a pronounced double-peak structure emerges. By Brownian dynamics simulations it is confirmed that this

[6] The physical meaning of this effective self-propulsion force is discussed in detail in chapter 8.

observation reflects the orientational degrees of freedom of the Janus particles. Apparently, for fast moving particles at high H_2O_2 concentration the rotational Brownian motion is largely restricted. In principle, one can also predict the probability distribution function of a particle with orientation inside the plane of motion analytically based on the Fokker-Planck equation by using Fourier transforms [238]. Thus, one obtains an infinite system of coupled ordinary differential equations for the various Fourier modes. By taking into account a sufficient number of modes, it is possible to derive an analytical solution which is arbitrarily close to the exact one [238].

While the theoretical description in chapter 2 follows the widely applied approach of assuming a constant propulsion for the motion of self-propelled particles, in **CHAPTER 3** we generalize this coarse-grained model toward a more detailed description by considering an explicit time dependence of the propulsion velocity.

This more fine-grained model is important both for biological microswimmers and for artificial self-propelled particles. In the former case it is required in particular in order to describe the time dependence in the motion resulting from the individual swimming strokes. This is relevant to the flagellar locomotion of sperm cells [63, 240] or *Chlamydomonas* [241–244] as well as to larger microorganisms such as *Daphnia*. These planktonic crustaceans perform a hopping motion which is powered by their antennae. It has been characterized experimentally [245, 246] and described theoretically by random walk [247] and continuous models [248, 249]. Clearly, for such types of motion a simple model based on a constant self-propulsion is insufficient for the description and understanding of the underlying dynamics. On the other hand, time-dependent swimming may also occur on much longer time scales. This is often the case when the motion of microorganisms is influenced by chemical [38, 132] or light gradients [163, 164]. In the context of flagellar-driven bacteria, changes of the environmental conditions or variability in the energy supply may affect the level of flagellar activity [250]. Examples include the increase of the swimming velocity as a consequence of food supply [251] or a reduction of the propulsion speed in poisoned environments [252].

With regard to artificial self-propelled particles, the theoretical study of a time-dependent propulsion velocity is interesting from a fundamental point of view as well as for many applications of micro- or nanomachines. The extensively studied catalytic propulsion mechanism based on the decomposition of hydrogen peroxide [83, 93] is intrinsically time-dependent as the fuel is consumed by the catalytic reaction so that its concentration is reduced. This leads to a decrease of the propulsion speed over time. Although many experiments are performed at low particle density and in the stationary regime during the time period shortly after the beginning of the reaction [213], in particular for clustering experiments at high density the fuel consumption and the related velocity reduction are important issues.
In other systems of artificial microswimmers, recent progress opens new ways of tuning the particles in a desired way (see also section 1.3). On the one hand, this is possible for spe-

cially designed catalytic Janus particles by applying spatially varying magnetic fields [117]. A controlled modification of the swimming velocity and direction can even be achieved by tuning only the magnitude of an applied homogeneous magnetic field [118]. Another particularly promising realization of self-propelled particles with a regulated time dependence of the swimming motion are systems where the propulsion velocity depends on the intensity of an illuminating light field. If the propulsion mechanism is based on the catalytic decomposition of hydrogen peroxide, light-activated bimaterial colloids consisting of a polymer sphere with a protruding hematite cube enable a switching between active and passive motion [113]. The active particles studied in reference [113] are also slightly magnetic so that they can additionally be steered by applying an external magnetic field.
The propulsion mechanism based on the local demixing of a slightly subcritical binary liquid mixture (see references [115] and [116]) even allows for a continuous regulation of the propulsion velocity by just modifying the intensity of an illuminating light field. Thus, by choosing an appropriate propulsion protocol all kinds of time-dependent swimming behavior can be realized. This is especially useful for optimizing micromachines such as for cargo transportation [72,73].

In its simplest form, temporal variations of the self-propulsion can be included in the Brownian equations of motion (1.3) by considering a time-dependent effective force $F(t)$ instead of the constant force F. Chapter 3 provides detailed calculations for three different realizations of such time-dependent self-propulsion. First, a periodic piecewise constant effective driving force is considered. This situation reflects the run-and-tumble motion, which is observed for many microorganisms [38,55,253]. Theoretically, in addition to single-particle models [253], interacting run-and-tumble particles [254, 255] and systems confined by gravity, traps, or walls [256] have been addressed. Secondly, we study the motional patterns resulting from a sinusoidal self-propulsion. Such a continuous periodic modeling is most appropriate when the temporal variations in the swimming velocity on the time scale of an individual swimming stroke [244,257] shall be taken into account. The third example we consider in chapter 3 is a power-law type of propulsion force. Such a behavior may be relevant in the context of growing clusters of active particles [258] or when the swimming speed of a predator is determined by the prey gradient [259]. A power-law scaling in time t as $\propto t^{\alpha}$ (with $\alpha = 0, 1, 2, \ldots$) leads to superdiffusive behavior in the long-time regime, which is characterized by a mean square displacement $\propto t^{2\alpha+1}$. Prior to this long-time scaling an intermediate regime with a $\propto t^{2\alpha+2}$ time dependence is identified. For the special case $\alpha = 0$, this transition is experimentally verified in chapter 2.

On top of the different types of time-dependent self-propulsion, we also study the influence of an additional torque on the swimming path statistics. An effective torque leading to a deterministic rotational motion is often induced by particle imperfections [260] or by asymmetric shapes [218]. The latter case will be considered in detail in chapter 7 of this thesis. In chapter 3, we exemplarily present theoretical calculations for a sinusoidal propulsion force in combination with a constant torque. Analytical results are provided both for the noise-free and the noise-averaged trajectories as well as for the mean square displacement. When fluctuations are neglected, closed swimming paths are obtained which can be quite complicated depending on the magnitude of the torque and the frequency of the self-propulsion force. In the presence of translational and rotational noise, interestingly the mean trajectories are self-similar curves which still bear the characteristics of the corresponding noise-free swimming paths for the same set of parameters. The concept of self-similarity is well established in the context of fractals [261, 262], networks [263, 264], growth processes [265, 266], and other areas of statistical physics [267, 268]. In chapter 3, we show that self-similarity is also an important property in the context of mean microswimmer trajectories. While the mean swimming path for the case of a constant propulsion force in combination with an additional constant torque is a logarithmic spiral [269], which is one of the simplest realizations of a self-similar curve, this feature is also characteristic for microswimmers with time-periodic self-propulsion as revealed by our explicit calculations.

Apart from **CHAPTER 4**, this thesis focuses on self-propelled particles with a fixed stable shape. Such rigid shapes are usually found in the context of artificial microswimmers, and in many situations they also provide a valid approximation when studying the swimming behavior of living microorganisms. In the latter case, the propulsion often originates from the nonreciprocal motion of flagella or cilia. Then the cell body itself is normally not significantly deformed so that a theoretical description based on rigid particle shapes is possible. However, in particular for some unicellular eukaryotic organisms such as amoebae [270, 271], surface deformations are of fundamental importance also with regard to cell migration. While it had previously been thought that amoebae only migrate across solid surfaces, recently it has been observed that *Dictyostelium* amoebae can also swim freely by membrane deformations with surprisingly high speeds [272, 273]. This has consequences on the general understanding of amoeboid cell migration [274]. A new model for amoeboid swimming has been proposed in which the cell is represented by a closed incompressible membrane [275]. A similar swimming behavior as for *Dictyostelium* amoebae has also been observed for neutrophils [272], a type of white blood cells in mammals. Moreover, related deformations of the entire body are important for euglenoid movement [276]. While swimming based on surface deformations is mainly relevant regarding eukaryotic cells, cyanobacteria are an example of prokaryotic organisms that move by a similar mechanism [23]. More precisely, longitudinal or transverse traveling surface waves have been suggested as the origin of the propulsion [24, 277].

Deformability is also essential for active droplets, which are sometimes used as a model system to characterize the dynamics of real microorganisms. Usually, the spontaneous motion (self-propulsion) of such droplets is based on the Marangoni effect [278]. In many experimental realizations, oil droplets propelling on fluid interfaces due to nonequilibrium chemical conditions have been considered [279, 280]. Various types of regular motion have been observed for self-motile oil droplets on a glass substrate under suitable boundary conditions [279]. As shown recently, under spatially isotropic conditions regular motion is enabled by breaking the fore-rear symmetry [280]. Further examples of active droplets include alcohol droplets moving on an aqueous solution [281], oil drops driven by actin polymerization [282], and oil droplets swimming through a liquid medium [283]. In the latter case, the self-propulsion originates from a chemical reaction occurring at the interface between the droplet and the surrounding fluid. Recent theoretical modeling suggests that the motion of active droplets can be considered as pusherlike or pullerlike, depending on the relation between propulsion and deformation [284, 285]. Besides the dynamics of single droplets, their collective behavior in active emulsions has also been studied [286, 287].

Because of the relevance of shape changes in some biological and artificial microswimmers, in chapter 4 we consider deformable active particles. The theoretical approach is similar to that originally proposed by Ohta and Ohkuma in reference [288]. It includes a coupling between the active particle motion and shape deformations and has been extensively studied in the past few years (see, e.g., references [289–298]). Related models, which also include shape deformations, have been published in references [299–305]. While all of these papers focus on the motion of deformable particles in a quiescent solvent, in chapter 4 of this thesis we propose a theoretical model for the dynamics of an active deformable particle exposed to a linear shear flow. Previously, only rigid self-propelled particles have been considered in shear geometries such as Couette flow [306, 307] and Poiseuille flow [308–311]. Very recently, the effect of a shear flow has also been analyzed for the specific situation of an active particle propelled by self-diffusiophoresis [312]. Under these conditions, the self-generated solute concentration gradients are usually distorted by the ambient flow so that the active motion of the self-propelled particle is affected on a very basic level. However, shape deformations have only been addressed for *passive* particles in shear flow so far [313].
The influence of fluid velocity gradients on the motion of biological microswimmers is often referred to as rheotaxis (see section 1.4). It is an important phenomenon in many natural systems and has recently been studied for spermatozoa [175, 179], bacteria [176, 177], and microalgae [314].

The nonlinear dynamical equations proposed in chapter 4 are based on symmetry considerations and describe the time evolution of the particle's position, its translational and angular velocities, and its deformation. Both shear-induced passive rotations and active rotational motion are taken into account. Although we mostly focus on the two-dimensional case for the detailed calculations, the equations generally also hold for a setup in three dimensions. The model combines two previously studied setups: deformable active particles in a quiescent solvent on the one hand [288, 294], and rigid self-propelled particles in linear shear flow on the other [306]. These situations are still contained as special cases in our more comprehensive description. The general equations are solved numerically for different conditions. Particular interest is directed at the interplay between the strength of the self-propulsion and the strength of the imposed shear flow. Depending on these parameters, a great variety of straight, periodic, quasi-periodic, and chaotic types of motion arises. Our findings for a deformable self-propelled particle under linear shear flow have inspired a subsequent study of a deformable microswimmer in a swirl flow [315].

CHAPTER 5 provides some important preparations for the study of rodlike microswimmers in asymmetric environments in chapter 6 and, especially, for the investigation of asymmetric self-propelled particles in chapters 7–9. A precondition for the understanding of the dynamics of asymmetric microswimmers is a detailed knowledge of the Brownian motion of passive colloidal particles with a similarly complex shape. For this purpose, a combined experimental and theoretical study of such particles is provided in chapter 5. In particular, it is discussed how the grand resistance matrix[7] of anisotropic colloidal particles with a given shape can be determined. This matrix is essential in the context of asymmetric microswimmers as it plays an important role for the coupling between translational and rotational propulsion.

While the Brownian motion of spherical colloidal particles has been analyzed in detail since the seminal works of A. Einstein [316] and M. J. Perrin [317], asymmetric model colloids with complex but well-defined shapes are much more difficult to access experimentally. Only recently such particles have become available as a consequence of substantially improved experimental synthesis techniques [318–321]. These enable quantitative measurements not only of the translational and rotational diffusion but also of the coupling between them.

A general theory for the Brownian motion of asymmetric particles has already been developed by H. Brenner nearly half a century ago [322, 323]. While he first concentrated on the coupling between translational and rotational Brownian motion of screwlike particles [322], two years after that he presented a general theory for arbitrary particle shapes [323]. In the following decades, translational and rotational diffusion were mostly considered separately [324–326]. For uniaxial nonspherical particles, such as ellipsoids, translational and rotational Brownian displacements are still decoupled in the body frame although this is not the case in the laboratory frame as indicated by the non-Gaussian behavior of the probability distribution function [326]. A coupling in the body frame as demonstrated by H. Brenner for screwlike particles [322] has only been addressed in purely theoretical studies [327–330] without applications to specific experimental systems. The explicit calculation of the grand resistance matrix for a given complex particle shape remains difficult as analytical solutions are only available for some simple shapes such as spheres [316], ellipsoids [331, 332], and dimers consisting of two spherical colloids that touch each other [333]. Therefore, in more complicated situations the grand resistance matrix has to be calculated numerically, which can be achieved by bead models.

[7]In chapter 5, the grand resistance matrix is referred to as "hydrodynamic friction tensor."

The new experimental possibilities have lead to a significantly increased interest in the Brownian motion of anisotropic particles in the last few years. One particular example that has been studied in detail by Chakrabarty et al. are boomerang-shaped particles which are confined to a quasi-two-dimensional setup and tracked by video microscopy [334–336]. In reference [334], the importance of the choice of the tracking point used for the analysis of experimental trajectories is illustrated. It is shown that this does not only affect the measured mean square displacement but even leads to nonzero mean displacements if the tracking point does not coincide with the center of hydrodynamic stress.[8] In reference [335], the image-processing algorithm used for the high-precision tracking of the boomerang-shaped particles is discussed in more detail. The accuracies for particle position and orientation are determined. In addition to that, diffusion coefficients are also measured for particles with various apex angles.
While in references [334] and [335] particles with equal arm lengths are considered, in reference [336] the asymmetric case with arms of different lengths is studied. In particular this latter study is very relevant with regard to the circular motion of L-shaped self-propelled particles as discussed in chapter 7 of this thesis. Indeed, some aspects of the technique that we already used to determine the diffusion coefficients from experimental trajectories [218] are described in detail in reference [336].

While customary video microscopy is only practicable in quasi-two-dimensional setups [338], there are also a few techniques which can be used to obtain three-dimensional trajectories of colloidal particles. In chapter 5, a method based on real space confocal microscopy [339, 340] is applied. It enables the three-dimensional tracking of the particle motion including full orientational resolution [216, 341]. As a consequence of recent technological progress, the achievable imaging speed is high enough so that a sufficiently large number of stacks in z direction can be scanned in a given time (70 stacks in about $0.8\,\mathrm{s}$ for the discussed setup). Alternatively, digital holographic microscopy [342, 343] can be used to track the three-dimensional dynamics of colloidal particles [344–346]. In that technique, coherent light is scattered from the particle to be tracked and subsequently interferes with a reference beam. The resulting interference pattern is digitally recorded as a hologram. Subsequently, this hologram is analyzed in order to determine the three-dimensional particle positions and orientations. Holographic microscopy is usually faster than confocal microscopy, whereas one important advantage of the latter technique is the possibility of applying it also in crowded environments.

What makes the analysis of general biaxial particles so difficult is the complexity of the 6×6-dimensional grand resistance matrix $\mathcal{H}$, which relates the translational and angular velocities to the hydrodynamic drag force and torque. In spite of the symmetry of $\mathcal{H}$, for a general biaxial particle there are still 21 independent entries. This leads to a complexity which is completely different from systems with simple particle shapes such as spheres or ellipsoids. When spherical

[8] The center of hydrodynamic stress is defined as the reference point for which all coupling coefficients vanish [337]. It exists for all nonskewed particles, where it is also identical to the center of diffusion and the center of reaction.

particles are concerned, there are only two independent entries of the grand resistance matrix, one for the translational and one for the rotational motion. For ellipsoidal or rodlike particles the friction coefficients become orientation-dependent, but still there are only four different nonzero entries of $\mathcal{H}$.
In chapter 5, we explicitly analyze the Brownian motion of three specific particle shapes: a regular trimer, a regular tetramer, and an irregular trimer, produced in an oil-water emulsion based on an evaporation process [347]. While the two former examples are characterized by a diagonal grand resistance matrix, for the third particle also several nonzero off-diagonal entries indicating a coupling between translational and rotational Brownian motion exist. From the three-dimensional trajectories measured by confocal microscopy, all independent friction coefficients are extracted by means of appropriate short-time correlation functions. In addition to this experimental determination, the grand resistance matrix is calculated numerically based on the Stokes equation for the flow field around a particle. For this purpose, stick boundary conditions are assumed on the particle surface. More precisely, a bead model as implemented in the software HYDROSUB [348] is applied. The exact shape and size of the particles are determined experimentally by confocal and scanning electron microscopy. Idealized by fused spheres, the obtained particle shapes serve as an input for the software HYDROSUB and are represented by a large number of rigidly connected beads. Further details on the implementation of this low-Reynolds-number hydrodynamic calculation based on the various programs of the HYDRO group [349] are given in the appendix.

While the main intention of this work is to elucidate important features of the single-particle dynamics of microswimmers, in **CHAPTER 6** a multi-particle system consisting of a passive object in a bath of self-propelled particles is investigated. Although many details of the motion of single microswimmers are not fully understood yet, there are already numerous publications in which collective phenomena of self-propelled particles have been addressed [28, 113, 350–359]. Most of these studies are based on computer simulations and present purely theoretical results, but in some cases simulation data have also been compared to real experimental systems. Important examples of collective effects in suspensions of self-propelled particles are clustering [50, 360–367], swarming [368–371], jamming [372, 373], lane formation [374, 375], and active turbulence [51, 371].

In chapter 6, we study the motion of a mesoscopic wedgelike carrier embedded in a bath of swimming rodlike bacteria. Experiments on a *Bacillus subtilis* suspension are compared to Brownian dynamics simulations. To take particle-particle interactions into account, a Yukawa segment model is used in combination with the overdamped equations of motion. The rod-shaped particles represent an intermediate stage between spherical rigid self-propelled particles as considered in chapters 2 and 3 and the asymmetric L-shaped microswimmers studied in chapters 7–9. To characterize the dynamics of the system, the friction coefficients of the involved particles have to be determined. For the rodlike swimmers this can be achieved by using the formulas provided in reference [376]. With regard to the specific shape of the wedgelike carrier, there are no corresponding formulas. Therefore, the coefficients are calculated numerically by means of an appropriate bead model as described in detail in the appendix. While the calculations in chapter 5 are based on the software HYDROSUB [348], this time the software HYDRO++ [349, 377] is used because it is more convenient when specific particle shapes which do not consist of certain predefined subunits are considered.

As a main finding, our results demonstrate that bacteria can power directed transport of otherwise passive objects. This poses new opportunities for the design of micromotors which extract their energy from a suspension of swimming bacteria. A recent, somewhat similar approach has shown that a bacterial bath can also induce a unidirectional rotation of objects on the microscale [378]. In that study, the energy was provided by *Escherichia coli* bacteria that self-assemble along the rotor boundaries.
While, on the one hand, the specific shape of a microwedge can be used to obtain a directed motion out of a bacterial bath, on the other hand, a V-shaped object is also very useful in the context of capturing active particles [379, 380]. Moreover, V-shaped asymmetric barriers have already been used for rectification [381] or separation purposes [382, 383]. Instead of a nondeformable V-shaped carrier, in a subsequent study the motion of a flexible polymer in a bacterial bath has been investigated [384]. Furthermore, the work presented in chapter 6 has inspired a detailed study on the influence of different kinds of interactions between the swimmers [385]. In particular, the special cases of weak and strong alignment have been considered.

Most of the earlier theoretical work on microswimmers focuses on highly symmetric particles with either spherical or rodlike shapes, which significantly facilitates the description. In many situations this assumption is a useful approximation which can, at least to some extent, provide valuable insights into the dynamics of self-propelled particles. However, despite the successful utilization of such idealized shapes for various modeling tasks, one has to keep in mind that they are often still a rigorous simplification. On a more fine-grained level, they are an exception rather than the rule. Especially in biological systems one usually encounters particles with much more complex shapes. Therefore, it is necessary to include such asymmetry in the theoretical modeling as well. When specific aspects of the motion of swimming microorganisms are studied, oversimplified models based on spherical particle shapes might fail completely. As an example, we refer to the different swimming modes contained in the run-and-tumble strategy used by many flagella-driven microorganisms [38, 55, 253]. To some extent, the motion can be phenomenologically described if time-dependent modifications of the translational and angular velocities of a spherical particle are included, similar to the model discussed in chapter 3. But this kind of modeling is not able to reveal the physical origin of the observed changes in the motile behavior. For that purpose, it is necessary to explicitly include the flagella in the theoretical description. Then, one can also understand the role of polymorphic transformations [386, 387] during tumbling of flagellated bacteria [388–390].

After having studied both the motion of spherical self-propelled particles and the Brownian dynamics of biaxial passive colloidal particles, in **CHAPTER 7** we are in a position to develop a general model describing the motion of rigid asymmetric microswimmers. Our intention is to directly apply the theory to a corresponding experimental setup and to compare our theoretical findings with experimental data of self-propelled particles. In principle, such asymmetric particles can be produced in a similar way as described in chapter 5 by assembling spheres of equal or different size. If the single building blocks are self-propelled Janus particles, various types of artificial microswimmers can be obtained [110, 391, 392]. One possibility of designing particles which do not only perform an active translational motion but also rotational self-propulsion are doublets consisting of two spherical Janus particles that are rigidly connected to each other [110, 392]. Generally, such Janus doublets move on circular trajectories whose radius depends on the relative orientation of the two assembled particles. In a similar way, it is also possible to prepare larger assemblies consisting of more than two particles. By incorporating not only active Janus spheres but also passive components one can create a variety of artificial microswimmers with different self-propulsion properties [391]. Instead of using Janus particles as building blocks, another option is to utilize assemblies of symmetrically coated colloids if the single particles have different surface properties. The active behavior of such assembled colloidal molecules is largely determined by the surface activities and mobilities of the involved particles [393].

The experiments discussed in chapter 7 are not performed with self-propelled particles consisting of spherical subunits but with L-shaped particles whose production process is based on photolithography [394], as described in reference [218]. The measurements were carried out in the group of C. Bechinger in Stuttgart. The well-defined L-shape of the particles is advantageous because of the flat surfaces. These allow for a regular coating of specific parts of the particles with a thin Au layer, which is required for the propulsion. For the particles under study, such a several-nm-thick Au layer is created at the bottom side of the short arms by thermal evaporation. With regard to asymmetric particles containing spherical building blocks, it would be much more difficult to obtain particles with certain well-defined propulsive properties. In particular, the coupling between translational and rotational propulsion could not easily be tuned quantitatively.
When the Au-coated L-shaped particles are suspended in a binary mixture consisting of water and 2,6-lutidine at critical concentration, a propulsive motion based on self-diffusiophoresis can be triggered by illumination with light. This specific self-propulsion mechanism was first proposed by Volpe et al. in reference [115] and can be explained as follows. Before illumination, the temperature of the solvent is slightly below the lower critical point of the water-lutidine mixture [395, 396]. Under these conditions, only passive Brownian motion occurs. When

the illumination is switched on, the temperature of the solvent is locally increased near the coated region of a particle because of the specifically chosen wavelength which is optimized for the absorption by gold [116]. Thus, the water-lutidine mixture is locally demixed in the vicinity of the Au-coated region. As a consequence of different surface properties of the Au layer and the uncoated regions of the particle regarding the water and lutidine components, spatial concentration gradients are generated. These chemical gradients across the surface of the particle induce a self-propelled motion based on diffusiophoresis [115, 116].
As shown by the experiments discussed in chapter 7, this self-propulsion mechanism can be utilized for the active motion of asymmetric artificial microswimmers. However, it was first characterized in detail for spherical particle shapes. In reference [116], the influence of several parameters, namely the illumination intensity, the particle size, and the functionalization of the gold cap, were studied. Furthermore, the dynamic behavior in a spatial light gradient was addressed. One main advantage of the specific self-propulsion mechanism is that the swimming speed can easily be tuned by altering the illumination intensity [115]. Moreover, the chemical demixing is reversible so that the experimental conditions do not change significantly, even if measurements are performed for a very long time. This latter aspect is particularly important for the statistical analysis in chapter 7 because it guarantees high accuracy for the quantitative comparison between theoretical predictions and experimental data.
Besides the investigation of single self-propelled particles in patterned surroundings such as walls, pores, and obstacles arranged in a periodical setup [115], the same propulsion mechanism has also been applied to study the collective behavior of suspensions of spherical self-propelled colloidal particles [362]. For that purpose, the gold layer was replaced by a carbon coating, but the self-propulsion mechanism is equivalent. While at low densities dynamical clusters are formed, at higher densities a phase separation into a clustered phase and a dilute phase of free particles has been observed [362]. The reversibility of the demixing of the critical solvent is an important requirement for experiments with such high concentrations of self-propelled particles.
In this thesis, we analyze experimental data for L-shaped microswimmers based on this type of self-propulsion in order to understand the effect of shape asymmetry on the particle motion. Moreover, we use this experimental system for the investigation of the gravitactic behavior of asymmetric self-propelled particles. This interesting phenomenon is studied in chapter 9.

The most striking difference between ideal spherical or rodlike self-propelled particles and microswimmers with an asymmetry around the propulsion axis is the fact that the former generally move on straight lines, while the latter move on characteristic circular trajectories. Spiral swimming paths have been known as an important feature of swimming biological microorganisms since more than a century [397]. As an example, we refer to the motion of sperm cells, which usually swim in circles or along helical paths in the absence of chemical gradients [134, 398]. Often the presence of a chemoattractant is required in order to achieve a deterministic drift motion. Circular motion is also observed for bacteria such as *Escherichia coli*, in particular near

surfaces or interfaces [36, 399]. While the sense of rotation is usually clockwise near planar solid surfaces [400], it is counterclockwise at liquid-air interfaces [399, 401].
For the L-shaped microswimmers studied in chapter 7, the circular motion is due to a velocity-dependent torque which arises as a consequence of the asymmetric particle shape in combination with the viscous forces mediated by the surrounding solvent. In previous theoretical studies, a related circular motion has been obtained for symmetric particles if an additional constant torque is considered [269, 402–404]. Interest has also been directed at the collective behavior of circle swimmers, which has been addressed in computer simulations [405, 406]. Furthermore, purely theoretical results are available for biaxial particles with prescribed constant forces and torques in three dimensions, where helical trajectories have been found [194].
In chapter 7 of this thesis, we present a theoretical concept which accurately describes and explains experimental findings on asymmetric L-shaped self-propelled particles. Here, an effective torque intrinsically evolves from the specific shape of the particle. Therefore, the effective propulsion forces and torques are not independent anymore as usually assumed in previous theoretical studies. We characterize the circular motion of the L-shaped microswimmers in detail by analyzing experimental data. Particular attention is paid to the relation between the angular and translational velocities, to the dependence of the radius of the circular trajectories on the swimming speed, and to the probability distribution of the angle between the particle orientation and the displacement vectors. These quantities are studied theoretically based on a set of coupled Langevin equations for the overdamped motion of an L-shaped self-propelled particle. The equations are in principle obtained as a generalization of equations (1.3) and explicitly read

$$\begin{aligned} \dot{\mathbf{r}} &= \beta F\big(\boldsymbol{\mathcal{D}}_{\mathrm{T}}\hat{\mathbf{u}}_{\perp} + l\mathbf{D}_{\mathrm{C}}\big) + \boldsymbol{\zeta}_{\mathbf{r}}\,, \\ \dot{\phi} &= \beta F\big(lD_{\mathrm{R}} + \mathbf{D}_{\mathrm{C}}\!\cdot\!\hat{\mathbf{u}}_{\perp}\big) + \zeta_{\phi}\,. \end{aligned} \tag{1.4}$$

Here, $\hat{\mathbf{u}}_{\perp}$ gives the direction of the effective self-propulsion force, which coincides with the orientation of the long arm of the L-shaped particle. To include the active rotational motion, which is characteristic for asymmetric microswimmers, the effective lever arm l enters the equations. It relates the propulsion force F to the effective intrinsic torque $M = lF$ (see chapter 7 for a more detailed explanation). Furthermore, the Langevin equations depend on the generalized diffusion tensor for the specific particle shape. For the two-dimensional motion studied here, the rotational part is characterized by a single rotational diffusion constant D_{R}. In contrast, for the translational part the 2×2-dimensional short-time diffusion tensor $\boldsymbol{\mathcal{D}}_{\mathrm{T}}(\phi) = D_{\parallel}\hat{\mathbf{u}}_{\parallel}\otimes\hat{\mathbf{u}}_{\parallel} + D_{\parallel}^{\perp}(\hat{\mathbf{u}}_{\parallel}\otimes\hat{\mathbf{u}}_{\perp} + \hat{\mathbf{u}}_{\perp}\otimes\hat{\mathbf{u}}_{\parallel}) + D_{\perp}\hat{\mathbf{u}}_{\perp}\otimes\hat{\mathbf{u}}_{\perp}$ with the outer product $\otimes$ is required. As opposed to highly symmetric particle shapes, in the general asymmetric case additional terms in the equations of motion result from the nonzero coupling coefficients $D_{\mathrm{C}}^{\parallel}$ and $D_{\mathrm{C}}^{\perp}$, which are contained in the translation-rotation coupling vector $\mathbf{D}_{\mathrm{C}}(\phi) = D_{\mathrm{C}}^{\parallel}\hat{\mathbf{u}}_{\parallel} + D_{\mathrm{C}}^{\perp}\hat{\mathbf{u}}_{\perp}$. With regard to the Gaussian noise terms, the main difference as compared to the case of spherical particle shapes in equations (1.3) is the coupling between translational and rotational noise, indicated by the nonzero correlation function $\langle \boldsymbol{\zeta}_{\mathbf{r}}(t_1)\zeta_{\phi}(t_2)\rangle = 2\mathbf{D}_{\mathrm{C}}\delta(t_1 - t_2)$. Moreover, instead of the dependence on only one diffusion coefficient D_{T}, now the translational correlation functions are determined by the translational short-time diffusion tensor $\boldsymbol{\mathcal{D}}_{\mathrm{T}}$, which contains the diffusion coefficients $D_{\parallel}$, $D_{\parallel}^{\perp}$, and $D_{\perp}$. We note that equations (1.4) can alternatively be written by using the resistance matrix formalism [19,407]. However, in the context of this thesis the present form is more appropriate as the diffusion coefficients can be directly determined from the experimental data.

Our theoretical model provides analytical expressions for various quantities characterizing the circular motion of the L-shaped particles. In particular, we show that the mean radius of the circular trajectories does not depend on the propulsion strength but only on the various diffusion and coupling coefficients and the effective lever arm. Thus, from the specific shape of a particle the noise-free swimming path can be predicted. Furthermore, the theory yields a linear relationship between the translational and angular velocities of the particle. Both this proportionality and the velocity-independent radius of curvature are verified in the experiments. In addition to the characterization of the noise-free swimming paths, the model gives a theoretical prediction for the noise-averaged trajectory. It is shown to be a logarithmic spiral, also known as *spira mirabilis* [269]. This specific curve is also obtained from the experimental data by dividing the measured trajectories into segments of equal length and subsequent averaging. All of the theoretical formulas depend on some of the diffusion and coupling coefficients of the considered particle. Hence, for the comparison between theory and experiment it is important to determine these parameters as precisely as possible. This is done both theoretically and experimentally in a similar way as for the anisotropic colloidal particles in chapter 5. The accuracy of the experimental determination based on short-time correlation functions is even much better here because the motion of the particles is restricted to two dimensions. In some respects, our technique is very similar to that later studied in detail by Chakrabarty et al. [335,336]. The theoretical method basically corresponds to the one used in chapters 5 and 6. However, the difference is that the substrate is explicitly taken into account this time. This is achieved by calculating the hydrodynamic interactions between the single beads representing the L-shaped particle based on the Stokeslet in the vicinity of a no-slip boundary [408]. The experimental and theoretical results for the diffusion coefficients are in very good agreement. We use the experimentally obtained numerical values in order to also quantitatively verify the theoretical prediction for the noise-averaged trajectory.

In addition to a freely moving L-shaped self-propelled particle, we briefly address the types of motion which occur when such a particle interacts with a solid wall. In the experiments, two different types of interaction are observed. The particle is either reflected by the wall, or it performs a stable sliding motion. Although our theoretical model does not include hydrodynamic particle-wall interactions [407,409], it is not only in qualitative agreement with the experimental observations but also provides a quantitative criterion which determines whether a particle approaching the wall at a certain angle enters the sliding or the reflection regime.

In most parts of this thesis, we use *effective* forces and torques to describe the active motion of self-propelled particles. However, it is important to keep in mind that the motion of a microswimmer is force-free and torque-free as long as no external forces or torques are present. In the second part of chapter 7, we already briefly explain how the effective forces and torques have to be interpreted and provide a physical justification of the equations of motion. This topic is covered in much more detail in the subsequent **CHAPTER 8**.

Although the concept of effective forces and torques is widely used in the context of microswimmer motion, there are still scientists who question whether this kind of modeling is appropriate [410]. This confusion probably arises because of the fact that a swimmer is force-free and torque-free [19]. Therefore, some people argue that the term "force" cannot be used at all. While it is true that force monopoles are not appropriate to model the solvent flow field around a microswimmer [411,412], effective forces are still a very useful concept to describe the resulting dynamics and to explain related experimental observations. In chapter 8, the applicability of this concept is studied in detail. The equations of motion are derived on a fundamental level by means of explicit hydrodynamic calculations.

We follow the nomenclature commonly used in literature and do not distinguish between microswimmers and self-propelled particles. Both terms refer to micron-sized objects that are able to perform an autonomous directed motion. However, in some papers [410,413] the following distinction is made. Rigid autonomously moving objects are referred to as "self-propelled particles" and the term "microswimmer" is only used if the active motion is based on shape changes. The latter can be realized by mutually moving parts such as in the three-sphere swimmer [414,415] or by surface distortions [277], for example. We do not use this special nomenclature but consider microswimmers as one class of self-propelled particles. However, they have to be distinguished from passive particles which can only move deterministically in externally applied fields or when they are carried by fluid currents. Examples of related numerical studies on Brownian particles driven across various arrangements of obstacles can be found in references [416–418]. Such particles that are exposed to an external body force typically induce a long-range flow pattern in the surrounding solvent. When r is the distance from the particle, the solvent velocity decays as $1/r$ in the far field. The corresponding modeling is based on force monopoles. This is different when self-propelled particles are concerned. As these are force-free, the far-field behavior is characterized by the force dipole contribution, which decays as $1/r^2$. The force dipole approach is well established for models of self-propelled particles which explicitly include the solvent [419–424].

One possibility of defining a self-propulsion force for microswimmers which create a force dipole in the surrounding solvent is to take the ratio of the dipole strength and a characteristic length scale of the particle [409]. Although this explicit interpretation as thrust force is not necessary for the effective propulsion forces in our theoretical model, the magnitudes agree well with experimental data of flagella thrust forces for swimming bacteria [218, 425, 426]. Such measurements can be performed, e.g., by using optical tweezers [427, 428]. Thus, one can determine the force that is needed in order to restrain the microswimmer from moving. This force is also referred to as "stall force" [54, 429, 430].

However, the main intention of our theoretical model is to describe and elucidate the interesting motional patterns of self-propelled particles. For that purpose, the effective quantities can be viewed as formally external forces and torques which are required to guide a corresponding passive particle with the same shape along exactly the same trajectory as the self-propelled particle. In chapter 8, we show that such a mapping of the equations of motion is possible. In order to do so, we explicitly consider two different models for mechanical and diffusiophoretic self-propulsion.

Mechanical self-propulsion at low Reynolds number is usually based on some kind of non-reciprocal periodic motion [19, 431]. A ubiquitous realization in nature is flagellar locomotion [407, 432], which is responsible for the propulsion of most bacteria, algae, and spermatozoa. The details of this propulsion strategy have also been addressed based on artificial setups [124, 125, 433]. Concerning artificial microswimmers, usually phoretic propulsion mechanisms are implemented (see section 1.3). These are typically modeled by prescribing a slip velocity on the particle surface [78, 79]. In chapter 8, we explicitly consider self-diffusiophoretic microswimmers, but the reasoning is similar for other kinds of self-phoretic propulsion.

The concept of effective forces and torques is particularly useful when both an active swimming motion and external fields are present [220, 434]. Any additional body force can directly be included in the equations of motion (1.4) without changing the structure. Only when the flow and pressure fields in the surrounding fluid become relevant and govern the dynamics, the theory has to be adapted. For this purpose, experiments with passive tracer particles [435–439] are helpful in order to identify the solvent flow patterns in a given situation. This will probably provide additional information about how the theoretical modeling has to be modified for particles in the vicinity of system boundaries or in suspensions at high density.

In **CHAPTER 9**, we study the dynamics of L-shaped self-propelled particles under gravity. This is one example of an external field affecting the motion of microswimmers. While under laboratory conditions it is possible to study the influence of one single parameter, biological microorganisms in nature usually respond to their environmental conditions in many different ways. This so-called tactic behavior has been described in detail in section 1.4. In chemotaxis, phototaxis, and thermotaxis, microorganisms actively change their behavior in response to the stimulus. In addition to these active types of taxes, several passive mechanisms based on physical phenomena exist. Important examples are magnetotaxis, rheotaxis, and gyrotaxis. All of these passive tactic mechanisms can also be studied in systems of artificial self-propelled particles.

Moreover, additional effects such as electrotaxis[9] [442] and tribotaxis [443], which are generally less relevant to the motion of biological microswimmers, can be addressed based on artificial setups. In tribotaxis, the motion is determined by spatially varying friction coefficients, which has been utilized to design an artificial microscopic walker consisting of two magnetically assembled colloids [443]. Nonconstant friction coefficients are especially relevant in situations far from equilibrium [444].

Most types of tactic behavior of biological microswimmers can be traced back to either physiological or passive physical mechanisms. However, as already discussed in section 1.4, this is not the case for gravitaxis. Rather, its origin is still highly controversially discussed. While some scientists strongly support the hypothesis of the existence of a physiological gravity sensor [189, 200], others argue that all observations can be explained by purely physical effects [190, 207, 210]. According to the current state of knowledge, a physiological mechanism is likely to contribute to the gravitactic behavior of some microorganisms, in particular *Euglena gracilis* [159, 204]. However, no conclusive evidence exists that such a gravity receptor is also available in other species such as *Chlamydomonas* [211, 212] and *Paramecium* [190].

Two different passive alignment mechanisms have been proposed to explain the orientation of microorganisms under gravity. The first explanation is based on an inhomogeneous mass distribution. If one part of a microorganism is heavier than the remaining body, the heavy side will point downward in the gravitational field. This effect causes the gravitational torque in gyrotaxis, which has been addressed in detailed theoretical studies. Using the Fokker-Planck equation as a starting point, a continuum model has been formulated [445], and in a more fine-grained approach resistive force theory has been used to include the contribution of the flagella [446]. Sometimes, bottom heaviness is also referred to as "gravity-buoyancy model" [210].
Alternatively, mechanical alignment may also occur due to an asymmetric body shape. In that case, viscous torques orient the particle. Both mechanisms are visualized in figure 1.1 based on a spherical particle with an inhomogeneous mass distribution and a T-shaped object with constant density. When these particles are submersed in a hypo- or hyperdensity medium, qualitatively different behavior is found under gravity. Independent of the difference between the averaged particle density $\bar{\rho}_{\mathrm{p}}$ and the density ρ_{s} of the surrounding solvent, the heavy part of the spherical particle is at the bottom. In a hypodensity medium with $\rho_{\mathrm{s}} < \bar{\rho}_{\mathrm{p}}$ the particle sediments downward [see figure 1.1(a)], while the direction of motion is upward in a hyperdensity solvent with $\rho_{\mathrm{s}} > \bar{\rho}_{\mathrm{p}}$ [see figure 1.1(b)]. Figure 1.1(c) shows the situation of a T-shaped particle with constant mass density ρ_{p} in a hypodensity fluid. In that case, the particle sediments downward with an upside-down orientation. Interestingly, in a hyperdensity medium the T-shaped particle does not only reverse its direction of motion but also its orientation [see figure 1.1(d)]. This can be explained by viscous torques which arise as a consequence of the

[9] Electrotaxis describes the response of motile cells to electric fields. It is also called galvanotaxis [440, 441].

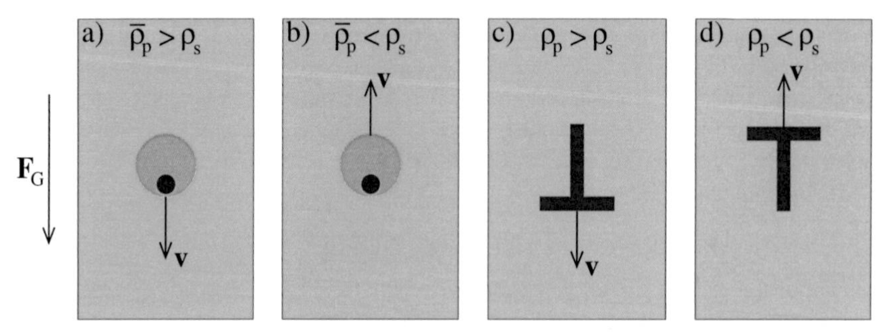

Figure 1.1: *Visualization of the passive sedimentation behavior of a spherical particle with an inhomogeneous mass distribution and an asymmetric T-shaped particle. The dark region inside the spherical particle indicates a higher mass density. When exposed to a gravitational force* $\mathbf{F}_G$*, the particle always orients itself with the heavy side pointing downward. In a hypodensity solvent with density* ρ_s *smaller than the mean particle density* $\bar{\rho}_p$*, the particle moves downward (a), while it moves upward in a hyperdensity medium with* $\rho_s > \bar{\rho}_p$ *(b). However, the particle orientation in both situations is the same. This is different for a T-shaped particle with a homogeneous mass distribution, where the orientation is upside-down in a hypodensity medium (c) and upright in a hyperdensity solvent (d).*

deviation between the center of mass and the center of hydrodynamic stress. The described behavior is characteristic of asymmetric particles. It has also been observed in experiments with *Paramecium* and gastrula larvae of sea urchins [210], which indicates the importance of shape asymmetry for gravitaxis of these microorganisms.

The role of a fore-rear asymmetry can also be illustrated by comparing the sedimentation behavior of rodlike and T-shaped particles. Due to the symmetry, for a rod the center of mass and the center of hydrodynamic stress coincide. Therefore, no torque is caused by gravity, and there is no preferred sedimentation angle. The rod may in principle move downward with any given orientation. The only difference between passive rods aligned parallel to the direction of gravity and others oriented perpendicular is that in the former case the particles sediment faster as a consequence of the different friction coefficients along the symmetry axes of the rod.

An unequivocal evidence of the role of shape asymmetry for gravitactic behavior can hardly be given based on experiments with microorganisms. By analyzing active swimming paths, it is possible to identify features which support one or the other explanation [189, 190, 202]. However, neither a physiological gravity sensor nor a dominant physical mechanism can be definitely ruled out. When immobilized cells are considered [210], it is not evident to what extent the immobilizing treatment itself alters the shape and the mass distribution [207]. This is clearly illustrated by the contradictory conclusions drawn in references [209] and [210] regarding the alignment mechanism in *Paramecium*. For these reasons, in chapter 9 we study the influence of gravity on the motion of artificial asymmetric microswimmers with a well-defined shape and homogeneous mass distribution. Thus, both a gravity sensor and bottom heaviness can be excluded. The experimental and theoretical results show that shape anisotropy alone can cause a swimming motion opposed to gravity.

As we consider L-shaped particles, which do not only have a fore-rear asymmetry but also an asymmetry around their propulsion axis, a versatile dynamical behavior under gravity is observed. The motion is governed by the interplay between the rotation resulting from the propulsion and the gravitational torque. Most importantly, this leads to a transition from a regime with straight trajectories to characteristic circling types of motion. As obtained from our theory, this transition does not occur for particles which only have a fore-rear asymmetry such as T-shaped particles. In that case, for vanishing noise only straight downward or upward trajectories are possible.

Our results suggest that for many particle shapes gravitactic behavior can be much more sophisticated than just swimming in a direction opposed to gravity.

CHAPTER 2

Non-Gaussian statistics for the motion of self-propelled Janus particles: experiment versus theory

The content of this chapter has been published in a similar form in *Physical Review E* **88**, 032304 (2013) by Xu Zheng, Borge ten Hagen, Andreas Kaiser, Meiling Wu, Haihang Cui, Zhanhua Silber-Li, and Hartmut Löwen (see reference [213]).

Non-Gaussian statistics for the motion of self-propelled Janus particles: Experiment versus theory

Xu Zheng,[1] Borge ten Hagen,[2,*] Andreas Kaiser,[2] Meiling Wu,[3]
Haihang Cui,[3] Zhanhua Silber-Li,[1,†] and Hartmut Löwen[2]

[1]*State Key Laboratory of Nonlinear Mechanics, Institute of Mechanics, CAS, Beijing 100190, People's Republic of China*

[2]*Institut für Theoretische Physik II: Weiche Materie, Heinrich-Heine-Universität Düsseldorf, D-40225 Düsseldorf, Germany*

[3]*Xi'an University of Architecture and Technology, Xi'an, 710055, People's Republic of China*

Spherical Janus particles are one of the most prominent examples for active Brownian objects. Here, we study the diffusiophoretic motion of such microswimmers in experiment and in theory. Three stages are found: simple Brownian motion at short times, superdiffusion at intermediate times, and finally diffusive behavior again at long times. These three regimes observed in the experiments are compared with a theoretical model for the Langevin dynamics of self-propelled particles with coupled translational and rotational motion. Besides the mean square displacement also higher displacement moments are addressed. In particular, theoretical predictions regarding the non-Gaussian behavior of self-propelled particles are verified in the experiments. Furthermore, the full displacement probability distribution is analyzed, where in agreement with Brownian dynamics simulations either an extremely broadened peak or a pronounced double-peak structure is found depending on the experimental conditions.

PACS numbers: 82.70.Dd, 05.40.Jc

* bhagen@thphy.uni-duesseldorf.de

† lili@imech.ac.cn

I. INTRODUCTION

Recently, the single and collective properties of self-propelled particles have been studied intensely [1–3]. Examples are found in quite different areas of physics and involve bacteria [4–10], spermatozoa [11–15], and even fish, birds, and mammals including humans [16–19]. Furthermore, various types of micron-sized man-made active particles have been developed [20–27]. One of the by now most popular artificial realizations of colloidal microswimmers is mesoscopic Janus particles which are put into motion by a chemical reaction in the surrounding solvent [28, 29]. In detail, this reaction is catalyzed at one surface of the Janus particle such that an asymmetric gradient field arises, which self-propels the particle by diffusiophoresis [30–33]. Several features of the resulting swimming behavior such as the direction of motion [34, 35], the dependence of the propulsion velocity on the particle size [36], and the swimming efficiency [37] have been investigated recently. Focus has also been directed at the flow pattern in the vicinity of a heated Janus particle [38], clustering in suspensions of self-propelled colloids [39–42], and controlling the locomotion of single Janus micromotors [43] by an external magnetic field [44, 45]. Experiments with self-propelled spherical Janus particles in periodical arrangements of obstacles [46] have inspired theoretical studies on possible applications for the sorting of chiral active particles [47] or separation purposes in binary mixtures of passive colloids [48]. Very recent simulations of microswimmers moving in a ratchet channel also suggest their applicability for pumping processes [49].

In general, the orientation of a Janus particle is fluctuating and therefore the particle performs a persistent random walk [50]. The mean square displacement hence crosses over from a ballistic regime, where the particle on average is self-propelled along its orientation, to a long-time diffusive behavior. The transition between these two regimes basically occurs at a time scale corresponding to the inverse rotational diffusion constant, i.e., when the particle has lost the memory of its initial orientation. However, while the mean square displacement is the standard quantity to characterize modes of propagation in self-propelled systems, there are only a few studies for the non-Gaussian behavior as revealed in the higher moments and, in particular, in the excess kurtosis. Pure theoretical calculations [51, 52] have addressed higher moments, but an analysis has never been performed based on experimental data for microswimmers. Non-Gaussianity is an important feature also in other disciplines of statistical physics including, e.g., the glass transition [53–56] and the analysis of rare events (such as earthquakes and stock crashes) [57]. Therefore, it is relevant from a fundamental point of view to get insight into the non-Gaussian statistics for microswimmer motion.

Here, we analyze higher moments characterizing non-Gaussianity in experimental trajectories of self-motile Janus particles and compare them to the theoretical predictions of a model based on the Langevin equations for the coupled translational and rotational motion of active Brownian particles. Moreover, we elucidate the interplay between the random and deterministic components of the particle displacements at very short times. We show that the crossover from diffusive short-time motion to superdiffusive motion at intermediate times [52] can also be verified experimentally, which supports the theoretical description of microswimmers by active Brownian models. Additional insights regarding the non-Gaussianity are obtained by analyzing the time evolution of the full probability distribution of particle displacements. Here, the experimental data show that the initial Gaussian curve transforms into a shape with a significantly broadened peak if Janus particles in solutions with low hydrogen peroxide (H_2O_2) concentrations are considered. In contrast, for high H_2O_2 concentration a pronounced double-peak structure is found. These fundamentally different features result from a restriction of the rotational Brownian motion in the case of strongly driven Janus particles. Our observations are confirmed by Brownian dynamics simulations, where the particle orientation is either freely diffusing on a unit sphere or restrained to a two-dimensional plane.

This paper is organized as follows: Section II introduces the methods used in experiment, theory, and simulation. The experimental observations are presented in Sec. III, where also a detailed discussion and interpretation in the context of the theoretical model is given. Finally, we conclude in Sec. IV.

II. METHODS

A. Experiment

In our experiments, we study the motion of spherical Pt-silica Janus particles. The fabrication method is similar to that illustrated in Ref. [31]. By electron beam evaporation, a layer of Pt (thickness about 7 nm) is deposited on the surface of one hemisphere of the particles (see Sec. A 1 in the Appendix for further details). After that the Janus particles are resuspended in distilled water (18.2 MΩ cm). Most of the experiments are performed with Janus spheres with diameter $d_1 = 2.08 \pm 0.05$ μm (measured by scanning electron microscopy). Whenever additional results for smaller particles with diameter $d_2 = 0.96 \pm 0.03$ μm are included for comparison, this is appropriately indicated.

The particle trajectories in water and in H_2O_2 solutions with different concentrations (1.25%–15%) are observed by video microscopy with an image field of view of 512×512 pixels (approximately 80×80 μm). To be able to observe also the particle dynamics at very short times, the time interval Δt between two images was reduced to 10 ms. After the preparation of the solutions, a 70 μl droplet with specified H_2O_2 concentration was put on a cover slip. Image series consisting of 600–1000 frames were captured in one position located about 2–5 μm above the glass substrate.

In the same droplet, five movies were taken in five different locations in the same horizontal plane. The measurements for each H_2O_2 concentration were repeated in 12–15 droplets independently. In order to have a good measurement reproducibility and to limit the influence of temperature and concentration fluctuations induced by the chemical reaction, a fresh test solution for each droplet was reprepared. The experiments were performed in the stationary regime from 1 to 9 min after the beginning of the catalytic reaction in the H_2O_2 solution. In this period the fuel concentration does not change significantly since the used particle density is very low. The displacements of the Janus particles were measured by trajectory tracking from the movies. To reach the requirements of the statistical analysis, for each concentration more than 1000 particles were considered.

In the images the Janus particles appear half bright (the silica side) and half dark (Pt coating side). In order to determine the exact center of each Janus particle, a two-step method using "find edge" and "Gaussian blur" was performed (see Sec. A 2 in the Appendix for details). Thus, the center of the Janus particles could be determined with a ± 0.5 pixel accuracy. After this preprocessing, the trajectories of individual particles can be tracked from the video material.

The dynamics of the particles in our system is strongly influenced by their translational and rotational Brownian motion. Thus, before investigating the self-propulsion on top of it, the diffusion coefficients D_t for translation and D_r for rotation have to be addressed. The translational diffusion coefficient is in principle given by the Stokes-Einstein equation

$$D_\mathrm{t} = \frac{k_\mathrm{B}T}{3\pi\eta d}, \tag{1}$$

where $k_\mathrm{B}T$ is the thermal energy and η is the viscosity of the solvent. Alternatively, D_t can also be directly determined from the two-dimensional mean square displacement $\langle(\Delta\mathbf{r})^2\rangle$ of passive Brownian particles via $D_\mathrm{t} = \langle(\Delta\mathbf{r})^2\rangle/(4\Delta t)$. Following this standard method the experimental data yield $D_\mathrm{t} = 0.175\ \mu\mathrm{m}^2\,\mathrm{s}^{-1}$ for the particles with diameter $d_1 = 2.08\ \mu\mathrm{m}$ [theoretical prediction based on Eq. (1): $D_\mathrm{t} = 0.211\ \mu\mathrm{m}^2\,\mathrm{s}^{-1}$]. In the case of the smaller particles ($d_2 = 0.96\ \mu\mathrm{m}$) the measurements give $D_\mathrm{t} = 0.416\ \mu\mathrm{m}^2\,\mathrm{s}^{-1}$ as compared to the theoretical value $D_\mathrm{t} = 0.456\ \mu\mathrm{m}^2\,\mathrm{s}^{-1}$. The small deviations between the measured and the predicted values are clearly due to hydrodynamic interactions with the glass substrate [58, 59], which slightly reduce the mobility of the particles. We estimate the rotational diffusion coefficient from the relation $D_\mathrm{r} = 3D_\mathrm{t}/d^2$, which directly follows from Eq. (1) and its analogon

$$D_\mathrm{r} = \frac{k_\mathrm{B}T}{\pi\eta d^3} \tag{2}$$

for rotational diffusion [60]. Using the experimentally determined values for D_t, one obtains $D_\mathrm{r} = 0.121\ \mathrm{s}^{-1}$ for the larger particles and $D_\mathrm{r} = 1.35\ \mathrm{s}^{-1}$ for the smaller ones.

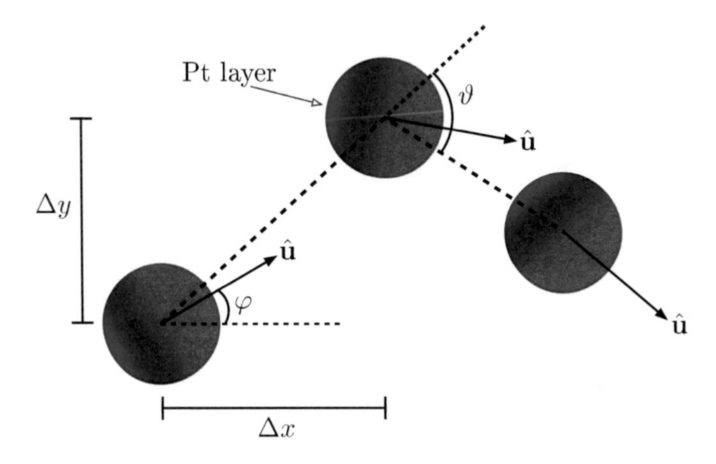

Figure 1. Schematic of the particle motion for two subsequent time steps and definition of several parameters used for its characterization. The translational motion is determined by the displacements Δx and Δy of the center-of-mass postion of the particle. The orientation vector $\hat{\mathbf{u}} = (\sin\theta\cos\varphi, \sin\theta\sin\varphi, \cos\theta)$ coincides with the direction of self-propulsion. Note that $\theta = 90°$ in the figure for the sake of clarity. While θ and φ define the particle orientation, ϑ is the angle between the directions of subsequent displacements. Due to the combination of Brownian motion and self-propulsion, $\hat{\mathbf{u}}$ is not necessarily collinear with the displacement direction.

B. Theory

In order to describe the dynamics of the Janus particles in our experiments, we use a theoretical model similar to that studied in detail in Ref. [52]. This general model for self-propelled Brownian particles is altered in a way such that it suits our experimental setup. Primarily, this means that the theoretical description is transferred from a one-particle situation to a dilute system with many, but not interacting particles as realized in our experiments. As the particles have different initial orientations that cannot easily be measured with sufficient accuracy, we always take an average and use corresponding theoretical results.

Starting with the Langevin equations for the overdamped motion of a Brownian particle, we include an effective driving force $\mathbf{F} = F\hat{\mathbf{u}}$, which accounts for the detailed self-propulsion mechanism of the active Janus particle on average and does not contradict the fact that the motion of a swimmer is force-free. $\mathbf{F}$ is parallel to a particle-fixed orientation vector $\hat{\mathbf{u}}$ that is defined by the position of the Pt layer (see Fig. 1).

The translational motion of the Janus spheres studied here is performed in two dimensions as gravity, in combination with electrostatic repulsion, keeps the particles close to the substrate, where the focal plane of the microscope is located. However, in principle the particles can rotate freely. This implies that the translational motion of one Janus particle is described by the two-dimensional projection of the Langevin equation

$$\frac{d\mathbf{r}}{dt} = \beta D_{\mathrm{t}} F \hat{\mathbf{u}} + \sqrt{2D_{\mathrm{t}}}\boldsymbol{\xi}_{\mathbf{r}} \tag{3}$$

for the center-of-mass position $\mathbf{r}(t) = [x(t), y(t)]$, where $\beta = 1/(k_{\mathrm{B}}T)$ is the inverse effective thermal energy. As the direction of the self-propulsion depends on the particle orientation $\hat{\mathbf{u}}$, Eq. (3) is coupled to the rotational Langevin equation

$$\frac{d\hat{\mathbf{u}}}{dt} = \sqrt{2D_{\mathrm{r}}}\boldsymbol{\xi}_{\hat{\mathbf{u}}} \times \hat{\mathbf{u}}\,. \tag{4}$$

The translational and rotational random motion due to the kicks of the solvent molecules is included by the Gaussian noise terms $\boldsymbol{\xi}_{\mathbf{r}}$ and $\boldsymbol{\xi}_{\hat{\mathbf{u}}}$ with zero mean and variances $\langle \boldsymbol{\xi}_{\mathbf{r}}(t_1) \otimes \boldsymbol{\xi}_{\mathbf{r}}(t_2)\rangle = \langle \boldsymbol{\xi}_{\hat{\mathbf{u}}}(t_1) \otimes \boldsymbol{\xi}_{\hat{\mathbf{u}}}(t_2)\rangle = \delta(t_1 - t_2)\mathbb{1}$, where $\mathbb{1}$ is the unit tensor. The corresponding orientational probability distribution for the freely diffusing orientation vector [60] is given by

$$P(\theta, \varphi, t) = \sum_{l=0}^{\infty} \sum_{m=-l}^{l} e^{-D_{\mathrm{r}} l(l+1)t}\, Y_l^{m*}(\theta_0, \varphi_0) Y_l^m(\theta, \varphi)\,, \tag{5}$$

where Y_l^m and Y_l^{m*} are the spherical harmonics and their complex conjugates. The spherical coordinates θ and φ define the particle orientation $\hat{\mathbf{u}} = (\sin\theta\cos\varphi, \sin\theta\sin\varphi, \cos\theta)$. Initial values at $t = 0$ are indicated by the index 0. For freely diffusing Janus particles with arbitrary initial orientation the analytical expressions for the different moments of the displacement probability distribution are given in Secs. III A and III B. These results are in good agreement with the experimental data for up to 5% H_2O_2 concentration. As discussed in detail in Sec. III E, our observations strongly suggest that for higher H_2O_2 concentration of the solvent the particle orientation is not homogeneously distributed on a unit sphere, but is to some extent restricted to the two-dimensional plane of translational motion. This requires an appropriate adaption of the theoretical model.

C. Simulation

While our model provides analytical expressions for the displacement moments, a corresponding Brownian dynamics simulation based on the same Langevin equations (3) and (4) allows us to also study the full distribution. Numerical results are obtained for 10^6 particle trajectories with arbitrary initial conditions and length 100 t_{r}, where $t_{\mathrm{r}} = 1/D_{\mathrm{r}}$ is the rotational diffusion time. The translational and rotational noise terms $\boldsymbol{\xi}_{\mathbf{r}}$ and $\boldsymbol{\xi}_{\hat{\mathbf{u}}}$ are implemented by independent Gaussian random numbers with zero mean and unit variance for each component. Simulation results are provided for the probability distributions for both the magnitude and the direction of displacements. The function $\Psi(\Delta x, t)$ gives the probability to find a particle at a certain distance Δx from its initial position after a specified time t (see schematic illustration in Fig. 1). The time evolution of $\Psi(\Delta x, t)$ is discussed in detail in Sec. III C. To elucidate the interplay between the random and the deterministic components of the particle motion, we also address the probability distribution $\Psi(\vartheta, t)$ of the angle ϑ between the directions of subsequent particle displacements (cf. Fig. 1) both in experiment and simulation (see Sec. III D).

III. RESULTS

A. Mean square displacement

To characterize the dynamics of the Janus particles, we first discuss the mean square displacement (MSD) $\langle(\Delta\mathbf{r})^2\rangle_{\hat{\mathbf{u}}_0}$. Here, $\Delta\mathbf{r} = \mathbf{r}(t) - \mathbf{r}_0$ is the two-dimensional translational displacement and the notation $\langle\cdots\rangle_{\hat{\mathbf{u}}_0}$ denotes a noise average with an additional averaging over the initial orientation $\hat{\mathbf{u}}_0$ of the particles. Figure 2 shows the experimental results for the MSD in a double logarithmic plot. We use dimensionless quantities $\langle(\Delta\mathbf{r})^2\rangle_{\hat{\mathbf{u}}_0}/d^2$ and $\tau = D_\mathrm{r}t$ as this is convenient for the discussion of the measurements in the context of our theoretical model. While the main figure and the left inset of Fig. 2 are based on measurements for particles with diameter $d_1 = 2.08$ µm, the right inset visualizes corresponding data for smaller particles ($d_2 = 0.96$ µm). Due to the different rotational diffusion coefficients ($D_\mathrm{r} = 0.121\ \mathrm{s}^{-1}$ for d_1 and $D_\mathrm{r} = 1.35\ \mathrm{s}^{-1}$ for d_2), the larger particles are more appropriate to also study the behavior at small values of τ. Therefore, we focus on these particles for our detailed statistical analysis. In the experiments, images were usually recorded with a frame rate of 10 frames per second (fps). Corresponding results for the MSD in water and in H_2O_2 solutions with different concentrations ranging from 1.25% to 5% are visualized by the solid lines in Fig. 2. However, to be able to resolve the very early time regime, additional measurements with a frame rate of 100 fps are included as well (see dashed lines in Fig. 2). In water the Janus particles undergo simple Brownian motion resulting in a linear time dependence of the MSD (see lowermost curve in Fig. 2). This changes when the particles are embedded in H_2O_2 solutions. A chemical reaction catalyzed by the Pt coated Janus particles is induced in the solvent [50], which triggers the self-propulsion and leads to three different regimes of motion.

At short times ($\tau < 10^{-2}$ for 2.5% H_2O_2 concentration), the particles undergo simple Brownian motion. The behavior corresponds to that of passive Brownian particles as the deterministic displacements due to the self-propulsion are not relevant at this early stage.

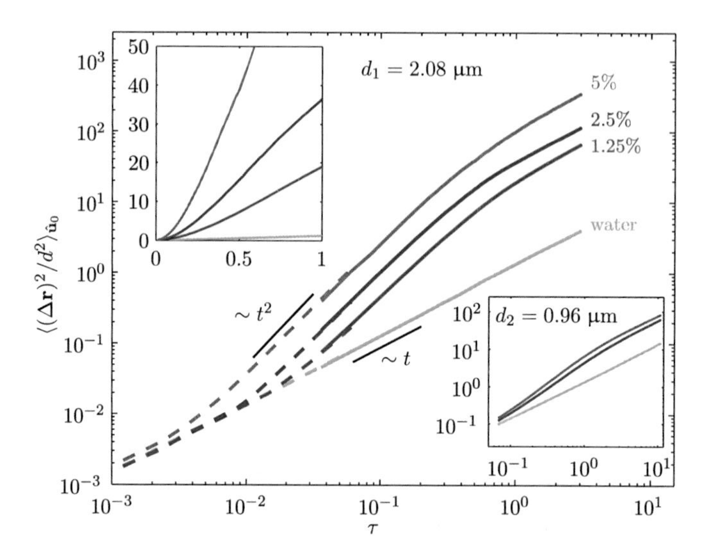

Figure 2. Double logarithmic plot of the experimental results for the MSD of Janus particles with diameter $d_1 = 2.08$ µm in water and in H_2O_2 solutions with different concentrations as a function of the scaled time $\tau = D_r t$. Various regimes of motion are identified. Dashed and solid curves refer to different measurements for the same H_2O_2 concentrations. Left inset: visualization of the data in a linear plot. Right inset: experimental results for smaller particles with diameter $d_2 = 0.96$ µm in water and in H_2O_2 solutions with concentrations of 2.5% and 5%.

We introduce the characteristic time scale τ_1 to describe the transition to the intermediate regime, where directed (active) motion dominates. Physically, τ_1 is the time that is required for the chemical reaction to bring about a propulsive motion comparable to the Brownian random displacements. It clearly decreases with increasing H_2O_2 concentration of the solution and can be used to measure the strength of the self-propulsion of the investigated particles.

In the second regime, the MSD yields a superdiffusive behavior (approximately $\langle(\Delta\mathbf{r})^2\rangle_{\hat{\mathbf{u}}_0} \propto t^2$) as the motion is dominated by the directed propulsive component. Finally, at a second time scale τ_2 the dynamics becomes diffusive again with an enhanced diffusion coefficient [50]. The transition to this third regime is also obvious in the linear plot of the MSD (see left inset in Fig. 2), where the nonlinear (quadratic) dependence at short times becomes linear for longer times. The transition occurs near $\tau_2 = 1/2$, which corresponds to $t_2 = 4.1$ s. This is the time scale where the particles lose their memory of the initial orientation due to

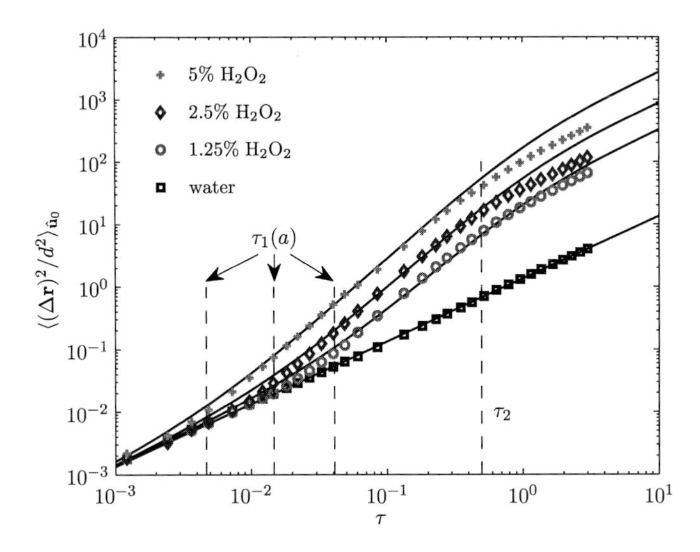

Figure 3. Comparison of the measured MSD (symbols) with the theoretical prediction (solid curves). The fitting parameter a is given in Table I. Dashed lines indicate the transition times $\tau_1(a) = 18/a^2$ and $\tau_2 = 1/2$ between the different regimes of motion.

rotational Brownian motion. Note that τ_2 is largely independent of the H_2O_2 concentration as opposed to the transition time τ_1. Previous experiments [34, 50] have observed the time scale τ_2. However, an experimental investigation of the time scale τ_1 has not been reported yet.

As visualized in Fig. 3, the experimental data show good agreement with our theoretical model. The solid curves represent best fits for short and intermediate times based on the prediction

$$\left\langle \frac{(\Delta \mathbf{r})^2}{d^2} \right\rangle_{\hat{\mathbf{u}}_0} = \frac{4}{3}\tau + \frac{1}{27}a^2 \left[2\tau - 1 + e^{-2\tau}\right] \tag{6}$$

for the two-dimensional MSD, which is obtained from Eqs. (3) and (5). Here and in the following the dimensionless parameter $a = \beta dF$ is used to characterize the strength of the self-propulsion. The fit curves in Fig. 3 are based on the values of a specified in Table I. We attribute the slight deviations at long times to small particle imperfections, in particular with regard to the Pt layer. These might induce a noncentral effective driving force, which leads to a tiny, but deterministic rotation of the particle and thus reduces the measured MSD for long times. This effect could be included in the theoretical model either by means of a renormalized rotational diffusion coefficient [50] or by explicitly considering an internal

Table I. Dimensionless self-propulsion force $a = \beta dF$ of the Janus particles as a function of the H_2O_2 concentration of the solvent. The values for a are obtained by fitting Eq. (6) to the experimental data for the MSD (see Fig. 3).

H_2O_2 concentration (%)	Scaled self-propulsion a
0	0
1.25	21
2.5	35
5	62

torque generated by the asymmetry of the particle [61]. Another source of deviations might be remnants of long-ranged hydrodynamic effects. Short-ranged particle-particle interactions can be excluded due to our tracking algorithm, where only particles with a specified minimum distance from each other are considered (for further details see Sec. A 2 in the Appendix). From Eq. (6) one also obtains a prediction for the transition time τ_1 between the initial diffusive and the superdiffusive regime. Equating the Brownian and the propulsive contributions yields $\tau_1 = 18/a^2$. In agreement with the experimental observations, τ_1 is antiproportional to the square of the self-propulsion force.

B. Excess kurtosis

On top of the analysis of the MSD, here we also address skewness S and excess kurtosis γ, which serve to quantify the non-Gaussian behavior of self-propelled particles [51]. They are given by

$$S = \frac{\langle(\Delta x)^3\rangle_{\hat{\mathbf{u}}_0}}{\langle(\Delta x)^2\rangle_{\hat{\mathbf{u}}_0}^{3/2}} \tag{7}$$

and

$$\gamma = \frac{\langle(\Delta x)^4\rangle_{\hat{\mathbf{u}}_0}}{\langle(\Delta x)^2\rangle_{\hat{\mathbf{u}}_0}^{2}} - 3\,, \tag{8}$$

respectively. Note that Eqs. (7) and (8) are only valid because $\langle\Delta x\rangle_{\hat{\mathbf{u}}_0} = 0$ in our system. Otherwise, the moments have to be replaced by the respective central moments. As the third moment $\langle(\Delta x)^3\rangle_{\hat{\mathbf{u}}_0}$ trivially vanishes due to the symmetry of $\Psi(\Delta x, t)$, resulting from the averaging over the initial orientation $\hat{\mathbf{u}}_0$ of the Janus particles, the skewness S is zero; though, our measurements in H_2O_2 solution clearly yield nonzero values for the excess kurtosis γ, which directly indicates non-Gaussian behavior. The curves in Fig. 4(a) are calculated from the experimental displacement data based on Eq. (8). Results are shown for pure water and H_2O_2 concentrations of 1.25%, 2.5%, and 5% corresponding to the analysis of the MSD in

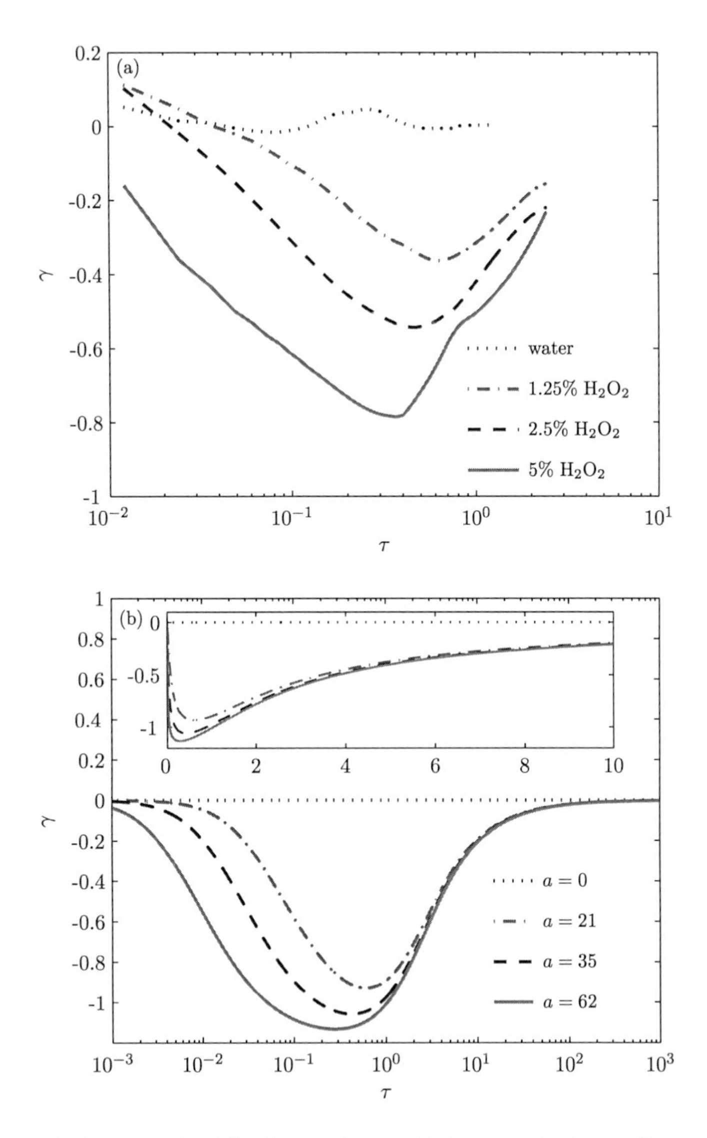

Figure 4. (a) Experimental and (b) theoretical results for the excess kurtosis γ. The theoretical curves are calculated for the values of the self-propulsion force extracted from the MSD fits in Fig. 3 (see Table I). Inset in (b): Visualization of the theoretical data in a linear plot.

Figs. 2 and 3. As expected, the reference measurements in water yield a nearly vanishing excess kurtosis γ, which indicates largely Gaussian behavior. The slight deviations from zero can be induced by a not perfectly symmetric particle shape. This leads to a situation similar to the Brownian motion of passive ellipsoids, where small positive values for the non-Gaussian parameter are observed as well [62]. However, the time dependence of the excess kurtosis changes drastically, when active Janus particles in H_2O_2 solutions are considered. The measured curves turn negative and present a minimum located between $\tau = 0.4$ and $\tau = 0.8$ depending on the H_2O_2 concentration. If the latter is increased, the position of the minimum is shifted to shorter times and it becomes more pronounced ($\gamma_{\text{min}} \approx -0.35$ for 1.25% and $\gamma_{\text{min}} \approx -0.8$ for 5% H_2O_2). This corresponds to the general observation that for all times a higher H_2O_2 concentration leads to more negative values for γ.

To derive the analytical expression for the excess kurtosis from our theoretical model, in addition to the MSD the fourth moment $\langle(\Delta x)^4\rangle_{\hat{\mathbf{u}}_0}$ of $\Psi(\Delta x, t)$ is also required. For the situation in our experiments, where the particles undergo three-dimensional rotational Brownian motion, one obtains

$$\begin{aligned}\left\langle\frac{(\Delta x)^4}{d^4}\right\rangle_{\hat{\mathbf{u}}_0} &= \frac{4}{3}\tau^2 + \frac{2}{27}a^2\tau\left[2\tau - 1 + e^{-2\tau}\right] \\ &+ \frac{1}{21870}a^4\left[90\tau^2 - 156\tau + 107 - 54\tau e^{-2\tau} - 108e^{-2\tau} + e^{-6\tau}\right].\end{aligned} \tag{9}$$

The final result for the excess kurtosis γ directly follows from Eq. (8) by inserting Eq. (9) and $\langle(\Delta x)^2\rangle_{\hat{\mathbf{u}}_0} = (1/2)\,\langle(\Delta \mathbf{r})^2\rangle_{\hat{\mathbf{u}}_0}$ [see Eq. (6)]. In Fig. 4(b) theoretical curves are plotted for a as determined for pure water and the various H_2O_2 concentrations from the analysis of the MSD (see Fig. 3 and Table I). The linear plot in the inset visualizes the pronounced negative long-time tail [52].

Basically, the theoretical results show the same tendency as discussed for the experimental curves in Fig. 4(a). In particular with regard to the general behavior and the position of the minimum the agreement is very good, although the experimental values for γ are usually less negative than the theoretical predictions. Slightly positive values as measured for very short times can again be ascribed to small deviations from an ideal isotropic particle shape, similar to our observations in pure water.

C. Displacement probability distribution

After the discussion of the displacement moments, in a next step we study the full probability distribution function $\Psi(\Delta x, t)$ for one-dimensional displacements, which reveals further details of the statistical characteristics of the particle motion. In water, the Janus particles show a simple diffusive behavior corresponding to Gaussian probability distributions at all times. However, for self-propelled particles the curves for $\Psi(\Delta x, t)$ significantly deviate from a Gaussian shape. In Fig. 5(a) exemplarily the experimental results for 5% H_2O_2 concentration are given. Data points are plotted for each pixel, corresponding to an interval of 0.16 μm. At the beginning ($t = 0.1$ s), $\Psi(\Delta x, t)$ is still nearly Gaussian. After $t = 0.5$ s, a broadening of the peak is observed, which further intensifies until $t = 2$ s. Furthermore, the wings of the distribution become steeper as time proceeds.

A theoretical prediction for $\Psi(\Delta x, t)$ is obtained numerically from the model equations (3) and (4). As opposed to the analytical results for the MSD and the excess kurtosis presented in Secs. III A and III B, the full displacement probability distribution is only accessible via a Brownian dynamics simulation (see Sec. II C). Figure 5(b) gives the simulation results calculated for $a = \beta dF = 62$. They show the same characteristic features—such as the broadened peak and the steep wings—as the experimental plots.

The shape of the displacement probability distribution curves is closely related to the particle dynamics in the different regimes of motion (see Sec. III A). At short times, when the random translational motion still dominates, $\Psi(\Delta x, t)$ is nearly Gaussian. In the intermediate regime, where the self-propulsion dictates the particle motion, the broadening of the peak emerges [see Fig. 5, plots for $t = 0.5$ s ($\tau = 0.0605$) and $t = 2$ s ($\tau = 0.242$)]. Thus, the appearance of the broadened peak accompanied by the steep wings is due to the active component of the motion. This shape also provides the explanation for the negative values of the excess kurtosis γ (see Sec. III B), which could only be suspected in earlier theoretical calculations [51, 52].

As we assume that the Janus particles undergo free rotational Brownian motion in three dimensions, their initial orientations are homogeneously distributed on a unit sphere [63]. This implies that the projections of all possible initial orientation vectors $\hat{\mathbf{u}}_0$ on the x axis are evenly spread between -1 and 1. Consequently, the contribution to the deterministic particle displacement in x direction is uniformly distributed as well, which explains the kind of rectangular shape of $\Psi(\Delta x, t)$ in the intermediate regime. Although the measured MSD (see Fig. 2) already indicates diffusive behavior again for $\tau > 1/2$, the non-Gaussian structure of $\Psi(\Delta x, t)$ still persists [see Fig. 5, plots for $t = 10$ s ($\tau = 1.21$)]. This yields that the displacement probability distribution is less sensitive to changes in the type of motion than the MSD. The prolonged presence of the broadened peak is consistent with the negative long-time tail observed for the excess kurtosis (see Fig. 4) and explains its origin. For very long times, $\Psi(\Delta x, t)$ is expected to become Gaussian again. While the experiments cannot be performed long enough to show this tendency clearly, it is confirmed by our simulation. The conversion back to a Gaussian shape occurs at τ on the order of 10^2, when also the excess kurtosis, which is a direct measure for the non-Gaussianity, approaches zero again.

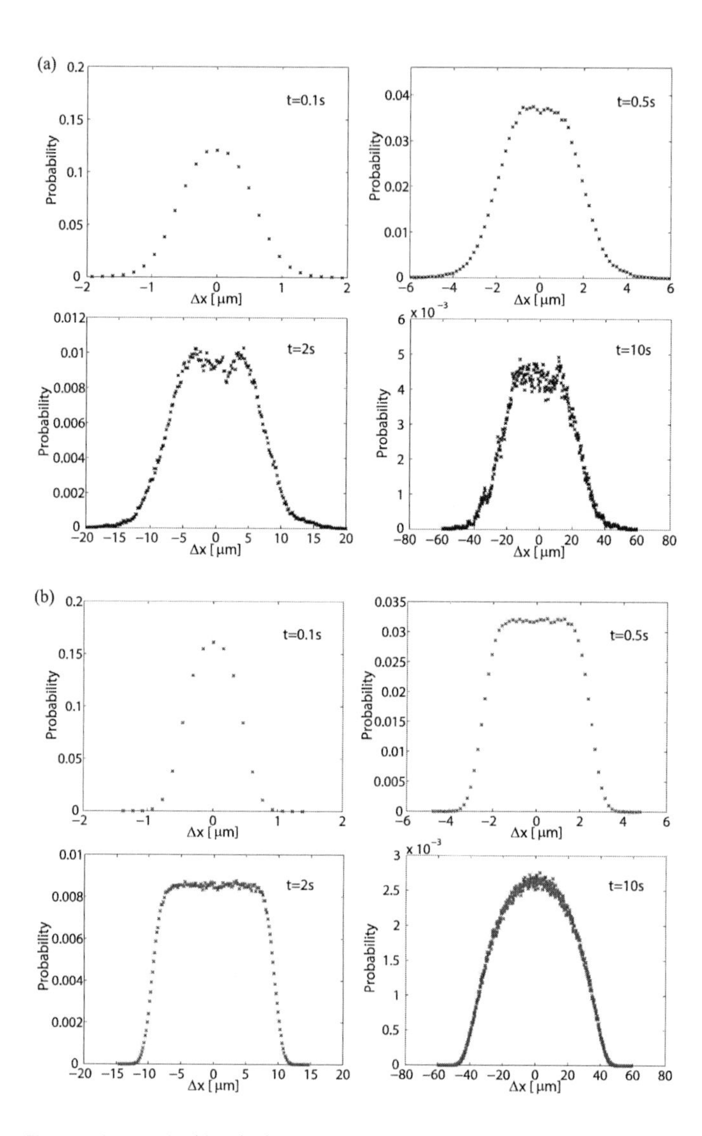

Figure 5. Time evolution of $\Psi(\Delta x, t)$: (a) experimental results for 5% H_2O_2 concentration; (b) corresponding simulation for $a = 62$.

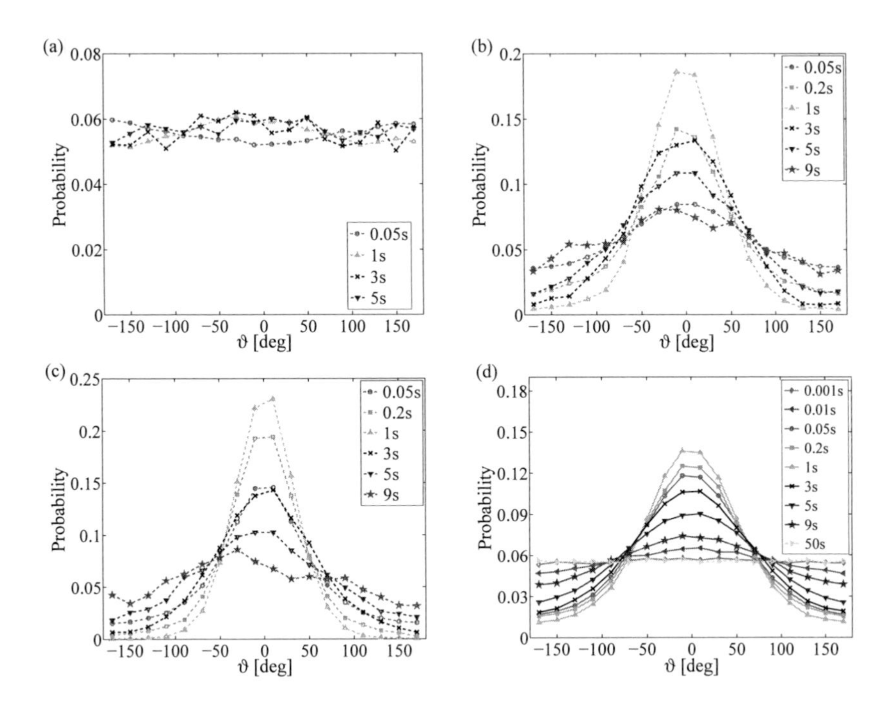

Figure 6. Time evolution of $\Psi(\vartheta, t)$: (a)–(c) experimental data for (a) water, (b) 2.5%, and (c) 5% H_2O_2 concentration. (d) Simulation results for $a = 62$.

D. Directional probability distribution

In this section, we briefly discuss an alternative approach to visualize the relative importance of the random and the deterministic contributions to the particle motion. It is based on the probability distribution function $\Psi(\vartheta, t)$ for the angle ϑ between the directions of subsequent particle displacements (see Fig. 1). While the Brownian noise induces arbitrary displacement directions (corresponding to a homogeneous distribution of ϑ between $-\pi$ and π), the self-propulsive motion is always collinear with the particle orientation $\hat{\mathbf{u}}$, determined by the position of the Pt layer, and thus favors values of ϑ near zero.

The experimental results for the time evolution of $\Psi(\vartheta, t)$ in solutions with different H_2O_2 concentrations are shown in Figs. 6(a)–6(c). In water [see Fig. 6(a)] the distribution is uniform at all times due to the random Brownian motion. However, for nonzero H_2O_2 concentration a peaked behavior of $\Psi(\vartheta, t)$ occurs [see Figs. 6(b) for 2.5% and 6(c) for 5%

H_2O_2]. Here, the peak height increases for short times until it reaches its maximum value at about 1 s. After that the curves become flatter again when the displacement directions decorrelate due to rotational Brownian motion. With increasing H_2O_2 concentration the peak attains higher maximum values and it becomes more pronounced at intermediate times. At long times there is no significant difference between the curves for 2.5% and 5% H_2O_2 concentration.

Figure 6(d) gives the simulation results for the time evolution of $\Psi(\vartheta, t)$ for $a = 62$. It is in good agreement with the corresponding experimental data and shows additional curves for very short and very long times that are not directly accessible in experiment. The three regimes (short-time diffusive, intermediate ballistic, and long-time diffusive) discussed in detail in the previous sections can also be extracted from the plots of $\Psi(\vartheta, t)$. For very short times [see curves for $t = 0.001$ s and $t = 0.01$ s in Fig. 6(d)] the directions of the particle displacements in two adjacent time intervals are completely uncorrelated. This yields that the passive Brownian motion is dominant in this regime. The pronounced peaks for intermediate values of t [see, in particular, curves for $t = 0.5$ s and $t = 1$ s in Fig. 6(d)] clearly show that the particle dynamics is largely influenced by the directed self-propelling component of the motion. Finally, for very long times the angular probability distribution becomes homogeneous again, indicating the long-time diffusive regime.

E. Orientational symmetry breaking for high H_2O_2 concentration

The previous discussion focused on results for up to 5% H_2O_2 concentration. We have also performed experiments with 10% and 15% solutions. Here, our video observation of the Janus particles strongly indicates that their orientation is not freely diffusing on a unit sphere any more, but is largely restricted to the x-y plane [64]. This symmetry breaking in the rotational motion directly affects the structure of the probability distribution function $\Psi(\Delta x, t)$ and also leads to different analytical expressions for the displacement moments. Assuming that the orientation vector of the particle always lies inside the two-dimensional plane of motion, the evolution of the single orientational angle ϕ is given by [52]

$$P(\phi, t) = \frac{1}{\sqrt{4\pi D_\mathrm{r} t}} \exp\left(-\frac{(\phi - \phi_0)^2}{4 D_\mathrm{r} t}\right) . \tag{10}$$

Consequently, from Eqs. (3) and (10) one obtains the orientation-averaged MSD

$$\left\langle \frac{(\Delta \mathbf{r})^2}{d^2} \right\rangle_{\hat{\mathbf{u}}_0} = \frac{4}{3}\tau + \frac{2}{9}a^2 \left[\tau - 1 + e^{-\tau}\right] \tag{11}$$

and the fourth moment

$$\begin{aligned}\left\langle \frac{(\Delta x)^4}{d^4} \right\rangle_{\hat{\mathbf{u}}_0} &= \frac{4}{3}\tau^2 + \frac{4}{9}a^2\tau\left[\tau - 1 + e^{-\tau}\right] \\ &+ \frac{1}{3888}a^4\Big[144\tau^2 - 540\tau + 783 - 240\tau e^{-\tau} - 784 e^{-\tau} + e^{-4\tau}\Big]\end{aligned} \tag{12}$$

determining the excess kurtosis. At first sight, these results seem to be very similar to their counterparts for free three-dimensional rotational Brownian motion as presented in Eqs. (6) and (9), respectively. Technically, they only differ in the prefactors of the various terms and in the arguments of the exponential functions. The larger absolute values of the latter for three-dimensional orientation indicate that the particles lose their orientational memory earlier than in the case with two-dimensional rotational Brownian motion. Despite the formal analogy of the analytical expressions for the displacement moments, both the experimental data and the simulation results reveal striking differences with regard to the full probability distribution function (see Fig. 7). While an extremely broadened peak is observed for isotropic rotational diffusion (see Fig. 5), a characteristic double peak occurs due to the symmetry breaking that restricts the particle orientations to the two-dimensional plane of translational motion (see Fig. 7). It is most pronounced after times on the order of several seconds. The origin of the double peak can be understood by considering the initial orientations of the Janus particles. If these are homogeneously distributed on a unit circle (and not on a unit sphere), the corresponding projections on the x axis are not evenly spread between -1 and 1. Instead of that, values close to the extrema have a higher statistical weight than values around zero. Consequently, the majority of the Janus particles carry out a significant directed displacement during the superdiffusive regime where the self-propulsion is dominant. Only few particles stay close to their initial position. This explains the characteristic double-peak structure observed in our experiments and verified by a corresponding computer simulation (see Fig. 7).

Figure 8 directly visualizes the dependence of $\Psi(\Delta x, t)$ on the H_2O_2 concentration. For this purpose, snapshots of the distributions after 2 s are shown. These reveal Gaussian behavior for pure water, a broadened peak for low, and a double peak for high H_2O_2 concentration. The existence of the double peak in the latter case is a second independent indicator for the orientational symmetry breaking, in addition to our video observation. We surmise that the limitation of the rotational freedom is due to hydrodynamic effects [65, 66]. In solutions with higher H_2O_2 concentration, the chemical reaction generates a stronger self-propulsion. Thus, the flow pattern in the vicinity of a Janus particle [38] might have increasing influence on its rotational motion. However, clearly more work is needed to fully understand the origin of the observed orientational symmetry breaking.

Finally, our theoretical description including limited rotational freedom could also explain the seemingly contradicting experimental results presented in Refs. [34] and [50]. In

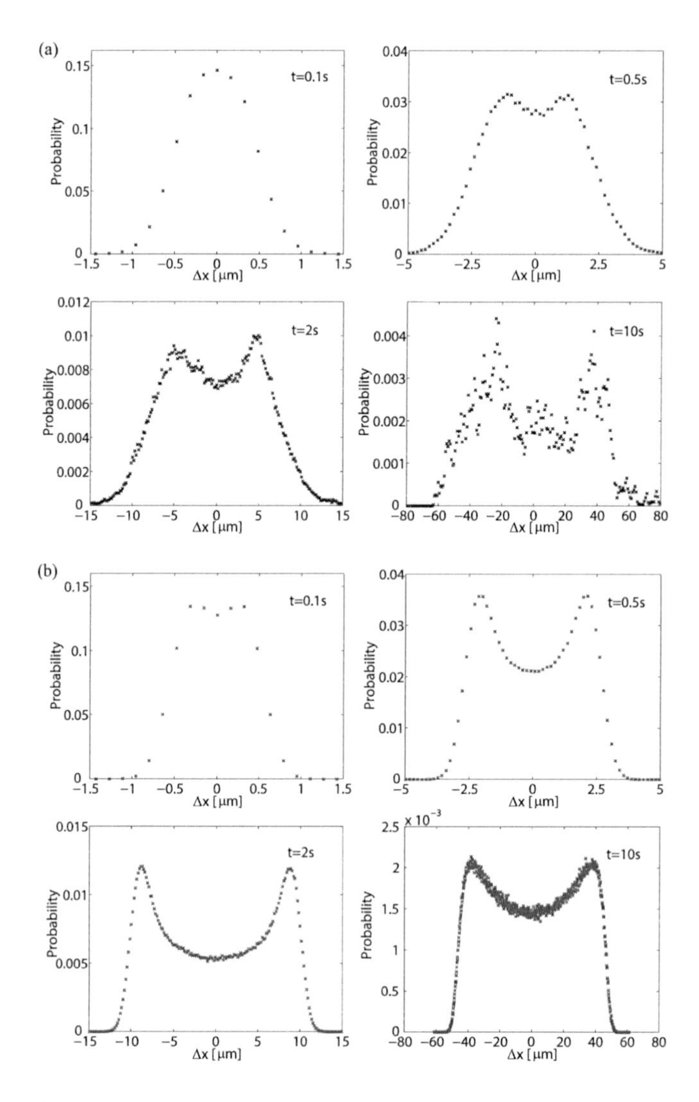

Figure 7. (a) Time evolution of the measured probability distribution $\Psi(\Delta x, t)$ for Janus particles in a 10% H_2O_2 solution. The occurrence of the double peak indicates that the particle orientation does not diffuse freely on a unit sphere for high H_2O_2 concentrations. (b) Reference simulation for particles whose orientation is restricted to the x-y plane.

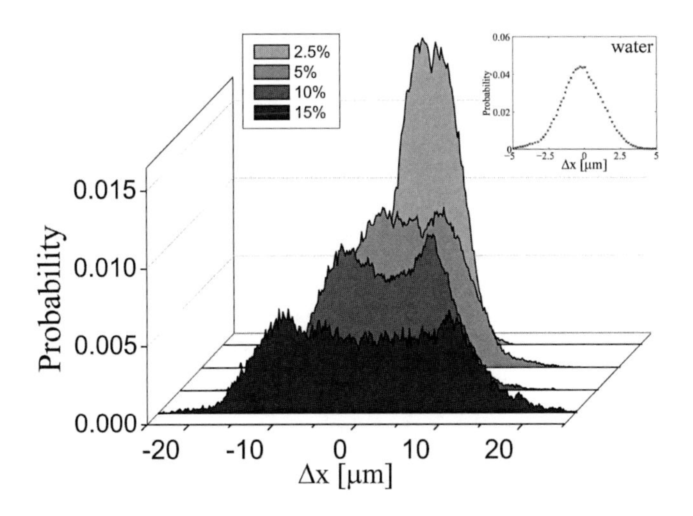

Figure 8. Experimental results for $\Psi(\Delta x, t)$ after 2 s for different H_2O_2 concentrations. The inset shows the Gaussian distribution measured in water.

Ref. [50] the rotational diffusion time $t_r = 1/D_r$ is measured to decrease as a function of the H_2O_2 concentration, which is attributed to an asymmetric Pt coverage leading to a deterministic rotation of the particles. On the contrary, a slight increase of t_r with higher H_2O_2 concentration is reported in Ref. [34], where t_r is estimated from the transition between the superdiffusive and the long-time diffusive regime. Following the argument of limited rotational freedom, this increase is not due to a change of the rotational diffusion coefficient, but could directly be explained by the different prefactors of τ in the exponents of Eqs. (6) and (11). The real situation in experiments with active Janus particles is most likely always somewhere in between free rotational diffusion and full restriction to two dimensions. While the good agreement between theory and experiment for low H_2O_2 concentrations (up to 5%)—as discussed in Secs. III A–III D—implies that the orientational limitation plays a minor role in those cases, a modified description is required for higher H_2O_2 concentrations.

IV. CONCLUSIONS

In summary, we have studied the non-Gaussian characteristics of the diffusiophoretic motion of self-propelled Pt-silica Janus spheres both in experiment and in theory. The propulsion strength is varied by means of different concentrations of the H_2O_2 solution, in which the particles are embedded. The good agreement between theory and experiment shows that in spite of the rather complicated underlying propulsion mechanism all the main features of the motion including the higher displacement moments can be understood by our model based on the translational and orientational Langevin equations [52, 60]. The analytical predictions have been experimentally verified not only for the mean square displacement, but also for the excess kurtosis characterizing the non-Gaussian behavior. This promises the applicability of our model to a broad range of experimental systems as the detailed propulsion mechanism can be accounted for by the implementation of an effective driving force. As illustrated here, the excess kurtosis is a helpful tool beyond the standard mean square displacement approach in order to understand the interplay between the deterministic and the random components of the dynamics of active Brownian systems. The characteristic non-Gaussian superdiffusive intermediate regime is enframed by two diffusive regimes: simple (passive) Brownian motion at short times and enhanced diffusion with a significantly increased diffusion constant [50] due to the active part of the motion at long times.

A deeper understanding of the non-Gaussianity is provided by the full probability distributions for the magnitude and the direction of displacements as obtained from the experiments in good agreement with a corresponding Brownian dynamics simulation. Concerning the magnitude of the displacements, the respective probability distribution for low H_2O_2 concentration reveals a significantly broadened peak at intermediate times, which is induced by the self-propulsion of the Janus particles. In agreement with the negative long-time tail of the excess kurtosis, the broadened peak is still observable when the particle dynamics has already changed to the enhanced diffusive regime. This phenomenon can be traced back to the superdiffusive regime, where a large number of particles performed significant deterministic displacements. In the experiments with high H_2O_2 concentration, a symmetry breaking manifested in a limitation of the rotational Brownian motion is found. It induces a pronounced double-peak structure of the displacement probability distribution and requires a modification of the theoretical description.

In order to generalize the presented results for spherical Janus particles, in a next step, it is interesting to analyze the non-Gaussian behavior of asymmetric particles. These can either be axisymmetric such as rods [67] and ellipsoids [68], or they can have an even more complicated anisotropic shape [69, 70]. While some results for the non-Gaussian behavior of passive [62] and active [52] axisymmetric particles are already available, an open question addresses the influence of more complicated particle shapes on the characteristic features of the particle dynamics beyond simple Brownian motion. In particular, an additional torque [71]—as automatically induced by an asymmetry around the propulsion axis [61]—significantly affects the motional behavior and leads to a modified displacement probability distribution. Another interesting aspect for future experimental studies are solvent flow effects [72, 73] which accelerate the displacement of microswimmers drastically [74]. In the present work the non-Gaussianity is already caused by the presence of the self-propulsion of the active particles. Thus, here it is a single particle phenomenon as dilute systems, where particle interactions are negligible, are investigated in our experiments. However, for situations with higher particle density [75–77], the interplay between hydrodynamic effects [78] and the self-propulsion of the particles is expected to give rise to new physical phenomena manifested also in the excess kurtosis of the displacement probability distribution and its higher moments.

ACKNOWLEDGMENTS

We thank Jiang Lei for using electron beam evaporation to prepare the Janus particles. This work was financially supported by the National Natural Science Foundation of China (Grants No. 11272322, No. 11202219, and No. 21005058), the ERC Advanced Grant INTERCOCOS (Grant No. 267499), and by the DFG within SFB TR6 (project C3).

Appendix A: Experimental apparatus and methods

1. Preparation of the Janus particles

The silica particles used in the experiments were produced by the University of Petroleum in China. The diameters of the two considered particle sizes are $d_1 = 2.08 \pm 0.05$ µm and $d_2 = 0.96 \pm 0.03$ µm measured by scanning electron microscopy (SEM) (see Fig. 9).

To fabricate the Janus particles, an aqueous suspension of silica particles is first deposited on a 4-inch silicon wafer by spin coating at low speed (800 rpm). After evaporating the water, a single layer of particles is formed on the wafer. Then, using electron beam evaporation (by an Innotec e-beam evaporator in the Institute of Semiconductors, Chinese Academy of

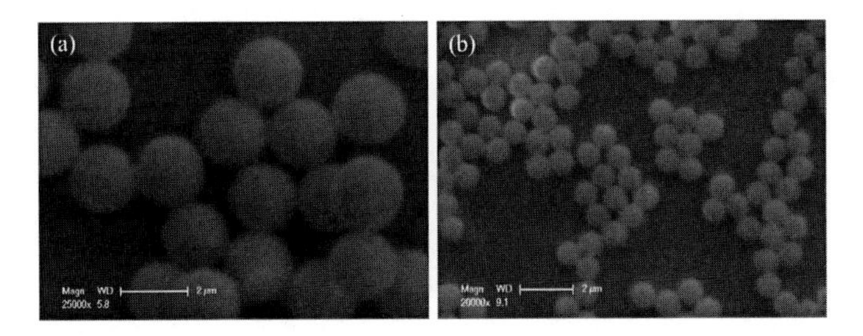

Figure 9. SEM images of the silica particles with diameters (a) $d_1 = 2.08 \pm 0.05$ µm and (b) $d_2 = 0.96 \pm 0.03$ µm.

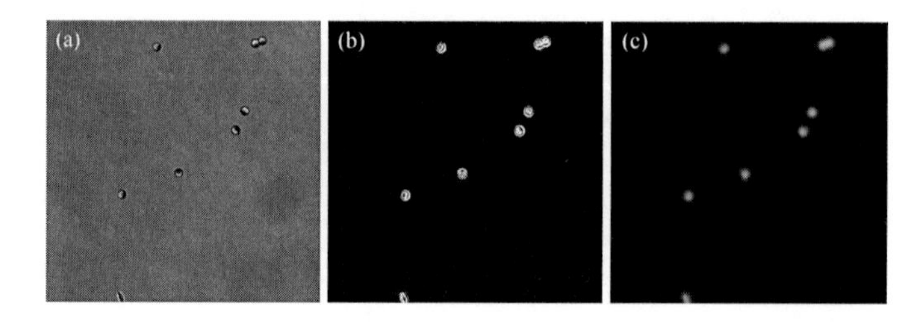

Figure 10. Image preprocessing with the program IMAGEJ: (a) the original image directly obtained by video microscopy from the experiments, (b) image after using the find edge function, and (c) image after using the Gaussian blur function.

Sciences), a layer of Pt (thickness about 7 nm) is deposited on the upper half surfaces of the particles. Finally, the half-coated Janus particles are collected from the silicon wafer using a razor blade and resuspended in distilled water (18.2 MΩ cm). The volumetric concentration of the Janus particle suspension is approximately 5×10^{-3}.

2. Image processing

We apply the following method to determine the exact center of the Janus particles, which appear half bright and half dark in the images (see Fig. 10). First, the find edge function of the program IMAGEJ is used, which highlights sharp intensity changes. As the sharpest changes occur at the particle edges, this function offers a way to reconstruct the round shape of the particle. Secondly, using the Gaussian blur function of IMAGEJ, the gray scale value distribution in the particle domain is determined. The point with the maximum gray scale value is considered to be the center of the particle. This method has a ± 0.5 pixel accuracy.

After this preprocessing, the particle positions (x, y) can be tracked by the software VIDEO SPOT TRACKER (V07.02). To guarantee that only individual particles are tracked, we omit aggregated particles and use a "dead zone" function, by which the region of approximately one diameter around the particle is monitored. If other particles enter into this zone, the tracking of the respective particles is stopped. Therefore, particle aggregation as well as particle-particle collisions and interactions can be excluded from our investigation.

---

[1] P. Romanczuk, M. Bär, W. Ebeling, B. Lindner, and L. Schimansky-Geier, Eur. Phys. J. Spec. Top. **202**, 1 (2012).

[2] M. E. Cates, Rep. Prog. Phys. **75**, 042601 (2012).

[3] M. C. Marchetti, J. F. Joanny, S. Ramaswamy, T. B. Liverpool, J. Prost, M. Rao, and R. A. Simha, Rev. Mod. Phys. **85**, 1143 (2013).

[4] X. Chen, X. Dong, A. Be'er, H. L. Swinney, and H. P. Zhang, Phys. Rev. Lett. **108**, 148101 (2012).

[5] A. Sokolov, I. S. Aranson, J. O. Kessler, and R. E. Goldstein, Phys. Rev. Lett. **98**, 158102 (2007).

[6] C. Dombrowski, L. Cisneros, S. Chatkaew, R. E. Goldstein, and J. O. Kessler, Phys. Rev. Lett. **93**, 098103 (2004).

[7] T. Ishikawa, N. Yoshida, H. Ueno, M. Wiedeman, Y. Imai, and T. Yamaguchi, Phys. Rev. Lett. **107**, 028102 (2011).

[8] H. H. Wensink, J. Dunkel, S. Heidenreich, K. Drescher, R. E. Goldstein, H. Löwen, and J. M. Yeomans, Proc. Natl. Acad. Sci. USA **109**, 14308 (2012).

[9] G. Miño, T. E. Mallouk, T. Darnige, M. Hoyos, J. Dauchet, J. Dunstan, R. Soto, Y. Wang, A. Rousselet, and E. Clement, Phys. Rev. Lett. **106**, 048102 (2011).

[10] J. Schwarz-Linek, C. Valeriani, A. Cacciuto, M. E. Cates, D. Marenduzzo, A. N. Morozov, and W. C. K. Poon, Proc. Natl. Acad. Sci. USA **109**, 4052 (2012).

[11] I. H. Riedel, K. Kruse, and J. Howard, Science **309**, 300 (2005).

[12] V. Kantsler, J. Dunkel, M. Polin, and R. E. Goldstein, Proc. Natl. Acad. Sci. USA **110**, 1187 (2013).

[13] D. M. Woolley, Reproduction **216**, 259 (2003).

[14] B. M. Friedrich and F. Jülicher, New J. Phys. **10**, 123035 (2008).

[15] J. Elgeti, U. B. Kaupp, and G. Gompper, Biophys. J. **99**, 1018 (2010).

[16] T. Vicsek and A. Zafeiris, Phys. Rep. **517**, 71 (2012).

[17] D. Helbing, I. Farkas, and T. Vicsek, Nature (London) **407**, 487 (2000).

[18] J. Zhang, W. Klingsch, A. Schadschneider, and A. Seyfried, J. Stat. Mech. (2012) P02002.

[19] J. L. Silverberg, M. Bierbaum, J. P. Sethna, and I. Cohen, Phys. Rev. Lett. **110**, 228701 (2013).

[20] R. Dreyfus, J. Baudry, M. L. Roper, M. Fermigier, H. A. Stone, and J. Bibette, Nature (London) **437**, 862 (2005).

[21] S. Thutupalli, R. Seemann, and S. Herminghaus, New J. Phys. **13**, 073021 (2011).

[22] A. Reinmüller, H. J. Schöpe, and T. Palberg, Langmuir **29**, 1738 (2013).

[23] H.-R. Jiang, N. Yoshinaga, and M. Sano, Phys. Rev. Lett. **105**, 268302 (2010).

[24] I. Buttinoni, G. Volpe, F. Kümmel, G. Volpe, and C. Bechinger, J. Phys.: Condens. Matter **24**, 284129 (2012).

[25] A. Snezhko and I. S. Aranson, Nat. Mater. **10**, 698 (2011).

[26] G. Rückner and R. Kapral, Phys. Rev. Lett. **98**, 150603 (2007).

[27] P. Tierno, R. Golestanian, I. Pagonabarraga, and F. Sagués, J. Phys. Chem. B **112**, 16525 (2008).

[28] W. F. Paxton, A. Sen, and T. E. Mallouk, Chem. Eur. J. **11**, 6462 (2005).

[29] A. Erbe, M. Zientara, L. Baraban, C. Kreidler, and P. Leiderer, J. Phys.: Condens. Matter **20**, 404215 (2008).

[30] J. L. Anderson, Annu. Rev. Fluid Mech. **21**, 61 (1989).

[31] R. Golestanian, T. B. Liverpool, and A. Ajdari, New J. Phys. **9**, 126 (2007).

[32] R. F. Ismagilov, A. Schwartz, N. Bowden, and G. M. Whitesides, Angew. Chem. Int. Ed. **41**, 652 (2002).

[33] M. N. Popescu, S. Dietrich, and G. Oshanin, J. Chem. Phys. **130**, 194702 (2009).

[34] H. Ke, S.-R. Ye, R. L. Carroll, and K. Showalter, J. Phys. Chem. A. **114**, 5462 (2010).

[35] S. J. Ebbens and J. R. Howse, Langmuir **27**, 12293 (2011).

[36] S. Ebbens, M.-H. Tu, J. R. Howse, and R. Golestanian, Phys. Rev. E **85**, 020401(R) (2012).

[37] B. Sabass and U. Seifert, J. Chem. Phys. **136**, 064508 (2012).

[38] T. Bickel, A. Majee, and A. Würger, Phys. Rev. E **88**, 012301 (2013).

[39] I. Buttinoni, J. Bialké, F. Kümmel, H. Löwen, C. Bechinger, and T. Speck, Phys. Rev. Lett. **110**, 238301 (2013).

[40] I. Theurkauff, C. Cottin-Bizonne, J. Palacci, C. Ybert, and L. Bocquet, Phys. Rev. Lett. **108**, 268303 (2012).

[41] J. Palacci, S. Sacanna, A. P. Steinberg, D. J. Pine, and P. M. Chaikin, Science **339**, 936 (2013).

[42] G. S. Redner, M. F. Hagan, and A. Baskaran, Phys. Rev. Lett. **110**, 055701 (2013).

[43] L. Baraban, M. Tasinkevych, M. N. Popescu, S. Sanchez, S. Dietrich, and O. G. Schmidt, Soft Matter **8**, 48 (2012).

[44] L. Baraban, D. Makarov, R. Streubel, I. Mönch, D. Grimm, S. Sanchez, and O. G. Schmidt, ACS Nano **6**, 3383 (2012).

[45] L. Baraban, D. Makarov, O. Schmidt, G. Cuniberti, P. Leiderer, and A. Erbe, Nanoscale **5**, 1332 (2013).

[46] G. Volpe, I. Buttinoni, D. Vogt, H.-J. Kümmerer, and C. Bechinger, Soft Matter **7**, 8810 (2011).

[47] M. Mijalkov and G. Volpe, Soft Matter **9**, 6376 (2013).

[48] W. Yang, V. R. Misko, K. Nelissen, M. Kong, and F. M. Peeters, Soft Matter **8**, 5175 (2012).

[49] P. K. Ghosh, V. R. Misko, F. Marchesoni, and F. Nori, Phys. Rev. Lett. **110**, 268301 (2013).

[50] J. R. Howse, R. A. L. Jones, A. J. Ryan, T. Gough, R. Vafabakhsh, and R. Golestanian, Phys. Rev. Lett. **99**, 048102 (2007).

[51] B. ten Hagen, S. van Teeffelen, and H. Löwen, Condens. Matter Phys. **12**, 725 (2009).

[52] B. ten Hagen, S. van Teeffelen, and H. Löwen, J. Phys.: Condens. Matter **23**, 194119 (2011).

[53] W. Kob, C. Donati, S. J. Plimpton, P. H. Poole, and S. C. Glotzer, Phys. Rev. Lett. **79**, 2827 (1997).

[54] A. M. Puertas, M. Fuchs, and M. E. Cates, Phys. Rev. E **67**, 031406 (2003).

[55] K. Vollmayr-Lee, W. Kob, K. Binder, and A. Zippelius, J. Chem. Phys. **116**, 5158 (2002).

[56] A. Arbe, J. Colmenero, F. Alvarez, M. Monkenbusch, D. Richter, B. Farago, and B. Frick, Phys. Rev. Lett. **89**, 245701 (2002).

[57] J.-P. Bouchaud and M. Potters, *Theory of Financial Risk and Derivative Pricing: From Statistical Physics to Risk Management*, 2nd ed. (Cambridge University Press, Cambridge, 2009).

[58] J. Happel and H. Brenner, *Low Reynolds Number Hydrodynamics: With Special Applications to Particulate Media*, 2nd ed., Mechanics of Fluids and Transport Processes (Kluwer Academic Publishers, Dordrecht, 1991), Vol. 1.

[59] D. J. Jeffrey, Phys. Fluids A **4**, 16 (1992).

[60] J. K. G. Dhont, *An Introduction to Dynamics of Colloids* (Elsevier, Amsterdam, 1996).

[61] F. Kümmel, B. ten Hagen, R. Wittkowski, I. Buttinoni, R. Eichhorn, G. Volpe, H. Löwen, and C. Bechinger, Phys. Rev. Lett. **110**, 198302 (2013).

[62] Y. Han, A. M. Alsayed, M. Nobili, J. Zhang, T. C. Lubensky, and A. G. Yodh, Science **314**, 626 (2006).

[63] G. Marsaglia, Ann. Math. Stat. **43**, 645 (1972).

[64] See Supplemental Material at `http://link.aps.org/supplemental/10.1103/PhysRevE.88.032304` for movies of Janus particles in water and in solutions with 2.5% and 10% H_2O_2 concentration.

[65] E. Gauger and H. Stark, Phys. Rev. E **74**, 021907 (2006).

[66] I. O. Götze and G. Gompper, Phys. Rev. E **82**, 041921 (2010).

[67] F. Höfling, E. Frey, and T. Franosch, Phys. Rev. Lett. **101**, 120605 (2008).

[68] Y. Han, A. Alsayed, M. Nobili, and A. G. Yodh, Phys. Rev. E **80**, 011403 (2009).

[69] R. Wittkowski and H. Löwen, Phys. Rev. E **85**, 021406 (2012).

[70] D. J. Kraft, R. Wittkowski, B. ten Hagen, K. V. Edmond, D. J. Pine, and H. Löwen, Phys. Rev. E **88**, 050301(R) (2013).

[71] S. van Teeffelen and H. Löwen, Phys. Rev. E **78**, 020101(R) (2008).

[72] L. Holzer, J. Bammert, R. Rzehak, and W. Zimmermann, Phys. Rev. E **81**, 041124 (2010).

[73] A. Zöttl and H. Stark, Phys. Rev. Lett. **108**, 218104 (2012).

[74] B. ten Hagen, R. Wittkowski, and H. Löwen, Phys. Rev. E **84**, 031105 (2011).

[75] H. H. Wensink and H. Löwen, Phys. Rev. E **78**, 031409 (2008).

[76] H. H. Wensink and H. Löwen, J. Phys.: Condens. Matter **24**, 464130 (2012).

[77] A. Kaiser and H. Löwen, Phys. Rev. E **87**, 032712 (2013).

[78] E. Lauga and T. R. Powers, Rep. Prog. Phys. **72**, 096601 (2009).

Statement of the author: This chapter presents the results of a collaboration with the Chinese experimentalists Xu Zheng, Meiling Wu, Haihang Cui, and Zhanhua Silber-Li. The experiments were mainly performed by Xu Zheng at the State Key Laboratory of Nonlinear Mechanics in Beijing. Meiling Wu, who was supervised by Haihang Cui at Xi'an University of Architecture and Technology, helped to obtain some of the experimental data. I developed the theoretical model, performed the analytical calculations, and coordinated the experimental and theoretical work in consultation with Hartmut Löwen. Furthermore, I wrote most of the theoretical parts of the paper and rewrote the other parts. Andreas Kaiser provided the simulation data used for figures 5(b), 6(d), and 7(b) in the article.

The collaboration with the Chinese experimentalists was initiated by a visit of Zhanhua Silber-Li at our institute in Düsseldorf in December 2012. In the following year, I visited her group in Beijing, where I could see the experimental setup and conduct further discussions in addition to our regular conversations via Skype.

CHAPTER 3

Swimming path statistics of an active Brownian particle with time-dependent self-propulsion

The content of this chapter has been published in a similar form in *Journal of Statistical Mechanics: Theory and Experiment* (2014) P02011 by Sonja Babel, Borge ten Hagen, and Hartmut Löwen (see reference [214]).

Swimming path statistics of an active Brownian particle with time-dependent self-propulsion

S Babel, B ten Hagen, and H Löwen
Institut für Theoretische Physik II: Weiche Materie,
Heinrich-Heine-Universität Düsseldorf, Universitätsstraße 1,
D-40225 Düsseldorf, Germany

E-mail: sbabel@thphy.uni-duesseldorf.de,
bhagen@thphy.uni-duesseldorf.de, and hlowen@thphy.uni-duesseldorf.de

Abstract. Typically, in the description of active Brownian particles, a constant effective propulsion force is assumed, which is then subjected to fluctuations in orientation and translation, leading to a persistent random walk with an enlarged long-time diffusion coefficient. Here, we generalize previous results for the swimming path statistics to a time-dependent, and thus in many situations more realistic, propulsion which is a prescribed input. We analytically calculate both the noise-free and the noise-averaged trajectories for time-periodic propulsion under the action of an additional torque. In the deterministic case, such an oscillatory microswimmer moves on closed paths that can be highly more complicated than the commonly observed straight lines and circles. When exposed to random fluctuations, the mean trajectories turn out to be self-similar curves which bear the characteristics of their noise-free counterparts. Furthermore, we consider a propulsion force which scales in time t as $\propto t^{\alpha}$ (with $\alpha = 0, 1, 2, \ldots$) and analyze the resulting superdiffusive behavior. Our predictions are verifiable for diffusiophoretic artificial microswimmers with prescribed propulsion protocols.

1. Introduction

The description of self-propelled particles and microswimmers is a rapidly growing domain of statistical physics [1–3]. Even the motion of a single swimmer is non-trivial since this is already a non-equilibrium situation that requires new concepts of statistical mechanics. Due to the micron size of the swimmers, inertial effects are negligible so that the Reynolds number is small, but there are Brownian fluctuations as for passive colloidal particles [4–7]. In its simplest form, one can generalize the Brownian dynamics of passive colloidal particles to self-propelled particles by including an additional driving term which leads to a constant propagation speed along the particle orientation. The orientation, however, is subject to thermal fluctuations and therefore there is a non-trivial coupling between particle orientation and translation. For such an active Brownian particle, low-order moments of the time-dependent displacement distribution have been analytically calculated recently [8, 9]. Moreover, the full displacement probability distribution has been studied in experiment and simulation [10]. The established simple picture of a persistent random walk with a persistence generated by the self-propulsion gives a strongly enhanced long-time diffusion constant as compared to a passive particle. In the noise-free limit, the swimmer moves deterministically with a speed $\mathbf{v}$ along its orientation on a straight line.

These results have been generalized for swimmers that are subjected to an internal additional torque. When fluctuations are neglected, this leads to a motion on circles in two dimensions [11,12] and on helical paths in three dimensions [13]. The noise-averaged trajectories are a spira mirabilis [11] in two and a concho-spiral in three dimensions [13]. The former was recently confirmed by experiments on asymmetric self-diffusiophoretic swimmers [12].

In all of the previous work on active Brownian particles [14–20], the effective propulsion was assumed to be independent of time. In this paper, we consider an explicit time dependence of the propagation speed $\mathbf{v}(t)$, which serves as a given input for the Brownian equations of motion. We calculate moments of the displacement probability distribution analytically and thereby generalize results known from previous work [8, 9, 11]. Our motivation to do so is threefold: first, real swimmers usually do not move with a constant propagation speed. In particular, the swimming stroke itself induces variations in time [21]. Even in the simple Golestanian three-sphere swimmer [22] the net motion is time-dependent, not to speak about swimming strokes in real microorganisms such as Chlamydomonas [23–25] or larger swimmers as Daphnia [26–28]. A time dependence on the time scale of the individual swimming stroke is typical rather than an exception. In addition, time-dependent propagation can occur on much longer time scales if bacteria are exposed to chemical or light gradients [29]. Therefore, most importantly, the model considered in this paper generalizes the previous coarse-grained models with constant self-propulsion towards a more realisitic description of the propulsion mechanism itself. Second, artificial diffusiophoretic microswimmers [30–34] offer the fascinating possibility of tuning the propagation speed on demand by varying the laser power externally [35] such that any prescribed form of $\mathbf{v}(t)$ can be programmed and our model is realized. Third, an analytical solution is interesting in itself as it may serve as a simple test case for experimental and simulation data.

We provide analytical solutions for both the noise-free and the noise-averaged swimming paths for time-periodic propulsion under the action of an additional constant torque. When fluctuations are neglected, such an oscillatory swimmer moves on closed trajectories that can be much more complicated than the commonly observed straight lines and circles. In the presence of translational and rotational Brownian random motion, the mean swimming path turns out to be a *self-similar* curve that still bears the characteristics of the noise-free case under very general periodicity assumptions. Self-similarity is known from many other areas of statistical physics, such as fractals [36], growth processes [37], networks [38], and critical phenomena [39]. Therefore our findings introduce the concept of similarity into the world of mean microswimmer paths. As an example for a non-periodic realization of the self-propulsion, we consider a power-law time dependence and show that a propagation speed which scales in time t as $\propto t^{\alpha}$ (with $\alpha = 0, 1, 2, \ldots$) induces superdiffusive behavior characterized by an exponent $2\alpha + 1$ in the time-dependent mean square displacement.

The paper is organized as follows: in section 2 we describe the model equations of active Brownian particles. Results are presented in sections 3 and 4, where in the latter an additional constant torque is considered on top of the time-dependent self-propulsion. Finally, we conclude in section 5.

2. The model

In our model we consider colloidal particles in a dilute solution where particle–particle interactions can be neglected. The dynamics in the low-Reynolds number regime is governed by the Langevin equations for completely over-damped Brownian motion. We assume that the motion of the particles is constrained to a two-dimensional plane. Such a situation is often realized in experiments with microswimmers where gravity keeps them close to the substrate [10, 12, 40]. However, the generalization to three dimensions is straightforward when following the procedure presented in reference [9] for self-propelled particles with constant propagation speed.

The colloid itself is regarded as a sphere with a hydrodynamic radius R. Its swimming path is determined by the center-of-mass position $\mathbf{r}(t) = [x(t), y(t)]$. To account for the detailed self-propulsion mechanism, we consider an effective time-dependent driving force $\mathbf{F}(t) = F(t)\hat{\mathbf{u}}$, where $\hat{\mathbf{u}} = (\cos\phi, \sin\phi)$ is a particle-fixed orientation vector defined by the angle ϕ between the x axis and the direction of propulsion (see figure 1(a)). Thus, the corresponding translational and orientational Langevin equations are given by

$$\frac{dx(t)}{dt} = \beta D \left[F(t)\cos(\phi(t)) + f_x(t)\right] , \tag{1}$$

$$\frac{dy(t)}{dt} = \beta D \left[F(t)\sin(\phi(t)) + f_y(t)\right] , \tag{2}$$

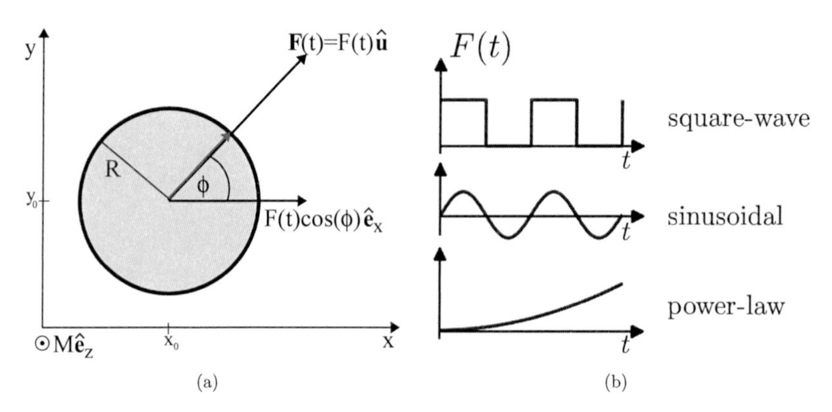

Figure 1. (a) Schematic view of a spherical self-propelled particle with hydrodynamic radius R as considered in our model. The motion is restricted to the two-dimensional x-y plane and characterized by the center-of-mass position $\mathbf{r}(t) = [x(t), y(t)]$ and the angle $\phi(t)$ representing the orientation $\hat{\mathbf{u}} = (\cos\phi, \sin\phi)$ of the particle relative to the x direction. The propulsion speed is determined by an effective time-dependent driving force $\mathbf{F}(t) = F(t)\hat{\mathbf{u}}$ and the particle may additionally be exposed to a constant torque $\mathbf{M} = M\hat{\mathbf{e}}_z$ (see section 4). (b) Overview of the types of self-propulsion with square-wave, sinusoidal, and power-law time dependence that are explicitly considered.

$$\frac{d\phi(t)}{dt} = \beta D_{\mathrm{r}}\, g(t) \tag{3}$$

with the inverse effective thermal energy $\beta = 1/(k_{\mathrm{B}}T)$. Brownian random fluctuations are implemented in equations (1)-(3) by means of zero-mean Gaussian noise terms $f_x(t)$, $f_y(t)$, and $g(t)$. The respective variances are given by $\langle f_x(t)f_x(t')\rangle = \langle f_y(t)f_y(t')\rangle = 2\delta(t-t')/(\beta^2 D)$ and $\langle g(t)g(t')\rangle = 2\delta(t-t')/(\beta^2 D_{\mathrm{r}})$, where angular brackets denote a noise average. The translational and rotational Brownian motion is characterized by the respective short-time diffusion constants D and D_{r} fulfilling $D/D_{\mathrm{r}} = 4R^2/3$ for a spherical particle. As equations (1) and (2) for the motion in x and y direction are formally identical for changed initial conditions, we will only present the results for the x component, but discuss trajectories in the full x-y plane. To solve the system of Langevin equations (1)-(3), first the angular equation (3) is considered. As the noise term $g(t)$ is Gaussian, following Wick's theorem the full angular probability distribution has to be Gaussian as well and can be obtained by calculating the first two moments of $\phi(t)$ (for more details see references [41, 42]). Using the orientational probability distribution as an input for the translational Langevin equations, analytical results for the mean position and the mean square displacement can be derived.

To account for the variable propagation speed which is often observed in the motion of real microswimmers, we study the influence of different types of time-dependent driving forces $F(t)$. Explicitly, we consider piecewise constant, sinusoidal, and power-law realizations of the self-propulsion (see figure 1(b)). A piecewise constant or "square-wave" self-propulsion force (see section 3.1) can mimic biological microorganisms which undergo a run-and-tumble motion [23,43–45], for example. When the swimming stroke itself leads to periodic variations in the propagation speed, a continuous description such as the sinusoidal driving force (see section 3.2) is the most appropriate one. Finally, a power-law type of self-propulsion (see section 3.3) may be relevant for organisms that enhance their swimming velocity by consuming food [46] or in situations where the velocity of a predator is determined by the prey gradient [47]. Furthermore, growing clusters of active particles [48] may require a power-law time dependence for the description of the propulsion.

Whereas some realizations of the different self-propulsion types can be directly studied experimentally with active particles in nature, such as the run-and-tumble motion of biological microorganisms [43, 49], recent progress in the field of artificial colloidal microswimmers makes it possible to tune man-made self-propelled objects in a way such that all considered kinds of swimming behavior are realized. This can be accomplished either by an external magnetic field [50] or in systems where the self-propulsion mechanism is triggered by a light source which can be switched on and off [51] or be regulated in a more sophisticated way [35,52]. Thus, any propulsion protocol can be achieved for diffusiophoretic artificial microswimmers.

3. Time-dependent self-propulsion

3.1. Square-wave self-propulsion force

First, we discuss the mean position and the mean square displacement of a particle propelling through a liquid as governed by the square-wave self-propulsion force

$$F(t) = \begin{cases} F_0 & \text{for } nT < t \leq (n+\frac{1}{2})T \\ 0 & \text{for } (n+\frac{1}{2})T < t \leq (n+1)T \end{cases} \quad \text{with } n = 0, 1, 2, \ldots, \tag{4}$$

where T ist the cycle duration (see inset in figure 2). Active and passive time intervals of equal length alternate. Here, we consider the case of a particle starting with the active regime (constant self-propulsion force F_0). Some of the statements below have to be modified if the particle starts in the exclusively diffusive regime.

3.1.1. Mean position As a result, the one-dimensional mean position of a particle with the self-propulsion force as defined in equation (4) is given by

$$\langle x(t) - x_0 \rangle = \frac{4}{3}\beta F_0 R^2 \cos(\phi_0) \times \begin{cases} \left[\dfrac{e^{D_r T} - e^{-D_r(n-1)T}}{e^{D_r T} + e^{D_r T/2}} + e^{-D_r n T} - e^{-D_r t}\right] & \text{for } nT < t \leq (n+\frac{1}{2})T \\ & \text{and } n = 0, 1, 2, \ldots \\ \left[\dfrac{e^{D_r T} - e^{-D_r n T}}{e^{D_r T} + e^{D_r T/2}}\right] & \text{for } (n+\frac{1}{2})T < t \leq (n+1)T \\ & \text{and } n = 0, 1, 2, \ldots\,. \end{cases} \tag{5}$$

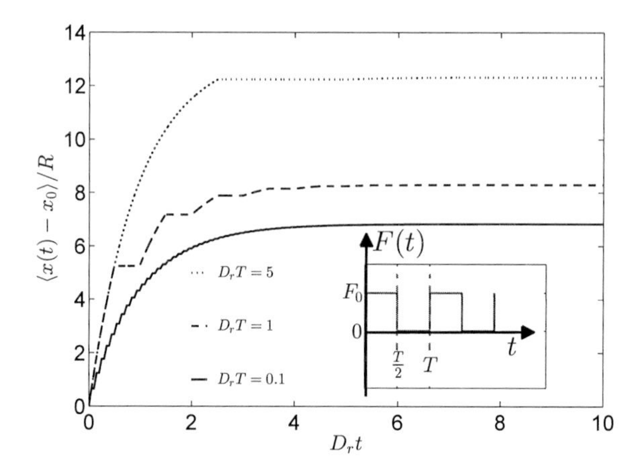

Figure 2. Mean position of a self-propelled particle with a square-wave propulsion force based on the analytical result in equation (5). Curves are shown for different values of the period T and fixed parameters $\beta R F_0 = 10$ and $\phi_0 = 0$. The stair-like form is due to the lack of an active contribution during every second time interval of length $T/2$ (as visualized in the sketch of the square-wave self-propulsion force in the inset). For long times all curves approach a maximum value which depends on the period length T.

Obviously, the mean position increases during a time interval of length $T/2$ and stays constant during the following time interval of the same length. This is also visualized in figure 2, where the dimensionless mean position $\langle x(t) - x_0\rangle/R$ is shown for different values of the scaled period $D_r T$. In all cases the curves exhibit a stair-like form, where the steps are smaller for larger times, and approach a constant value for long times t. This final mean position depends on the period T and is obtained as the asymptotic solution from equation (5):

$$\lim_{t,n\to\infty} \langle x(t) - x_0\rangle = \frac{4}{3}\beta F_0 R^2 \cos(\phi_0) \frac{1}{1 + \mathrm{e}^{-D_r T/2}} \,. \tag{6}$$

Explicitly, the limits $t \to \infty$ for very short ($T \to 0$) and very long ($T \to \infty$) periods are given by

$$\lim_{T\to 0} \lim_{t,n\to\infty} \langle x(t) - x_0\rangle = \frac{2}{3}\beta F_0 R^2 \cos(\phi_0) \tag{7}$$

and

$$\lim_{T\to\infty} \lim_{t,n\to\infty} \langle x(t) - x_0\rangle = \frac{4}{3}\beta F_0 R^2 \cos(\phi_0) \,, \tag{8}$$

respectively. Clearly, the result in equation (7) equals the case of a constant self-propulsion force $F = F_0/2 = \langle F(t)\rangle = (1/T)\int_0^T F(t)\mathrm{d}t$ and equation (8) corresponds to the case of a constant self-propulsion force F_0.

3.1.2. Mean square displacement While the mean position already elucidates some of the physics of microswimmers with time-dependent self-propulsion, usually the standard quantitiy for characterizing the particle dynamics is the mean square displacement $\langle (x(t)-x_0)^2 \rangle$.

Our analytical result is as follows:

$$\begin{aligned}\langle (x(t)-x_0)^2 \rangle = 2Dt + \frac{16}{9}\left(\beta F_0 R^2\right)^2 \Big[& D_\mathrm{r}\left(t - n\frac{T}{2}\right) - n\,\xi\left(\frac{1}{2}\right) - \frac{\cos(2\phi_0)}{12}\,\xi\,(2)\,\rho_{n-1}(4) \\ & + \frac{\cos(2\phi_0)}{3}\,\xi\left(\frac{1}{2}\right)\rho_{n-1}(4) + (n-1)\,\xi\left(\frac{1}{2}\right)\rho_{-1/2}(1)\mathrm{e}^{-D_\mathrm{r}T/2} \\ & + \xi\left(\frac{1}{2}\right)\rho_{-1/2}(1)\,\rho_{-n}(1)\mathrm{e}^{-D_\mathrm{r}(n+1/2)T} \\ & + \frac{\cos(2\phi_0)}{3}\,\xi\left(\frac{1}{2}\right)\rho_{-1/2}(3)\Big(\rho_{n-1}(1) - \rho_{n-1}(4)\Big) \\ & - \mathrm{e}^{-D_\mathrm{r}T/2}\,\xi\left(\frac{1}{2}\right)\rho_{n-1}(1)\mathrm{e}^{D_\mathrm{r}nT}\,\tilde{\xi}(1) - \frac{\cos(2\phi_0)}{3}\,\xi\left(\frac{3}{2}\right)\rho_{n-1}(3)\,\tilde{\xi}(1) \\ & + \mathrm{e}^{D_\mathrm{r}nT}\,\tilde{\xi}(1) + \frac{\cos(2\phi_0)}{12}\,\tilde{\xi}(4) - \frac{\cos(2\phi_0)}{3}\mathrm{e}^{-3D_\mathrm{r}nt}\,\tilde{\xi}(1)\Big] \end{aligned} \tag{9}$$

for $nT < t \leq (n+\frac{1}{2})T$ with $n = 0, 1, 2, \dots$ and, correspondingly,

$$\begin{aligned}\langle (x(t)-x_0)^2 \rangle = 2Dt + \frac{16}{9}\left(\beta F_0 R^2\right)^2 \Big[& \left(\frac{D_\mathrm{r}T}{2} - \xi\left(\frac{1}{2}\right)\right)(n+1) - \frac{\cos(2\phi_0)}{12}\,\xi\,(2)\;\rho_n(4) \\ & + \frac{\cos(2\phi_0)}{3}\,\xi\left(\frac{1}{2}\right)\;\rho_n\,(4) + n\,\xi\left(\frac{1}{2}\right)\;\rho_{-1/2}(1)\mathrm{e}^{-D_\mathrm{r}T/2} \\ & + \xi\left(\frac{1}{2}\right)\mathrm{e}^{-D_\mathrm{r}(n+3/2)T}\;\rho_{-1/2}(1)\;\rho_{-(n+1)}(1) \\ & + \frac{\cos(2\phi_0)}{3}\,\xi\left(\frac{1}{2}\right)\;\rho_{-1/2}(3)\Big(\rho_n(1) - \rho_n(4)\Big)\Big] \end{aligned} \tag{10}$$

for $(n+\frac{1}{2})T < t \leq (n+1)T$. Here, the notations

$$\rho_m(a) := \frac{\mathrm{e}^{aD_\mathrm{r}T} - \mathrm{e}^{-aD_\mathrm{r}Tm}}{\mathrm{e}^{aD_\mathrm{r}T} - 1}, \tag{11}$$

$$\xi(a) := 1 - \mathrm{e}^{-aD_\mathrm{r}T}, \tag{12}$$

$$\tilde{\xi}(a) := \mathrm{e}^{-aD_\mathrm{r}t} - \mathrm{e}^{-aD_\mathrm{r}Tn} \tag{13}$$

are used.

Figure 3 visualizes the mean square displacement for the same parameter combinations as in figure 2 for the mean position. The dominant feature is the stair-like pattern resulting from the square-wave force. As shown by the linear representation in figure 3(a), the steps are significantly more equally sized than with regard to the mean position, where the steps become rapidly flatter with increasing time. The transition from the first active to the first passive regime is most obvious in the logarithmic representation in figure 3(b). In general, a longer period T leads to larger values of the mean square displacement for intermediate and long times. The long-term diffusion coefficient D_l for

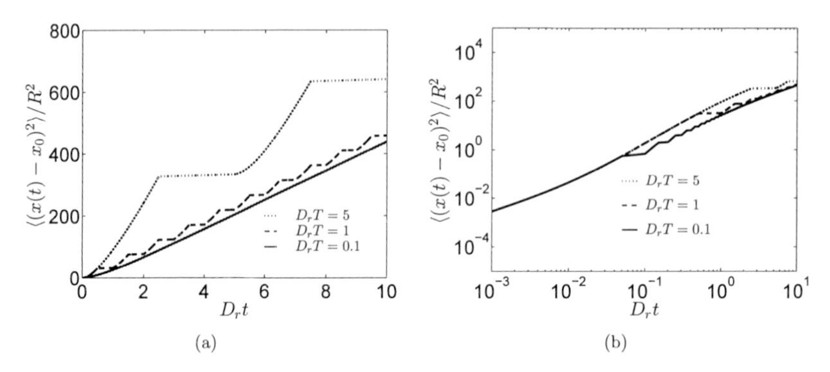

Figure 3. Analytically obtained mean square displacement of a self-propelled particle with a square-wave propulsion force: (a) linear and (b) logarithmic representation for the same situations as in figure 2. A longer period T leads to larger values of the mean square displacement.

$t \gg T$ is analytically calculated as

$$\begin{aligned} D_\mathrm{l} &= \lim_{t\to\infty} \frac{1}{2t}\langle(x(t)-x_0)^2\rangle \\ &= D + \frac{4}{9}\Big(\beta F_0 R^2\Big)^2 D_\mathrm{r} + \frac{8}{9}\Big(\beta F_0 R^2\Big)^2 \,\frac{1}{T}\,\frac{\mathrm{e}^{-D_\mathrm{r}T/2}-1}{1+\mathrm{e}^{-D_\mathrm{r}T/2}}\,. \end{aligned} \tag{14}$$

For a very short period T, equation (14) reduces to

$$\lim_{T\to 0} D_\mathrm{l} = D + \frac{2}{9}\Big(\beta F_0 R^2\Big)^2 D_\mathrm{r} = D_\mathrm{l}\Big|_{F=\langle F(t)\rangle = F_0/2=\mathrm{const.}}\,, \tag{15}$$

which corresponds to the case of a constant self-propulsion force $F = F_0/2$. On the other hand, the limit for a very long period is

$$\lim_{T\to\infty} D_\mathrm{l} = D + \frac{4}{9}\Big(\beta F_0 R^2\Big)^2 D_\mathrm{r}\,. \tag{16}$$

This result exhibits a factor $1/2$ in the second term as compared to the solution for a constant propulsion force F_0, which originates from the linear time dependence of the mean square displacement. During every second time interval of length $T/2$ the particle motion is completely passive so that no contribution resulting from the self-propulsion arises.

3.2. Sinusoidal self-propulsion force

To account for the effect of a continuous time-periodic propulsion, as often induced by the detailed swimming mechanism of biological microorganisms, we solve the Langevin equations (1)-(3) for a sinusoidal self-propulsion force

$$F(t) = F_0(\sin(\omega t) + c)\,. \tag{17}$$

It is characterized by the amplitude F_0, the frequency ω, and the offset cF_0 (see inset in figure 4(a)).

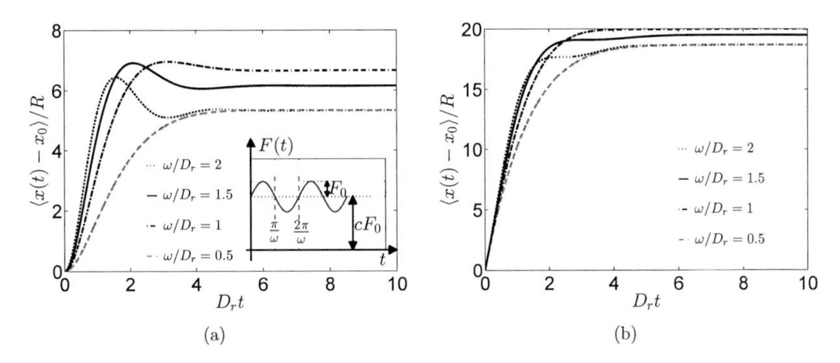

Figure 4. Mean position of a self-propelled particle with sinusoidal driving force for different values of the frequency ω. The offset is $c = 0$ in (a) and $c = 1$ in (b) for fixed values of the scaled amplitude $\beta R F_0 = 10$ and the initial orientation $\phi_0 = 0$. Inset: characterization of the sinusoidal self-propulsion force.

3.2.1. Mean position We obtain

$$\langle x(t) - x_0 \rangle = \frac{4}{3}\beta F_0 R^2 D_\mathrm{r} \cos(\phi_0) \left[\frac{\mathrm{e}^{-D_\mathrm{r} t}}{D_\mathrm{r}^2 + \omega^2} \left(-D_\mathrm{r} \sin(\omega t) - \omega \cos(\omega t) \right) + \frac{\omega}{D_\mathrm{r}^2 + \omega^2} + \frac{c}{D_\mathrm{r}} \left(1 - \mathrm{e}^{-D_\mathrm{r} t} \right) \right] \tag{18}$$

for the particle's mean position. The periodicity resulting from the driving force is washed out for long times. The last term in equation (18) vanishes if no constant contribution is considered in equation (17), i.e., if $c = 0$. In the limit $\omega \to 0$, as well as for $\omega \to \infty$, the mean position equals the solution obtained for a constant self-propulsion force $F = cF_0$.

The analytical expression for the mean position (equation (18)) is visualized in figure 4. The curves initially increase and reach a constant final value for long times, after a transient regime where the effect of the specific periodic type of the self-propulsion is visible. The existence and the position of one or more local maxima for intermediate times depend on the value of ω. For large ω the first maximum occurs earlier in time and is more distinct. At short times, higher values of ω lead in general to higher values of the mean position than observed for smaller ω. The final value for long times, which is analytically given by

$$\lim_{t \to \infty} \langle x(t) - x_0 \rangle = \frac{4}{3}\beta F_0 R^2 D_\mathrm{r} \cos(\phi_0) \left(\frac{\omega}{D_\mathrm{r}^2 + \omega^2} + \frac{c}{D_\mathrm{r}} \right), \tag{19}$$

is maximal for $\omega = D_\mathrm{r}$. For $\omega = \kappa D_\mathrm{r}$ with an arbitrary value of κ it is the same as for $\omega = D_\mathrm{r}/\kappa$.

3.2.2. Mean square displacement The mean square displacement of an active particle with sinusoidal self-propulsion force is

$$
\begin{aligned}
\langle(x(t)-x_0)^2\rangle = 2Dt + \frac{16}{9}\left(\beta F_0 R^2\right)^2 D_\mathrm{r}^2 \Bigg[& \frac{1}{D_\mathrm{r}^2+\omega^2}\left(\frac{D_\mathrm{r} t}{2} - \frac{D_\mathrm{r}\sin(2\omega t)}{4\omega} - \frac{1}{2}\sin^2(\omega t)\right) \\
& - \frac{\omega}{D_\mathrm{r}^2+\omega^2}\,\eta^-(1,1,1) + \frac{\cos(2\phi_0)}{9D_\mathrm{r}^2+\omega^2}\left(\frac{3}{2}\sin(\omega t)\,D_\mathrm{r}\,\eta^0(2,1,4)\right. \\
& \left. + \frac{3}{2}\,\frac{\omega^2}{16D_\mathrm{r}^2+4\omega^2}\left(\mathrm{e}^{-4D_\mathrm{r}t}-1\right) + \frac{\omega}{4}\,\eta_2^-(2,1,4) - \omega\,\eta^-(1,1,1)\right) \\
& + \frac{c}{D_\mathrm{r}}\left(\frac{1-\cos(\omega t)}{\omega} + \eta^-(1,1,1)\right. \\
& \left. + \frac{1}{3}\cos(2\phi_0)\left(\eta^-(4,1,4) - \eta^-(1,1,1)\right)\right) \\
& + \frac{c}{D_\mathrm{r}^2+\omega^2}\left(\frac{D_\mathrm{r}}{\omega}(1-\cos(\omega t)) - \sin(\omega t) + \frac{\omega}{D_\mathrm{r}}\left(1-\mathrm{e}^{-D_\mathrm{r}t}\right)\right) \\
& + \frac{c\,\cos(2\phi_0)}{\omega^2+9D_\mathrm{r}^2}\left(3D_\mathrm{r}\,\eta^-(4,1,4) - \omega\,\tilde{\eta}(-4,1,4)\right) \\
& + \frac{c\,\cos(2\phi_0)}{\omega^2+9D_\mathrm{r}^2}\,\frac{\omega}{D_\mathrm{r}}\left(1-\mathrm{e}^{-D_\mathrm{r}t}\right) + \frac{c^2 t}{D_\mathrm{r}} + \frac{c^2}{D_\mathrm{r}^2}\left(\mathrm{e}^{-D_\mathrm{r}t}-1\right) \\
& + \frac{c^2\,\cos(2\phi_0)}{3D_\mathrm{r}^2}\left(1-\mathrm{e}^{-D_\mathrm{r}t} + \frac{1}{4}\left(\mathrm{e}^{-4D_\mathrm{r}t}-1\right)\right)\Bigg]
\end{aligned} \tag{20}
$$

with the short notations

$$\eta_a^-(b,c,d) := \frac{(bD_\mathrm{r}\sin(a\omega t) + c\omega\cos(a\omega t))\,\mathrm{e}^{-dD_\mathrm{r}t} - c\omega}{(c\omega^2) + (bD_\mathrm{r})^2}\,, \tag{21}$$

$$\eta_a^0(b,c,d) := \frac{(bD_\mathrm{r}\sin(a\omega t) + c\omega\cos(a\omega t))\,\mathrm{e}^{-dD_\mathrm{r}t}}{(c\omega^2) + (bD_\mathrm{r})^2}\,, \tag{22}$$

$$\tilde{\eta}_a(b,c,d) := \frac{(bD_\mathrm{r}\cos(a\omega t) + c\omega\sin(a\omega t))\,\mathrm{e}^{-dD_\mathrm{r}t} - bD_\mathrm{r}}{(c\omega)^2 + (bD_\mathrm{r})^2}\,, \tag{23}$$

$$\eta^-(b,c,d) \equiv \eta_1^-(b,c,d)\,, \tag{24}$$

$$\eta^0(b,c,d) \equiv \eta_1^0(b,c,d)\,, \tag{25}$$

$$\tilde{\eta}(b,c,d) \equiv \tilde{\eta}_1(b,c,d)\,. \tag{26}$$

The result in equation (20) is illustrated in figure 5. Obviously, for $\phi_0 = 0$ and $c = 0$, smaller values of ω lead to a slower but longer initial increase due to the sine in equation (17). Consequently, for short times one obtains a larger mean square displacement for larger ω while for longer times more significant displacements result for smaller values of ω (see figure 5(a)). The curves for $\omega/D_\mathrm{r} = 5$ and $\omega/D_\mathrm{r} = 10$ in figure 5(b) yield some irregularities at short times: the first maximum of the oscillation is particularly large whereas the second one is much smaller than expected. This behavior is induced by the sign changes of the sine in equation (17). As the particle still has some memory of its initial orientation, the mean square displacement is significantly reduced when the propagation direction reverses. For long times, the rotational diffusion eliminates all

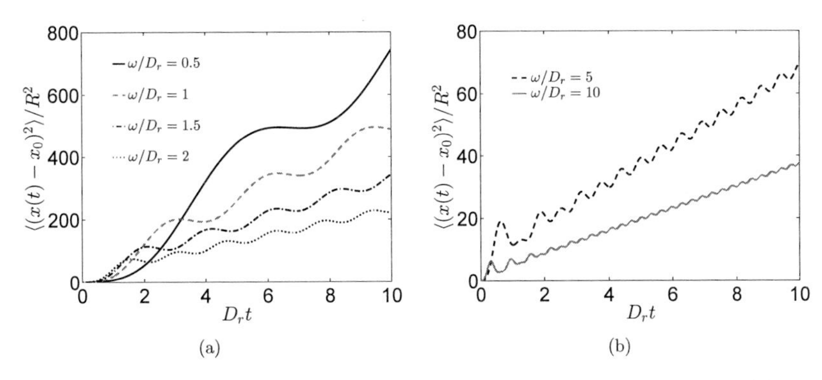

Figure 5. Mean square displacement of a self-propelled particle with sinusoidal driving force for different values of the frequency ω. The parameters are $\beta R F_0 = 10$, $\phi_0 = 0$, and $c = 0$. (a) Curves for low values of ω between 0.5 and 2. (b) For larger ω, a transition from a slowly oscillating initial regime to a regular periodicity with double frequency is observed.

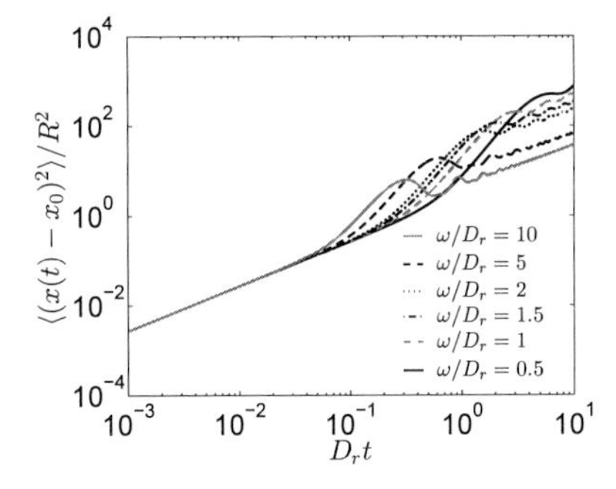

Figure 6. Same curves as in figures 5(a) and 5(b), but now in a logarithmic representation.

orientation-dependent effects. Thus, a periodic behavior with double frequency occurs. It is no longer possible to distinguish between sign changes of the propulsion force from + to − and from − to +. The logarithmic plots in figure 6 represent the particularly large first oscillation even more clearly. It is most obvious for larger values of ω and is followed by a leveled second peak.

By neglecting the exponentially decreasing terms, the long-time behavior of the mean square displacement is obtained as

$$\langle(x(t)-x_0)^2\rangle = 2Dt + \frac{16}{9}\Big(\beta F_0 R^2\Big)^2 D_\mathrm{r}^2\left[\frac{t}{2}\,\frac{D_\mathrm{r}}{D_\mathrm{r}^2+\omega^2} + \frac{tc^2}{D_\mathrm{r}} + \mathrm{const.}\,(\omega,\phi_0,c)\right.$$

$$- \sin(2\omega t)\left(\frac{D_\mathrm{r}}{4\omega}\frac{1}{D_\mathrm{r}^2+\omega^2}\right) - \sin^2(\omega t)\left(\frac{1}{2}\frac{1}{D_\mathrm{r}^2+\omega^2}\right)$$
$$- \sin(\omega t)\frac{c}{D_\mathrm{r}^2+\omega^2} - \cos(\omega t)\left(\frac{c}{D_\mathrm{r}\omega} + \frac{c\,D_\mathrm{r}}{\omega}\frac{1}{D_\mathrm{r}^2+\omega^2}\right)\Bigg]. \quad (27)$$

In the limit $\omega \to 0$ the solutions for a constant self-propulsion force are recovered. Otherwise, for $\omega \neq 0$, the result for the long-time diffusion coefficient D_l is

$$D_\mathrm{l} = \frac{1}{2t}\lim_{t\to\infty}\langle(x(t)-x_0)^2\rangle = D + \left(\beta F_0 R^2\right)^2 D_\mathrm{r}^2\left(\frac{4}{9}\frac{D_\mathrm{r}}{D_\mathrm{r}^2+\omega^2} + \frac{8}{9}\frac{c^2}{D_\mathrm{r}}\right), \quad (28)$$

corresponding to a situation with a constant force $F = cF_0$ if $\omega \to \infty$.

3.3. Power-law self-propulsion force

Finally, we consider a power-law time dependence

$$F(t) = F_0\,(D_\mathrm{r}t)^\alpha \qquad \text{with } \alpha = 0, 1, 2, \ldots \quad (29)$$

and an arbitrary but constant prefactor F_0. In principle, on large time scales the proportionality of the driving force to t^α with $\alpha > 0$ corresponds to a random walk with continuously increasing step size.

3.3.1. Mean position Solving equations (1)-(3) for a self-propulsion according to equation (29) gives the mean position

$$\langle x(t) - x_0\rangle = \frac{4}{3}\beta F_0 R^2 \cos(\phi_0)\left[\alpha! - \sum_{k=0}^{\alpha}(D_\mathrm{r}t)^{\alpha-k}\,\mathrm{e}^{-D_\mathrm{r}t}\frac{\alpha!}{(\alpha-k)!}\right] \quad (30)$$

with the long-time limit

$$\lim_{t\to\infty}\langle x(t) - x_0\rangle = \frac{4}{3}\beta F_0 R^2\cos(\phi_0)\,\alpha!\,. \quad (31)$$

For $\alpha = 1, 2, 3$ the mean position is visualized in figure 7. For short and intermediate times the curves increase until they reach a constant final value which depends on the specific exponent α. The larger the value of α the longer lasts the initial stage and the higher is the final mean displacement.

3.3.2. Mean square displacement The one-dimensional mean square displacement for a power-law self-propulsion force is

$$\langle(x(t)-x_0)^2\rangle = 2Dt + \frac{16}{9}\left(\beta F_0 R^2\right)^2\Bigg\{\sum_{k=0}^{\alpha}(-1)^k\frac{\alpha!}{(\alpha-k)!}\frac{1}{2\alpha-k+1}(D_\mathrm{r}t)^{2\alpha-k+1}$$
$$+ (-1)^\alpha\alpha!\left(\sum_{k=0}^{\alpha}(D_\mathrm{r}t)^{\alpha-k}\mathrm{e}^{-D_\mathrm{r}t}\frac{\alpha!}{(\alpha-k)!} - \alpha!\right)$$
$$+ \cos(2\phi_0)\Bigg[\sum_{k=0}^{\alpha}\frac{1}{3^{k+1}}\frac{\alpha!}{(\alpha-k)!}\left(\sum_{j=0}^{2\alpha-k}\frac{1}{4}(D_\mathrm{r}t)^{2\alpha-k-j}\mathrm{e}^{-4D_\mathrm{r}t}\frac{(2\alpha-k)!}{(2\alpha-k-j)!}\right.$$

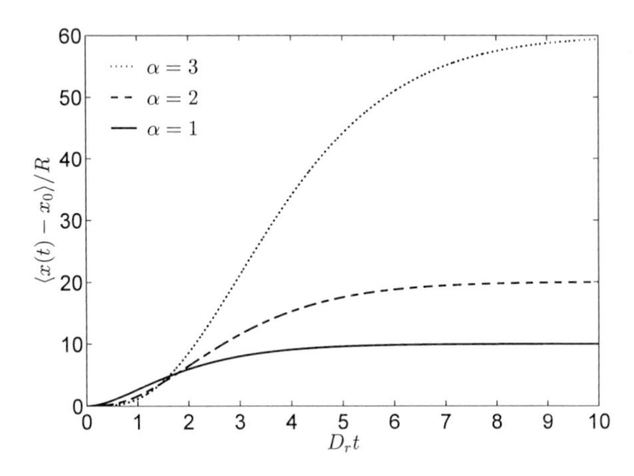

Figure 7. Analytically calculated mean position of a self-propelled particle with a power-law self-propulsion force for different exponents α. The constant parameters are $\beta R F_0 = 10$ and $\phi_0 = 0$. All curves approach a maximum value determined by equation (31).

$$ -\frac{1}{4}(2\alpha - k)!\bigg) - \frac{\alpha!}{3^{\alpha+1}}\bigg(\sum_{k=0}^{\alpha}(D_r t)^{\alpha-k}\mathrm{e}^{-D_r t}\frac{\alpha!}{(\alpha-k)!} - \alpha!\bigg)\bigg]\bigg\}. \tag{32} $$

While equation (32) depends on the initial orientation ϕ_0 of the particle, the mean square displacement can also be given in the two-dimensional version

$$ \langle(\mathbf{r}(t) - \mathbf{r}_0)^2\rangle = 4Dt + \frac{32}{9}\left(\beta F_0 R^2\right)^2\Bigg[\sum_{k=0}^{\alpha}(-1)^k\frac{\alpha!}{(\alpha-k)!\,(2\alpha-k+1)}(D_r t)^{2\alpha-k+1} + (-1)^\alpha\alpha!\bigg(\sum_{k=0}^{\alpha}(D_r t)^{\alpha-k}\mathrm{e}^{-D_r t}\frac{\alpha!}{(\alpha-k)!} - \alpha!\bigg)\Bigg], \tag{33} $$

which is independent of the initial conditions. The visualization of equation (33) in figure 8 exhibits three qualitatively different time regimes. A diffusive regime at short times is followed by a superdiffusive $\propto t^{2\alpha+2}$ regime. Finally, for $t > 1/D_r$, which corresponds to the characteristic time scale for the rotational Brownian motion, the curves enter another superdiffusive regime, where the scaling is $\propto t^{2\alpha+1}$. Whereas the time for this last transition does not depend on the exponent α of the self-propulsion force, the crossover from the diffusive to the first superdiffusive regime occurs at the time

$$ D_r t^*(\alpha) = \left[\frac{2}{3}(\beta F_0 R)^2\left(\sum_{k=0}^{\alpha}(-1)^k\frac{(\alpha!)^2}{(\alpha-k)!(\alpha+k+2)!}\right)\right]^{-1/(2\alpha+1)}, \tag{34} $$

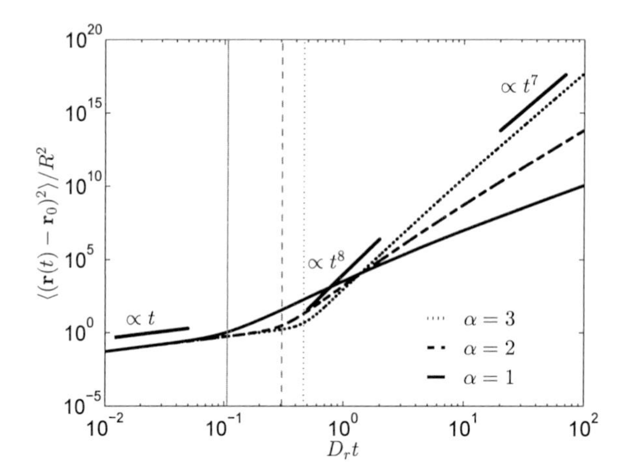

Figure 8. Mean square displacement of a self-propelled particle with a power-law driving force in a logarithmic representation. The strength of the self-propulsion is given by $\beta RF_0 = 100$. Three different regimes $\propto t$, $\propto t^{2\alpha+2}$, and $\propto t^{2\alpha+1}$ are identified and explicitly indicated for the exponent $\alpha = 3$.

which is determined by the specific type of power law and the propulsion strength. For the various curves in figure 8 the transition time according to equation (34) is indicated by vertical lines.

4. Results for an additional constant torque

For many experimental systems it is possible to describe the motion of the respective natural or artificial microswimmers by implementing only an effective self-propulsion force corresponding to a translational swimming velocity [8, 10, 53] in the Langevin equations. However, a more detailed investigation often yields that either particle imperfections or asymmetric shapes [12, 54–56] induce a deterministic rotational motion of the swimming object. To account for this, an additional torque $\mathbf{M} = M\hat{\mathbf{e}}_z$ (see figure 1) has to be considered in the orientational Langevin equation, while equations (1) and (2) stay the same. The updated version of equation (3) is given by

$$\frac{d\phi}{dt} = \beta D_\mathrm{r} \left[M + g(t)\right] , \tag{35}$$

which leads to $\langle\phi(t)\rangle = \phi_0 + \beta D_\mathrm{r} M t$ and $\langle(\phi(t) - \langle\phi(t)\rangle)^2\rangle = 2D_\mathrm{r} t$ for the first and second moments of the angular displacement distribution. As a constant torque does not destroy the Gaussianity of the orientational distribution [9], the Langevin equations can be solved similarly to the torque-free case discussed in section 3. In the following, this is done exemplarily for the sinusoidal self-propulsion force (see section 3.2).

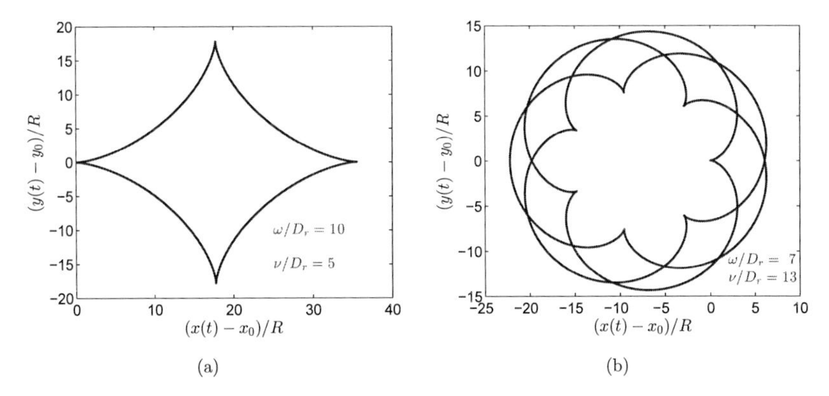

Figure 9. Noise-free trajectories of a self-propelled particle with sinusoidal self-propulsion and an additional constant torque. The plots are based on equation (37). Curves are shown for $\beta RF_0 = 100$, $\phi_0 = 0$, $c = 0$, and different values of ω and ν: (a) $\omega = 10D_\mathrm{r}$ and $\nu = 5D_\mathrm{r}$; (b) $\omega = 7D_\mathrm{r}$ and $\nu = 13D_\mathrm{r}$. For the case $c = 0$, closed trajectories are obtained as long as $\omega \neq \nu$.

4.1. Trajectories for vanishing noise

To gain a better understanding of the interplay between the oscillating driving force and an additional constant torque, we first consider the noise-free case by neglecting the random terms in equations (1), (2), and (35). With $\nu = \beta D_\mathrm{r} M$ this leads to

$$\phi(t) = \phi_0 + \nu t \tag{36}$$

for the rotational motion and

$$\begin{aligned} x(t) - x_0 = \beta D F_0 \Bigg\{ & \cos(\phi_0) \left[-\frac{1}{2} \left(\frac{\cos(t(\omega-\nu))}{\omega-\nu} + \frac{\cos(t(\omega+\nu))}{\omega+\nu} \right) + \frac{\omega}{\omega^2-\nu^2} \right] \\ & - \sin(\phi_0) \frac{1}{2} \left(\frac{\sin(t(\omega-\nu))}{\omega-\nu} - \frac{\sin(t(\omega+\nu))}{\omega+\nu} \right) \\ & + \frac{c}{\nu} (\sin(\phi_0 + \nu t) - \sin(\phi_0)) \Bigg\} \end{aligned} \tag{37}$$

for the translational particle displacement.

As illustrated in figure 9, for $c = 0$ the particle moves on closed trajectories which display a certain number of vertices in a regular pattern. They occur whenever the sign of the propulsion force changes, i.e., in time steps of π/ω. If $\omega > \nu$, the vertices face outward; for $\omega < \nu$ they face inward. The number N of vertices per closed loop depends on the frequencies ω and ν according to

$$\begin{aligned} N &= 2\omega \ \mathrm{lcm}\left(\frac{1}{\omega+\nu}, \frac{1}{|\omega-\nu|} \right) \\ &= \frac{2\omega}{(\omega+\nu)|\omega-\nu|} \ \mathrm{lcm}\left(|\omega-\nu|, \omega+\nu\right) . \end{aligned} \tag{38}$$

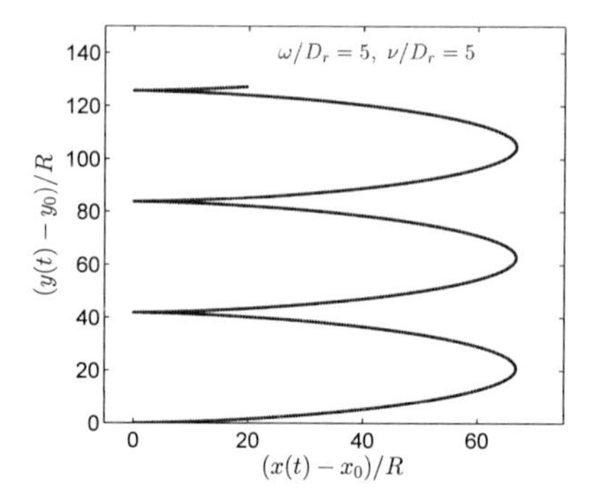

Figure 10. Same as in figure 9, but now for $\omega = \nu = 5D_{\mathrm{r}}$. The special case $\omega = \nu$ is the only situation for which the trajectories are not closed.

Here, lcm denotes the least common multiple as the product of the highest order of each prime factor. It is generalized to fractions by also allowing negative exponents.

Closed trajectories are always obtained as long as $\omega \neq \nu$. However, for the special case $\omega = \nu$ the propulsion direction reverses just at the moment when the orientation has changed by 180°. Thus, a trajectory with a continuously increasing displacement in one direction is established (see figure 10).

4.2. Mean position

For non-zero noise, the mean position of an active particle with self-propulsion as defined in equation (17) is a linear superposition of the contributions originating from a purely sinusoidal force $F(t) = F_0 \sin(\omega t)$ on the one hand and a constant force $F = cF_0$ on the other hand. As the latter case has already been considered in reference [42], here we present only the result for $c = 0$:

$$\begin{aligned}\langle x(t) - x_0\rangle = \beta D F_0 \Bigg[&\frac{\mathrm{e}^{-D_{\mathrm{r}}t}}{2}\Bigg(-\frac{D_{\mathrm{r}} \sin\left(t(\omega-\nu) - \phi_0\right) + (\omega-\nu)\cos\left(t(\omega-\nu)-\phi_0\right)}{(\omega-\nu)^2 + D_{\mathrm{r}}^2} \\ &- \frac{D_{\mathrm{r}} \sin\left(t(\omega+\nu) + \phi_0\right) + (\omega+\nu)\cos\left(t(\omega+\nu)+\phi_0\right)}{(\omega+\nu)^2 + D_{\mathrm{r}}^2}\Bigg) \\ &+ \frac{1}{2}\frac{(\omega-\nu)\cos(\phi_0) - D_{\mathrm{r}}\sin(\phi_0)}{(\omega-\nu)^2 + D_{\mathrm{r}}^2} + \frac{1}{2}\frac{(\omega+\nu)\cos(\phi_0) + D_{\mathrm{r}}\sin(\phi_0)}{(\omega+\nu)^2 + D_{\mathrm{r}}^2}\Bigg]\,. \end{aligned} \tag{39}$$

The corresponding mean trajectories are similar to the noise-free ones (see figures 11 and 12 as compared to figures 9 and 10, respectively). However, when taking the Brownian random terms into account, we do not obtain closed mean swimming paths. Instead of that, the size of the curves reduces exponentially. As can be seen clearly in figures 11 and 12, the mean trajectories are *self-similar*. This characteristic feature also follows directly from the analytical expression in equation (39). The scaling factor for the

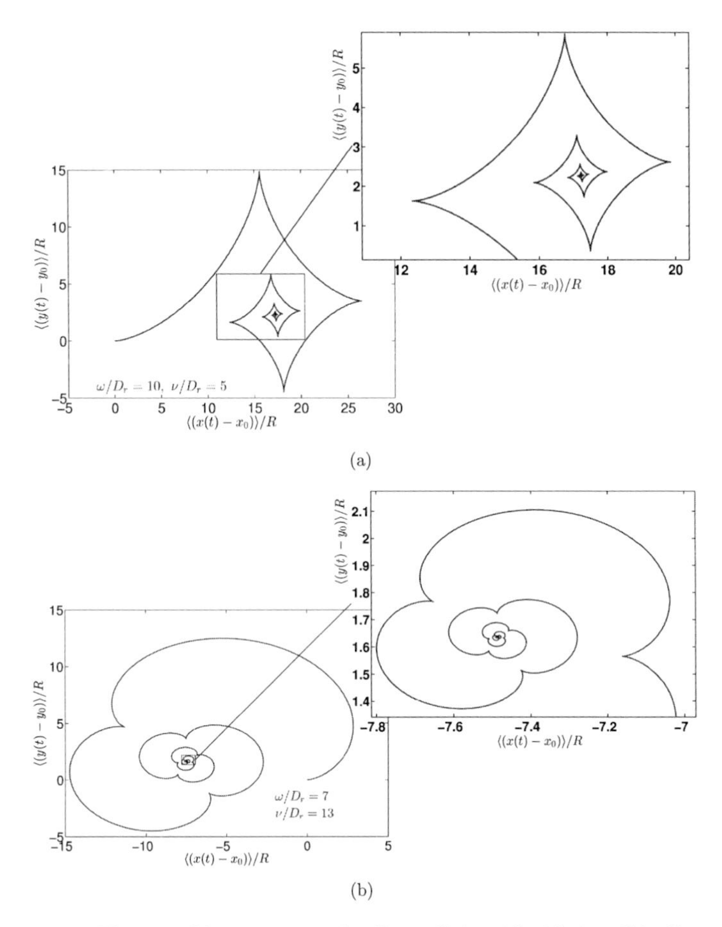

Figure 11. Mean trajectories of a self-propelled particle with sinusoidal self-propulsion and an additional constant torque for $\beta R F_0 = 100$, $\phi_0 = 0$, $c = 0$, and different values of ω and ν: (a) $\omega = 10D_{\mathrm{r}}$ and $\nu = 5D_{\mathrm{r}}$; (b) $\omega = 7D_{\mathrm{r}}$ and $\nu = 13D_{\mathrm{r}}$. The curves are self-similar, as illustrated by the closeups of the framed regions, and bear the same characteristics as their noise-free counterparts (see figure 9).

self-similarity is $\mathrm{e}^{-D_{\mathrm{r}}t}$. All other terms are periodic in time t with a period of either $T_1 = 2\pi/(\omega+\nu)$ or $T_2 = 2\pi/|\omega-\nu|$. Thus, after

$$T = 2\pi \ \mathrm{lcm}\left(\frac{1}{\omega+\nu}, \frac{1}{|\omega-\nu|}\right) \tag{40}$$

the scaled trajectory overlaps with itself.

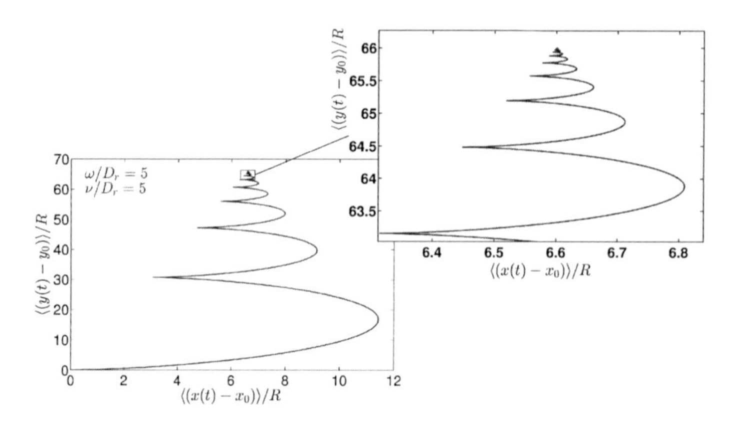

Figure 12. Noise-averaged trajectory of a self-propelled particle with sinusoidal self-propulsion and an additional constant torque for $\omega = \nu = 5D_\mathrm{r}$, $\beta RF_0 = 100$, $\phi_0 = 0$, and $c = 0$. Similar to the noise-free counterpart (see figure 10), one obtains a trajectory with a preferred direction of translation. The curve is self-similar, as visualized in the closeup of the framed area in the plot.

The self-similarity is an important property of mean microswimmer trajectories, also in the context of a comparison with the situation of a constant force, where the mean swimming path was shown to be a logarithmic spiral [11]. While the latter is one of the simplest realizations of a self-similar curve, it is not intuitive that this feature also survives when a sinusoidal self-propulsion force is considered.

4.3. Mean square displacement

The analytical expression for the mean square displacement of a self-propelled particle with sinusoidal self-propulsion and an additional constant torque is given in equation (A.1) in the appendix. Figure 13 shows the corresponding curves for different values of the frequencies ω and ν. At long times, the special case $\omega = \nu$ induces much larger values for the mean square displacement than obtained for $\omega \neq \nu$. This can easily be explained by comparing the mean trajectories in figure 11 with figure 12. Only the situation $\omega = \nu$, where the sign of the self-propulsion changes exactly after half a revolution of the particle, generates a motion primarily in one specific direction. This results in a much higher value for the mean square displacement.

The long-term diffusion coefficient is analytically given by

$$D_\mathrm{l} = D + \frac{D_\mathrm{r}}{8}\beta^2 D^2 F_0^2 \left[\frac{1}{(\omega-\nu)^2 + D_\mathrm{r}^2} + \frac{1}{(\omega+\nu)^2 + D_\mathrm{r}^2} + \frac{4c^2}{D_\mathrm{r}^2 + \nu^2}\right]. \quad (41)$$

For $\omega = \nu$ the first term inside the square brackets in equation (41) becomes maximal because only the squared rotational diffusion coefficient remains in the denominator.

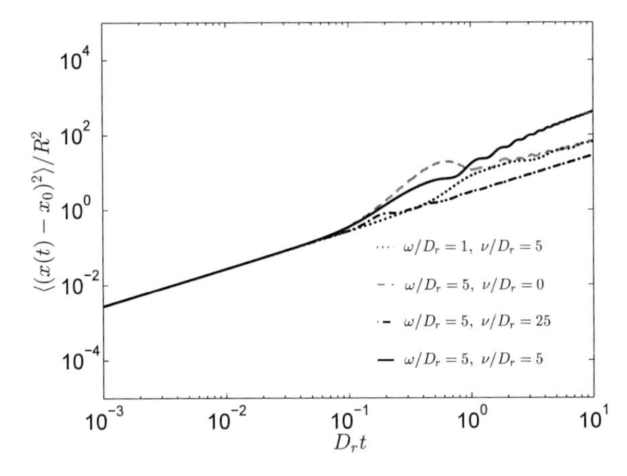

Figure 13. Mean square displacement of a self-propelled particle with sinusoidal self-propulsion force and an additional constant torque for different values of ω and ν. The fixed parameters are given by $\beta R F_0 = 10$, $\phi_0 = 0$, and $c = 0$. At long times, the mean square displacement is much larger for the special case $\omega = \nu$ than for $\omega \neq \nu$.

In contrast, a strong torque significantly reduces the mean square displacement, as illustrated by the solid curve in figure 13.

5. Conclusion

In conclusion, we have studied the influence of a time-dependent self-propulsion on the Brownian dynamics of an active colloidal particle. Our model based on the coupled translational and rotational Langevin equations provides analytical solutions for the mean position and the mean square displacement of swimmers with either time-periodic or continuously increasing propagation speed. Thus, previous coarse-grained models with constant self-propulsion are generalized towards a more realistic description of the propulsion mechanism, also on the time scale of an individual swimming stroke. The analysis yields that the noise-free path of a time-dependent swimmer can be quite complex, involving trajectories much more complicated than the commonly observed straight lines and circles. Moreover, we have analytically calculated the noise-averaged trajectories for time-periodic propulsion under the action of an additional torque. Interestingly, such an oscillatory microswimmer moves on average on a self-similar curve. If the effective self-propulsion force scales in time as $\propto t^{\alpha}$, superdiffusive behavior is found in the long-time regime where the mean square displacement reveals a $\propto t^{2\alpha+1}$ time dependence after an intermediate regime with a scaling $\propto t^{2\alpha+2}$. These new exponents are expected to also affect the non-Gaussian behavior of self-propelled particles [10,42].

An interesting next step would be to include a time dependence not only with regard to the self-propulsion force but also for the additional torque. Such a variation in time can either be externally prescribed by a magnetic field [50, 57], for example, or it can be of stochastic nature, as observed for slightly curved rods which undergo fluctuation-induced flipping leading to two equivalent stable states with an opposite sign of the torque [58].

Acknowledgment

This work was supported by the ERC Advanced Grant INTERCOCOS (Grant No. 267499).

Appendix

Here, we present the analytical result for the mean square displacement of a self-propelled particle with sinusoidal self-propulsion under the action of an additional constant torque. It is given by

$$\begin{aligned}\langle (x(t)-x_0)^2\rangle = 2Dt + \beta^2 D^2 F_0^2 \Bigg\{ & \frac{D_\mathrm{r}}{2}\,\omega_1^+ \left(\frac{t}{2} - \frac{\sin(2\omega t)}{4\omega} + \frac{c}{\omega}\,(1-\cos(\omega t))\right) \\ & - \frac{1}{2}\tilde{\omega}_1^+ \left(\frac{\sin^2(\omega t)}{2\omega} + \frac{c}{\omega}\sin(\omega t)\right) + \frac{c\,D_\mathrm{r}}{D_\mathrm{r}^2+\nu^2}\left(\frac{1}{\omega}\,(1-\cos(\omega t)) + ct\right) \\ & + \left(\frac{1}{2}\tilde{\omega}_1^+ - \frac{c\,D_\mathrm{r}}{D_\mathrm{r}^2+\nu^2}\right)(A_1+cB_1) + \left(\frac{1}{2}D_\mathrm{r}\,\omega_1^- + \frac{c\,\nu}{D_\mathrm{r}^2+\nu^2}\right)(A_2+cB_2) \\ & + \frac{1}{2}\sum_{k=\pm 1}\frac{1}{(\omega-k\nu)^2+9D_\mathrm{r}^2}\Big[-3D_\mathrm{r}\Big(Z_2(4D_\mathrm{r},-2\phi_0,\omega-2k\nu) \\ & + c\,Y_1(-4D_\mathrm{r},-2\phi_0,\omega-2k\nu)\Big) \\ & - (\omega-k\nu)\Big(Z_1(4D_\mathrm{r},-2\phi_0,\omega-2k\nu) + cY_2(-4D_\mathrm{r},-2\phi_0,\omega-2k\nu)\Big)\Big] \\ & + \frac{1}{2}\tilde{\omega}_3^+\Big(Z_1(D_\mathrm{r},2\phi_0,\nu) + c\,Y_2(-D_\mathrm{r},2\phi_0,\nu)\Big) \\ & + \frac{3}{2}D_\mathrm{r}\,\omega_3^-\Big(Z_2(D_\mathrm{r},2\phi_0,\nu) + c\,Y_1(-D_\mathrm{r},2\phi_0,\nu)\Big) \\ & + \frac{c}{9D_\mathrm{r}^2+\nu^2}\Big[-3D_\mathrm{r}\Big(Z_1(4D_\mathrm{r},2\phi_0,2\nu) + c\,Y_2(-4D_\mathrm{r},2\phi_0,2\nu) \\ & - Z_1(D_\mathrm{r},2\phi_0,\nu) - c\,Y_2(-D_\mathrm{r},2\phi_0,\nu)\Big) \\ & + \nu\,\Big(Z_2(4D_\mathrm{r},2\phi_0,2\nu) + c\,Y_1(-4D_\mathrm{r},2\phi_0,2\nu) \\ & - Z_2(D_\mathrm{r},2\phi_0,\nu) - c\,Y_1(-D_\mathrm{r},2\phi_0,\nu)\Big)\Big]\Bigg\}, \end{aligned} \tag{A.1}$$

where

$$\omega_a^+ = \frac{1}{(\omega+\nu)^2+(aD_\mathrm{r})^2} + \frac{1}{(\omega-\nu)^2+(aD_\mathrm{r})^2}, \tag{A.2}$$

$$\tilde{\omega}_a^+ = \frac{\omega+\nu}{(\omega+\nu)^2+(aD_\mathrm{r})^2} + \frac{\omega-\nu}{(\omega-\nu)^2+(aD_\mathrm{r})^2}, \tag{A.3}$$

$$\omega_a^- = \frac{1}{(\omega+\nu)^2+(aD_\mathrm{r})^2} - \frac{1}{(\omega-\nu)^2+(aD_\mathrm{r})^2}, \tag{A.4}$$

$$\begin{aligned} A_1 =& \frac{\mathrm{e}^{-D_\mathrm{r}t}}{2}\Bigg[-\frac{D_\mathrm{r}\sin\left(t\left(\omega-\nu\right)\right)}{(\omega-\nu)^2+D_\mathrm{r}^2} - \frac{(\omega-\nu)\cos\left(t(\omega-\nu)\right)}{(\omega-\nu)^2+D_\mathrm{r}^2} \\ &- \frac{D_\mathrm{r}\sin\left(t\left(\omega+\nu\right)\right)}{(\omega+\nu)^2+D_\mathrm{r}^2} - \frac{(\omega+\nu)\cos\left(t(\omega+\nu)\right)}{(\omega+\nu)^2+D_\mathrm{r}^2}\Bigg] \\ &+ \frac{1}{2}\frac{\omega-\nu}{(\omega-\nu)^2+D_\mathrm{r}^2} + \frac{1}{2}\frac{\omega+\nu}{(\omega+\nu)^2+D_\mathrm{r}^2}, \end{aligned} \tag{A.5}$$

$$\begin{aligned} A_2 =& \frac{\mathrm{e}^{-D_\mathrm{r}t}}{2}\Bigg[\frac{(\omega-\nu)\sin\left(t\left(\omega-\nu\right)\right)}{(\omega-\nu)^2+D_\mathrm{r}^2} - \frac{D_\mathrm{r}\cos\left(t(\omega-\nu)\right)}{(\omega-\nu)^2+D_\mathrm{r}^2} \\ &- \frac{(\omega+\nu)\sin\left(t\left(\omega+\nu\right)\right)}{(\omega+\nu)^2+D_\mathrm{r}^2} + \frac{D_\mathrm{r}\cos\left(t(\omega+\nu)\right)}{(\omega+\nu)^2+D_\mathrm{r}^2}\Bigg] \\ &+ \frac{1}{2}\frac{D_\mathrm{r}}{(\omega-\nu)^2+D_\mathrm{r}^2} - \frac{1}{2}\frac{D_\mathrm{r}}{(\omega+\nu)^2+D_\mathrm{r}^2}, \end{aligned} \tag{A.6}$$

$$B_1 = \frac{\mathrm{e}^{-D_\mathrm{r}t}}{D_\mathrm{r}^2+\nu^2}\Big(\nu\sin\left(\nu t\right) - D_\mathrm{r}\cos\left(\nu t\right)\Big) + \frac{D_\mathrm{r}}{D_\mathrm{r}^2+\nu^2}, \tag{A.7}$$

$$B_2 = \frac{\mathrm{e}^{-D_\mathrm{r}t}}{D_\mathrm{r}^2+\nu^2}\Big(-D_\mathrm{r}\sin\left(\nu t\right) - \nu\cos\left(\nu t\right)\Big) + \frac{\nu}{D_\mathrm{r}^2+\nu^2}, \tag{A.8}$$

$$Y_1(a,b,c) = \frac{\mathrm{e}^{at}}{a^2+c^2}\left(a\ \sin(ct+b) - c\cos(ct+b)\right) - \frac{a\ \sin(b) - c\cos(b)}{a^2+c^2}, \tag{A.9}$$

$$Y_2(a,b,c) = \frac{\mathrm{e}^{at}}{a^2+c^2}\left(a\ \cos(ct+b) + c\sin(ct+b)\right) - \frac{a\ \cos(b) + c\sin(b)}{a^2+c^2}, \tag{A.10}$$

$$\begin{aligned} Z_1(a,b,c) =& \frac{\mathrm{e}^{-at}}{2}\Bigg[\frac{-a\sin\left(t(\omega-c)-b\right)}{(\omega-c)^2+a^2} - \frac{(\omega-c)\cos\left(t(\omega-c)-b\right)}{(\omega-c)^2+a^2} \\ &- \frac{a\sin\left(t(\omega+c)+b\right)}{(\omega+c)^2+a^2} - \frac{(\omega+c)\cos\left(t(\omega+c)+b\right)}{(\omega+c)^2+a^2}\Bigg] \\ &+ \frac{1}{2}\frac{(\omega-c)\cos\left(b\right)}{(\omega-c)^2+a^2} + \frac{1}{2}\frac{(\omega+c)\cos\left(b\right)}{(\omega+c)^2+a^2} \\ &- \frac{1}{2}\frac{a\sin\left(b\right)}{(\omega-c)^2+a^2} + \frac{1}{2}\frac{a\sin\left(b\right)}{(\omega+c)^2+a^2}, \end{aligned} \tag{A.11}$$

$$\begin{aligned} Z_2(a,b,c) =& \frac{\mathrm{e}^{-at}}{2}\Bigg[\frac{-a\cos\left(t(\omega-c)-b\right)}{(\omega-c)^2+a^2} + \frac{(\omega-c)\sin\left(t(\omega-c)-b\right)}{(\omega-c)^2+a^2} \\ &+ \frac{a\cos\left(t(\omega+c)+b\right)}{(\omega+c)^2+a^2} - \frac{(\omega+c)\sin\left(t(\omega+c)+b\right)}{(\omega+c)^2+a^2}\Bigg] \\ &+ \frac{1}{2}\frac{(\omega-c)\sin\left(b\right)}{(\omega-c)^2+a^2} + \frac{1}{2}\frac{(\omega+c)\sin\left(b\right)}{(\omega+c)^2+a^2} \\ &+ \frac{1}{2}\frac{a\cos\left(b\right)}{(\omega-c)^2+a^2} - \frac{1}{2}\frac{a\cos\left(b\right)}{(\omega+c)^2+a^2}. \end{aligned} \tag{A.12}$$

References

[1] Cates M E, *Diffusive transport without detailed balance in motile bacteria: does microbiology need statistical physics?*, 2012 *Rep. Prog. Phys.* **75** 042601.

[2] Romanczuk P, Bär M, Ebeling W, Lindner B, and Schimansky-Geier L, *Active Brownian particles*, 2012 *Eur. Phys. J. Special Topics* **202** 1.

[3] Marchetti M C, Joanny J F, Ramaswamy S, Liverpool T B, Prost J, Rao M, and Simha R A, *Hydrodynamics of soft active matter*, 2013 *Rev. Mod. Phys.* **85** 1143.

[4] Ivlev A, Löwen H, Morfill G, and Royall C P, 2012 *Complex Plasmas and Colloidal Dispersions: Particle-resolved Studies of Classical Liquids and Solids (Series in Soft Condensed Matter* vol 5*)* (Hackensack: World Scientific).

[5] Löwen H, *Colloidal soft matter under external control*, 2001 *J. Phys.: Condens. Matter* **13** R415.

[6] Löwen H, *Brownian dynamics of hard spherocylinders*, 1994 *Phys. Rev. E* **50** 1232.

[7] Kirchhoff T, Löwen H, and Klein R, *Dynamical correlations in suspensions of charged rodlike macromolecules*, 1996 *Phys. Rev. E* **53** 5011.

[8] Howse J R, Jones R A L, Ryan A J, Gough T, Vafabakhsh R, and Golestanian R, *Self-motile colloidal particles: from directed propulsion to random walk*, 2007 *Phys. Rev. Lett.* **99** 048102.

[9] ten Hagen B, van Teeffelen S, and Löwen H, *Brownian motion of a self-propelled particle*, 2011 *J. Phys.: Condens. Matter* **23** 194119.

[10] Zheng X, ten Hagen B, Kaiser A, Wu M, Cui H, Silber-Li Z, and Löwen H, *Non-Gaussian statistics for the motion of self-propelled Janus particles: experiment versus theory*, 2013 *Phys. Rev. E* **88** 032304.

[11] van Teeffelen S and Löwen H, *Dynamics of a Brownian circle swimmer*, 2008 *Phys. Rev. E* **78** 020101(R).

[12] Kümmel F, ten Hagen B, Wittkowski R, Buttinoni I, Eichhorn R, Volpe G, Löwen H, and Bechinger C, *Circular motion of asymmetric self-propelling particles*, 2013 *Phys. Rev. Lett.* **110** 198302.

[13] Wittkowski R and Löwen H, *Self-propelled Brownian spinning top: dynamics of a biaxial swimmer at low Reynolds numbers*, 2012 *Phys. Rev. E* **85** 021406.

[14] Wensink H H, Dunkel J, Heidenreich S, Drescher K, Goldstein R E, Löwen H, and Yeomans J M, *Meso-scale turbulence in living fluids*, 2012 *Proc. Nat. Acad. Sci. USA* **109** 14308.

[15] Wensink H H and Löwen H, *Emergent states in dense systems of active rods: from swarming to turbulence*, 2012 *J. Phys.: Condens. Matter* **24** 464130.

[16] Kaiser A and Löwen H, *Vortex arrays as emergent collective phenomena for circle swimmers*, 2013 *Phys. Rev. E* **87** 032712.

[17] Redner G S, Hagan M F, and Baskaran A, *Structure and dynamics of a phase-separating active colloidal fluid*, 2013 *Phys. Rev. Lett.* **110** 055701.

[18] Cates M E and Tailleur J, *When are active Brownian particles and run-and-tumble particles equivalent? Consequences for motility-induced phase separation*, 2013 *Europhys. Lett.* **101** 20010.

[19] Bialké J, Speck T, and Löwen H, *Crystallization in a dense suspension of self-propelled particles*, 2012 *Phys. Rev. Lett.* **108** 168301.

[20] Buttinoni I, Bialké J, Kümmel F, Löwen H, Bechinger C, and Speck T, *Dynamical clustering and phase separation in suspensions of self-propelled colloidal particles*, 2013 *Phys. Rev. Lett.* **110** 238301.

[21] Friedrich B M, Riedel-Kruse I H, Howard J, and Jülicher F, *High-precision tracking of sperm swimming fine structure provides strong test of resistive force theory*, 2010 *J. Exp. Biol.* **213** 1226.

[22] Najafi A and Golestanian R, *Simple swimmer at low Reynolds number: three linked spheres*, 2004 *Phys. Rev. E* **69** 062901.

[23] Polin M, Tuval I, Drescher K, Gollub J P, and Goldstein R E, *Chlamydomonas swims with two 'gears' in a eukaryotic version of run-and-tumble locomotion*, 2009 *Science* **325** 487.

[24] Friedrich B M and Jülicher F, *Flagellar synchronization independent of hydrodynamic interactions*, 2012 *Phys. Rev. Lett.* **109** 138102.

[25] Bennett R R and Golestanian R, *Emergent run-and-tumble behavior in a simple model of Chlamydomonas with intrinsic noise*, 2013 *Phys. Rev. Lett.* **110** 148102.

[26] Ordemann A, Balazsi G, and Moss F, *Pattern formation and stochastic motion of the zooplankton Daphnia in a light field*, 2003 *Physica A.* **325** 260.

[27] Komin N, Erdmann U, and Schimansky-Geier L, *Random walk theory applied to Daphnia motion*, 2004 *Fluct. Noise Lett.* **4** L151.

[28] Vollmer J, Vegh A G, Lange C, and Eckhardt B, *Vortex formation by active agents as a model for Daphnia swarming*, 2006 *Phys. Rev. E* **73** 061924.

[29] Hoell C and Löwen H, *Theory of microbe motion in a poisoned environment*, 2011 *Phys. Rev. E* **84** 042903.

[30] Anderson J L, *Colloid transport by interfacial forces*, 1989 *Annu. Rev. Fluid Mech.* **21** 61.

[31] Ismagilov R F, Schwartz A, Bowden N, and Whitesides G M, *Autonomous movement and self-assembly*, 2002 *Angew. Chem. Int. Ed.* **41** 652.

[32] Rückner G and Kapral R, *Chemically powered nanodimers*, 2007 *Phys. Rev. Lett.* **98** 150603.

[33] Golestanian R, Liverpool T B, and Ajdari A, *Designing phoretic micro- and nano-swimmers*, 2007 *New J. Phys.* **9** 126.

[34] Popescu M N, Dietrich S, Tasinkevych M, and Ralston J, *Phoretic motion of spheroidal particles due to self-generated solute gradients*, 2010 *Eur. Phys. J. E* **31** 351.

[35] Buttinoni I, Volpe G, Kümmel F, Volpe G, and Bechinger C, *Active Brownian motion tunable by light*, 2012 *J. Phys.: Condens. Matter* **24** 284129.

[36] Peitgen H-O, Jürgens H, and Saupe D, 1992 *Chaos and Fractals: New Frontiers of Science* (New York: Springer).

[37] Meakin P, 1998 *Fractals, Scaling and Growth far from Equilibrium (Cambridge Nonlinear Science Series* vol 5*)* (Cambridge: Cambridge University Press).

[38] Song C M, Havlin S, and Makse H A, *Self-similarity of complex networks*, 2005 *Nature* **433** 392.

[39] Sornette D, 2000 *Critical Phenomena in Natural Sciences: Chaos, Fractals, Selforganization, and Disorder: Concepts and Tools* (Berlin: Springer).

[40] Theurkauff I, Cottin-Bizonne C, Palacci J, Ybert C, and Bocquet L, *Dynamic clustering in active colloidal suspensions with chemical signaling*, 2012 *Phys. Rev. Lett.* **108** 268303.

[41] Doi M and Edwards S F, 1986 *The Theory of Polymer Dynamics* (Oxford: Oxford Science Publications).

[42] ten Hagen B, van Teeffelen S, and Löwen H, *Non-Gaussian behaviour of a self-propelled particle on a substrate*, 2009 *Condens. Matter Phys.* **12** 725.

[43] Berg H C and Brown D A, *Chemotaxis in Escherichia coli analysed by three-dimensional tracking*, 1972 *Nature* **239** 500.

[44] Tailleur J and Cates M E, *Statistical mechanics of interacting run-and-tumble bacteria*, 2008 *Phys. Rev. Lett.* **100** 218103.

[45] Nash R W, Adhikari R, Tailleur J, and Cates M E, *Run-and-tumble particles with hydrodynamics: sedimentation, trapping, and upstream swimming*, 2010 *Phys. Rev. Lett.* **104** 258101.

[46] Strefler J, Ebeling W, Gudowska-Nowak E, and Schimansky-Geier L, *Dynamics of individuals and swarms with shot noise induced by stochastic food supply*, 2009 *Eur. Phys. J. B* **72** 597.

[47] Arditi R, Tyutyunov Y, Morgulis A, Govorukhin V, and Senina I, *Directed movement of predators and the emergence of density-dependence in predator-prey models*, 2001 *Theor. Popul. Biol.* **59** 207.

[48] Cremer P and Löwen H, *Scaling of cluster growth for coagulating active particles*, 2014 *Phys. Rev. E* **89** 022307.

[49] Rosser G, Baker R E, Armitage J P, and Fletcher A G, *Modelling and analysis of bacterial tracks suggest an active reorientation mechanism in Rhodobacter sphaeroides*, 2014 *J. R. Soc. Interface* **11** 20140320.

[50] Baraban L, Makarov D, Streubel R, Mönch I, Grimm D, Sanchez S, and Schmidt O G, *Catalytic Janus motors on microfluidic chip: deterministic motion for targeted cargo delivery*, 2012 *ACS Nano* **6** 3383.

[51] Palacci J, Sacanna S, Steinberg A P, Pine D J, and Chaikin P M, *Living crystals of light-activated colloidal surfers*, 2013 *Science* **339** 936.

[52] Volpe G, Buttinoni I, Vogt D, Kümmerer H-J, and Bechinger C, *Microswimmers in patterned environments*, 2011 *Soft Matter* **7** 8810.

[53] Ebbens S, Tu M-H, Howse J R, and Golestanian R, *Size dependence of the propulsion velocity for catalytic Janus-sphere swimmers*, 2012 *Phys. Rev. E* **85** 020401(R).

[54] Kraft D J, Wittkowski R, ten Hagen B, Edmond K V, Pine D J, and Löwen H, *Brownian motion and the hydrodynamic friction tensor for colloidal particles of complex shape*, 2013 *Phys. Rev. E* **88** 050301(R).

[55] Wagner J, Märkert C, Fischer B, and Müller L, *Direction dependent diffusion of aligned magnetic rods by means of X-ray photon correlation spectroscopy*, 2013 *Phys. Rev. Lett.* **110** 048301.

[56] Chakrabarty A, Konya A, Wang F, Selinger J V, Sun K, and Wei Q-H, *Brownian motion of boomerang colloidal particles*, 2013 *Phys. Rev. Lett.* **111** 160603.

[57] Baraban L, Makarov D, Schmidt O G, Cuniberti G, Leiderer P, and Erbe A, *Control over Janus micromotors by the strength of a magnetic field*, 2013 *Nanoscale* **5** 1332.

[58] Takagi D, Braunschweig A B, Zhang J, and Shelley M S, *Dispersion of self-propelled rods undergoing fluctuation-driven flips*, 2013 *Phys. Rev. Lett.* **110** 038301.

Statement of the author: This work was performed together with Sonja Babel in the context of her Bachelor thesis at the Institut für Theoretische Physik II, Heinrich-Heine-Universität Düsseldorf. Sonja Babel derived the bulk of the analytical results under the guidance of Hartmut Löwen and me, while I wrote most parts of the paper. I also checked the theoretical expressions and repeated some of the calculations.

The content of this chapter was published as an invited article for the proceedings of the International Conference on Statistical Physics STATPHYS25 in Seoul, South Korea.

CHAPTER 4

Dynamics of a deformable active particle under shear flow

The content of this chapter has been published in a similar form in *Journal of Chemical Physics* **139**, 104906 (2013) by Mitsusuke Tarama, Andreas M. Menzel, Borge ten Hagen, Raphael Wittkowski, Takao Ohta, and Hartmut Löwen (see reference [215]).

Dynamics of a deformable active particle under shear flow

Mitsusuke Tarama,[1,2,3,4,a)] Andreas M. Menzel,[1,2] Borge ten Hagen,[2] Raphael Wittkowski,[5] Takao Ohta,[1,4,6] and Hartmut Löwen[2]
[1)] *Department of Physics, Kyoto University, Kyoto 606-8502, Japan*
[2)] *Institut für Theoretische Physik II: Weiche Materie, Heinrich-Heine-Universität Düsseldorf, D-40225 Düsseldorf, Germany*
[3)] *Institute for Solid State Physics, The University of Tokyo, Kashiwa, Chiba 277-8581, Japan*
[4)] *Department of Physics, Graduate School of Science, The University of Tokyo, Tokyo 113-0033, Japan*
[5)] *SUPA, School of Physics and Astronomy, University of Edinburgh, Edinburgh, EH9 3JZ, United Kingdom*
[6)] *Soft Matter Center, Ochanomizu University, Tokyo 112-0012, Japan*

The motion of a deformable active particle in linear shear flow is explored theoretically. Based on symmetry considerations, we propose coupled nonlinear dynamical equations for the particle position, velocity, deformation, and rotation. In our model, both, passive rotations induced by the shear flow as well as active spinning motions, are taken into account. Our equations reduce to known models in the two limits of vanishing shear flow and vanishing particle deformability. For varied shear rate and particle propulsion speed, we solve the equations numerically in two spatial dimensions and obtain a manifold of different dynamical modes including active straight motion, periodic motions, motions on undulated cycloids, winding motions, as well as quasi-periodic and chaotic motions induced at high shear rates. The types of motion are distinguished by different characteristics in the real-space trajectories and in the dynamical behavior of the particle orientation and its deformation. Our predictions can be verified in experiments on self-propelled droplets exposed to a linear shear flow.

[a)] Electronic mail: tarama@scphys.kyoto-u.ac.jp

I. INTRODUCTION

In the last decade, the motion and modeling of active particles has attracted much attention in the field of nonequilibrium physics.[1–5] A major part of active particles are artificial colloidal microswimmers with fixed stable shapes,[6–8] but there are also cases in which the particles are deformable and do change shape during their motion. Such deformability is of basic importance for active droplets[9–11] but is also relevant for living swimmers like protozoa and other microorganisms.[12–15] Therefore a basic theoretical description for active deformable self-propelled particles and microswimmers is needed.

In a quiescent solvent, dynamical equations of motion were recently put forward which couple the particle position and deformability.[16–26] One of the unexpected results was a spontaneous circling motion due to the coupling of deformability and self-propulsion.[16–18] However, in most practical situations,[27–29] various external fields are present to influence the particle motion. They are, for instance, induced by a chemoattractant, phototaxis, and gravity,[30–35] or external walls.[36–39] An important particular case is a solvent flow field such as a Couette flow with a constant shear gradient or a Poiseuille flow through tubes. There are several studies of rigid self-propelled particles in various shear geometries.[40–43] However, despite its practical relevance, the motion of a deformable self-propelled particle in a solvent flow has not been considered theoretically yet. The corresponding modeling is expected to be complex since already rigid (undeformable) active particles have been shown to perform periodic motion on cycloids (rather than on straight lines) once they are exposed to a linear shear flow field.[42]

In this paper we close this gap and propose a theoretical model for the motion of an active deformable particle in shear flow. We use symmetry considerations to obtain coupled nonlinear dynamical equations for the particle position, velocity, deformation, and its active rotations. In our model, a passive rotation induced by the shear flow and an active spinning motion are both taken into account. On the one hand, for vanishing shear flow, our equations reduce to previous models for deformable particles.[16,19,20] On the other hand, for vanishing particle deformability, we obtain the cycloidal motion as embodied in previous investigations.[42] For varied shear rate and particle propulsion speed, we solve the equations numerically in two spatial dimensions and obtain a manifold of different dynamical modes including active straight motion, periodic motions, regular and undulated cycloidal motions, winding motions, as well as quasi-periodic and chaotic motions induced at high shear rates. The types of motion are distinguished by different characteristics in the dynamical behavior of the particle positions, velocity orientations, and its deformations. We are not aware of any experiments of deformable active particles in shear flow, but in principle these experiments are conceivable building upon recent analysis of self-propelled droplets[10,44] that can be subjected to an additional shear flow.

The organization of this paper is as follows. In Sec. II, we introduce time-evolution equations for an active deformable particle under an external flow. In Sec. III, we consider the special case of a round disk-shaped non-deformable active particle by eliminating the variable for deformation. We relate this case to previous work.[42] Next, in Sec. IV, we present numerical results for the dynamics of a deformable active particle under steady shear flow. In Sec. V, a spontaneous particle rotation for the dynamics is additionally included and numerical results are presented. Finally, Sec. VI is devoted to a summary and to conclusions.

II. COUPLED DYNAMICAL EQUATIONS

Based on symmetry considerations, we now derive a set of coupled nonlinear dynamical equations to describe the motion of a deformable active particle under an externally imposed flow field. These equations are first listed for the general case of three spatial dimensions and an unspecified flow field. Afterwards we will confine ourselves to a two-dimensional geometry and consider a simple shear flow. As a first approach, we only investigate the influence of the shear flow on the particle dynamics and do not consider the inverse effect.

In the following, we denote the prescribed externally imposed flow field that the particle is exposed to as $\boldsymbol{u}$. It is a given function of space. To proceed as normal,[45,46] the elongational part of the fluid flow is extracted by the symmetric second-rank tensor $\mathbf{A}$ with components

$$A_{ij} = \frac{1}{2}\left(\partial_i u_j + \partial_j u_i\right) . \tag{1}$$

Similarly, the rotational part is extracted via the anti-symmetric second-rank tensor $\mathbf{W}$ with components

$$W_{ij} = \frac{1}{2}\left(\partial_i u_j - \partial_j u_i\right) . \tag{2}$$

We denote the center-of-mass position of the particle at time t as $\mathbf{x}(t)$. In general, the total particle velocity $d\boldsymbol{x}/dt$ has two contributions. On the one hand, the particle is "passively" advected by the externally imposed prescribed flow field $\boldsymbol{u}$. On the other hand, the particle can "actively" self-propel with respect to the surrounding fluid. The corresponding "active" velocity measured relatively to the surrounding fluid flow is denoted as $\mathbf{v}$. Altogether, we obtain the equation of motion

$$\frac{dx_i}{dt} = u_i + v_i \, , \tag{3}$$

where the index $i = 1, 2, 3$ labels the Cartesian coordinates.

In our description, the active velocity $\mathbf{v}$ is one of our major dynamical variables describing the behavior of the particle. The other ones are its deformation that we characterize by the second-rank traceless symmetric tensor $\mathbf{S}$, and an "active" particle rotation described by the second-rank antisymmetric tensor $\boldsymbol{\Omega}$. Both of these tensors are briefly introduced in the following.

For simplicity, we only include elongational and flattening deformations of the particle. The tensor $\mathbf{S}$ represents these deformations.[47,48] We first consider the two-dimensional case, where orientations in the two-dimensional plane can be parameterized by a single angle ϖ.

This angle ϖ is now used to measure directions from the particle center of mass. The distance from the particle center to its boundary in the direction ϖ at time t is denoted as $R(\varpi, t)$. Large deformations are not taken into account, so that $R(\varpi, t)$ is single-valued with respect to the angle ϖ.

For a steady circular shape we have $R(\varpi, t) = R_0$, with R_0 the particle radius. We now consider deviations $\delta R(\varpi, t)$ from the circular shape, such that the distance from the particle center to its boundary becomes $R(\varpi, t) = R_0 + \delta R(\varpi, t)$. Next, the deviation from the circular shape $\delta R(\varpi, t)$ is expanded into a Fourier series

$$\delta R(\varpi, t) = \sum_{m=2}^{\infty} \left(z_m(t) e^{im\varpi} + z_{-m}(t) e^{-im\varpi} \right) . \tag{4}$$

In this expansion, the zeroth mode is excluded by assuming that the area of the particle is conserved. The first Fourier mode would represent a translation of the center of mass, which we already took into account by the velocity variable $\mathbf{v}$. Therefore, the lowest mode describing deformations is the second one, which actually represents an elliptical deformation. In two dimensions, we can define a symmetric tensor as $S_{11} = -S_{22} = z_2 + z_{-2} = s\cos 2\theta$ and $S_{12} = S_{21} = i(z_2 - z_{-2}) = s \sin 2\theta$, where we have defined $z_{\pm 2} = (s/2)e^{\mp 2i\theta}$. So the searched-for tensor $\mathbf{S}$ can be written in the form

$$\mathbf{S} = \begin{pmatrix} S_{11} & S_{12} \\ S_{12} & -S_{11} \end{pmatrix} = s \begin{pmatrix} \cos 2\theta & \sin 2\theta \\ \sin 2\theta & -\cos 2\theta \end{pmatrix} . \tag{5}$$

Here, s corresponds to the degree of deformation, and θ to the orientation of the long axis of deformation. Then, the distance from the particle center to its boundary for an elliptical deformation is given by

$$\begin{aligned} R(\varpi, t) &= R_0 + s(t) \cos 2 \left[\varpi - \theta(t) \right] \\ &= R_0 + S_{11}(t) \cos 2\varpi + S_{12}(t) \sin 2\varpi . \end{aligned} \tag{6}$$

In the case of three spatial dimensions, the deviation δR must be expanded into spherical harmonics $Y_{\ell m}(\tilde{\varpi})$ with coefficients $c_{\ell m}(t)$ and $\tilde{\varpi}$ the solid angle. Likewise, the minimum mode of the deformation, $\ell = 2$, represents an ellipsoidal deformation. See Ref. 18 for the relations between $c_{\ell m}$ and $\mathbf{S}$ in three spatial dimensions.

Finally, our last dynamic variable $\mathbf{\Omega}$ characterizes an "active" rotational motion of the particle around its center of mass. We call this a spinning motion.[19,20] This antisymmetric second-rank tensor $\mathbf{\Omega}$ can be obtained from the corresponding vector of angular velocity $\boldsymbol{\omega}$ via[20]

$$\Omega_{ij} = \epsilon_{ijk} \omega_k , \tag{7}$$

where ϵ_{ijk} denotes the components of the Levi-Civita tensor. Summation over repeated indices is implied, as throughout the remaining part of this paper.

The spinning motion $\mathbf{\Omega}$ occurs in addition to the rotational motion prescribed by the external flow field $\mathbf{u}$, see Eq. (2). In other words, the anti-symmetric tensor $\mathbf{\Omega}$ represents the relative rotation with respect to the rotational motion $\mathbf{W}$ of the surrounding fluid flow. Therefore, $\mathbf{W} + \mathbf{\Omega}$ describes the total angular velocity with respect to the laboratory frame from which the flow field is parameterized.

In total, we have introduced three central dynamical variables to characterize the state of a deformable active particle: $\mathbf{v}$ for the active propulsion velocity, $\mathbf{S}$ for the particle deformation, and $\mathbf{\Omega}$ for the active rotational motion. Based on symmetry arguments, our model for the dynamic evolution of these variables is derived. We consider the following set of coupled nonlinear equations:

$$\frac{dv_i}{dt} + a_2 \left(W_{ik} + \Omega_{ik}\right) v_k = \alpha v_i - (v_k v_k) v_i - a_1 S_{ik} v_k \,, \tag{8}$$

$$\begin{aligned} &\frac{dS_{ij}}{dt} - b_2 \left[S_{ik} \left(W_{kj} + \Omega_{kj}\right) - \left(W_{ik} + \Omega_{ik}\right) S_{kj}\right] \\ &\quad = -\kappa S_{ij} + b_1 \left[v_i v_j - \frac{\delta_{ij}}{d} \left(v_k v_k\right)\right] + b_3 \Omega_{ik} S_{k\ell} \Omega_{\ell j} + b_4 \Omega_{k\ell} \Omega_{k\ell} S_{ij} \\ &\quad + \nu_1 \left[A_{ij} - \frac{\delta_{ij}}{d} A_{kk}\right] + \nu_2 \left[A_{ik} S_{kj} + S_{ik} A_{kj} - \frac{2\delta_{ij}}{d} S_{k\ell} A_{\ell k}\right] , \end{aligned} \tag{9}$$

$$\frac{d\Omega_{ij}}{dt} = \zeta \Omega_{ij} + \Omega_{ik} \Omega_{k\ell} \Omega_{\ell j} + c_1 \left(S_{ik} \Omega_{kj} + \Omega_{ik} S_{kj}\right) + c_2 S_{ik} \Omega_{k\ell} S_{\ell j} \,. \tag{10}$$

Here δ_{ij} denotes the Kronecker delta, and d is the dimension of space. The coefficients α, κ, ζ, a_1, a_2, b_1, b_2, b_3, b_4, c_1, c_2, ν_1, and ν_2 are phenomenological coupling parameters. We now comment on each of the terms in this set of equations for the time evolution, Eqs. (8)–(10). In principle, more terms and higher-order couplings can be included, but the current model covers the main physical aspects that we intend to describe.

We start with the first two terms on the right-hand side of Eq. (8). They can be rewritten as

$$-\frac{\partial F}{\partial v_i} \quad \text{with} \quad F = -\frac{\alpha}{2} \left(v_k v_k\right) + \frac{1}{4} \left(v_k v_k\right)^2 , \tag{11}$$

where F is a Lyapunov function controlling the spontaneous self-propulsion. With increasing α, F describes a bifurcation at $\alpha = 0$ corresponding to the onset of active motion with $\mathbf{v} \neq \mathbf{0}$. In the same way, the first two terms on the right-hand side of Eq. (10) can be rewritten as

$$-\frac{\partial G}{\partial \Omega_{ij}} \quad \text{with} \quad G = \frac{\zeta}{2} \text{tr } \mathbf{\Omega}^2 + \frac{1}{4} \text{tr } \mathbf{\Omega}^4 , \tag{12}$$

introducing another Lyapunov function G. Likewise, this function characterizes the onset of the spontaneous rotation of the particle around its center of mass when ζ becomes positive. Together, the coefficients α and ζ characterize the strength of activity, for self-propulsion and for active rotation, respectively. In contrast to that, an active deformation of the particle is not considered. The first term on the right-hand side of Eq. (9) with $\kappa > 0$ always induces a relaxation of the deformation back to a spherical (circular) shape, at least when the coupling to $\mathbf{v}$, $\boldsymbol{\Omega}$, and to the surrounding flow field $\boldsymbol{u}$ allow it.

Next, we consider the terms with the coefficients a_2 in Eq. (8) and b_2 in Eq. (9). They have similar origin and include reorientations of the particle velocity and elongation axes due to the shear flow and due to the active spinning motion. In the passive case, they would contain the advective reorientation of the particle axes due to the fluid flow. Since an active particle can follow a prescribed rule on how to react to external rotational flow fields, the numerical values of the coefficients cannot be generally fixed at this point. We assume $a_2 > 0$. In principle, this contribution with $a_2 > 0$ can describe a sort of Magnus effect, with a force acting onto the particle in the direction perpendicular to its velocity and angular velocity. For a rigid particle – i.e. an undeformable particle – of spherical or ellipsoidal shape, rotational and translational motions do not couple to each other to linear order.[49] Accordingly, the coupling between the velocity $\mathbf{v}$ and the rotational part of the flow field $\mathbf{W}$ is nonlinear in the a_2-term.

The third term on the right-hand side of Eq. (8) with the coefficient a_1 and the second term on the right-hand side of Eq. (9) with the coefficient b_1 are the leading-order coupling terms between the velocity $\mathbf{v}$ and the deformation $\mathbf{S}$. Their influence was already extensively studied in previous investigations of deformable self-propelled particles.[16,18] On the one hand, deformations can reorient the particle velocity and change its speed via the a_1-term. Bended particle trajectories can result from this contribution. On the other hand, via the b_1-term, deformations can be induced when the particle self-propels.

In addition to that, we include further coupling contributions between the deformation $\mathbf{S}$ and the active rotation $\boldsymbol{\Omega}$.[19,20] These are the terms with the coefficients b_3 and b_4 on the right-hand side of Eq. (9), and the terms with the coefficients c_1 and c_2 on the right-hand side of Eq. (10). For $b_3 > 0$ and $b_4 > 0$, the self-driven active rotation in a two-dimensional space enhances the degree of deformation, while it reduces it for $b_3 < 0$ and $b_4 < 0$. When we confine ourselves to two spatial dimensions in Secs. IV and V, these two terms are equivalent for $b_3 = 2b_4$. We remark that, in three spatial dimensions, the term with the coefficient b_3 has an additional effect to rotate the particle, in contrast to the b_4-term. The third and fourth terms on the right-hand side of Eq. (10) include the analogous effects on the rotations $\boldsymbol{\Omega}$ induced by the deformation $\mathbf{S}$. Here, note that the c_1 term vanishes in two dimensions. Again, since different sorts of active particles may feature different coupling properties between their deformations and rotations, the values of these coefficients depend on the system under consideration.

Finally, the elongational part of the externally imposed flow field can lead to a deformation of the particle. This effect is included by the last two contributions on the right-hand side of Eq. (9). The corresponding coefficients ν_1 and ν_2 describe how the active particle reacts to the straining part of the flow field. Our terms are consistent with those of a previous study on the dynamics of a non-active liquid droplet in a fluid flow, where also elliptical shape deformations were considered.[46,50] We note that the contribution with the coefficient ν_2 vanishes for a two-dimensional geometry of incompressible flow. Such a case is studied below.

In contrast to Eq. (9), the tensor $\mathbf{A}$ containing the elongational part of the fluid flow does not enter Eq. (8) for the velocity $\mathbf{v}$. This is because $\mathbf{v}$ is defined as the *relative* velocity with respect to the fluid flow $\mathbf{u}$, see Eq. (3). Likewise, the rotational part of the fluid flow characterized by the tensor $\mathbf{W}$ is absent in Eq. (10): $\boldsymbol{\Omega}$ describes the *relative* rotation with respect to the surrounding flow field. In principle, coupling terms between $\boldsymbol{\Omega}$ and the tensor $\mathbf{A}$ are possible. This would mean that the active particle features a way of reacting to an elongational flow by adjusting its spinning motion. However, we do not consider such a process.

Generally our equations apply to a three-dimensional set-up. For simplicity, however, we confine ourselves to two spatial dimensions for the remaining part of this paper. Furthermore, we from now on specify the externally imposed flow field $\mathbf{u}$ to a linear steady shear flow,

$$\mathbf{u} = (\dot{\gamma} y, 0)\,, \tag{13}$$

with $\dot{\gamma}$ as the shear rate.

III. DYNAMICS WITHOUT DEFORMATION

The full set of dynamic equations (3), (8)–(10) is very complex. To get a first overview, we start by studying a reduced model. More precisely, we neglect deformability, i.e. we set $\mathbf{S} = \mathbf{0}$, and consider a circularly-shaped rigid particle. Under certain assumptions, an analytical solution can be obtained in this case.

Prescribing $\mathbf{S} = \mathbf{0}$, Eq. (9) is dropped from our system of equations. Eqs. (3), (8), and (10) reduce to

$$\frac{dx_i}{dt} = v_i + u_i\,, \tag{14}$$

$$\frac{dv_i}{dt} = \alpha v_i - (v_k v_k)\, v_i - a_2\,(W_{ik} + \Omega_{ik})\, v_k\,, \tag{15}$$

$$\frac{d\Omega_{ij}}{dt} = \zeta \Omega_{ij} + \Omega_{ik}\Omega_{kl}\Omega_{lj}\,. \tag{16}$$

We parameterize the vectors and tensors by $\mathbf{x} = (x, y)$,

$$\mathbf{v} = (v\cos\phi, v\sin\phi)\,, \tag{17}$$

$\Omega_{11} = \Omega_{22} = 0$, as well as $\Omega_{12} = -\Omega_{21} = \omega$, and insert $\mathbf{W}$ using Eqs. (2) and (13). Then Eqs. (14)–(16) become

$$\frac{dx}{dt} = v\cos\phi + \dot{\gamma}y\,, \tag{18}$$

$$\frac{dy}{dt} = v\sin\phi\,, \tag{19}$$

$$\frac{dv}{dt} = \alpha v - v^3\,, \tag{20}$$

$$\frac{d\phi}{dt} = a_2\left(-\frac{\dot{\gamma}}{2} + \omega\right), \tag{21}$$

$$\frac{d\omega}{dt} = \zeta\omega - \omega^3\,. \tag{22}$$

Next, we assume that the magnitudes of the velocity v and of the relative rotation ω relax quickly, so that they are given by the steady state solutions of Eqs. (20) and (22), respectively. In this situation, together with $\alpha > 0$ and $\zeta > 0$ implying self-propulsion and active spinning, we have

$$v = \sqrt{\alpha}\,, \tag{23}$$

$$\omega = \pm\sqrt{\zeta}\,, \tag{24}$$

where the positive and negative signs in Eq. (24) correspond to counter-clockwise and clockwise rotations, respectively. Using these solutions, Eq. (21) reads

$$\frac{d\phi}{dt} = a_2\left(-\frac{\dot{\gamma}}{2} \pm \sqrt{\zeta}\right). \tag{25}$$

From Eqs. (18), (19), (23), and (25), the trajectory of the center of mass can be calculated as

$$\begin{aligned} x(t) = {} & \frac{\sqrt{\alpha}\left\{a_2\left(-\frac{\dot{\gamma}}{2} \pm \sqrt{\zeta}\right) - \dot{\gamma}\right\}}{a_2^2\left(-\frac{\dot{\gamma}}{2} \pm \sqrt{\zeta}\right)^2}\left\{\sin[\phi(t)] - \sin\phi_0\right\} \\ & + \dot{\gamma}\left(\frac{\sqrt{\alpha}}{a_2\left(-\frac{\dot{\gamma}}{2} \pm \sqrt{\zeta}\right)}\cos\phi_0 + y_0\right)t + x_0\,, \end{aligned} \tag{26}$$

$$y(t) = \frac{-\sqrt{\alpha}}{a_2\left(-\frac{\dot{\gamma}}{2} \pm \sqrt{\zeta}\right)}\left\{\cos[\phi(t)] - \cos\phi_0\right\} + y_0\,, \tag{27}$$

$$\phi(t) = a_2\left(-\frac{\dot{\gamma}}{2} \pm \sqrt{\zeta}\right)t + \phi_0\,. \tag{28}$$

Here, (x_0, y_0) and ϕ_0 are the position of the center of mass and the direction of the velocity vector at $t = 0$, respectively. This set of solutions represents a cycloidal trajectory.

A similar cycloidal trajectory has previously been obtained for an active rigid circularly-shaped particle by some of the present authors.[42] In that case, the dynamics of the particle features a polarity axis,[51] the orientation of which in the two-dimensional plane can be characterized by the angle ϕ. It marks the direction of the self-propulsion that generates a relative velocity with respect to the surrounding flow field. The equations of motion introduced in Ref. 42 can be written in the form

$$\frac{dx}{dt} = \dot{\gamma} y + \tilde{\alpha}\left[\cos\phi + f_x\right], \tag{29}$$

$$\frac{dy}{dt} = \tilde{\alpha}\left[\sin\phi + f_y\right], \tag{30}$$

$$\frac{d\phi}{dt} = -\frac{\dot{\gamma}}{2} + \tilde{\mu}(1+g), \tag{31}$$

where $\tilde{\alpha}$ is a normalized effective self-propulsion force that is proportional to the self-propulsion velocity v in the over-damped regime considered in Ref. 42. $\tilde{\mu}$ accounts for an additional self-induced[52] or externally imposed[53] torque on the particle. Equations (29)–(31) contain Gaussian white noise terms f_x, f_y, and g, which are not included in the present approach. Our equations (18), (19), and (25), together with the asymptotic steady-state magnitudes of the translational and angular velocities in Eqs. (23) and (24), respectively, are consistent with Eqs. (29)–(31) when the noise terms are neglected. By solving this zero temperature limit for $\tilde{\alpha} = \alpha^{1/2}$ and $\tilde{\mu} = \pm\zeta^{1/2}$, one recovers the results presented in Eqs. (26)–(28) for $a_2 = 1$.

For the special case of $(\dot{\gamma}/2) = \pm\sqrt{\zeta}$, the solutions of Eqs. (18), (19), and (25), together with Eq. (23), read

$$x(t) = \left(\frac{\dot{\gamma}}{2}\sqrt{\alpha}\sin\phi_0\right) t^2 + \left(\sqrt{\alpha}\cos\phi_0 + \dot{\gamma} y_0\right) t + x_0\,, \tag{32}$$

$$y(t) = \left(\sqrt{\alpha}\sin\phi_0\right) t + y_0\,, \tag{33}$$

$$\phi(t) = \phi_0\,. \tag{34}$$

The physical meaning of this limit is that the spontaneous rotation compensates the rotation due to the surrounding flow field, i.e. $(\dot{\gamma}/2) = \pm\sqrt{\zeta}$ in Eq. (25) or correspondingly $(\dot{\gamma}/2) = \tilde{\mu}$ in Eq. (31). Also Eqs. (32) and (33) are consistent with the ones correspondingly obtained in Ref. 42.

IV. DYNAMICS WITHOUT ACTIVE ROTATION

In Sec. III, we studied a rigid non-deformable particle as a first step. We now include deformability, but do not consider an active spinning motion of the particle. Although various different dynamic states can be found in this situation, the dynamics is still much simpler than with active rotations included, as is shown later. The dynamic equations must be solved numerically.

Choosing $\zeta < 0$ hinders active spinning. To solve Eqs. (3) and (8)–(10), we use a fourth-order Runge-Kutta method. We checked the numerical accuracy by comparing results obtained for different time increments.

The full parameter space is far too complex to be exhaustively explored. We therefore concentrate on the impact of only two parameters that we consider central to the current problem. One of them is the strength of the self-propulsion of the particle characterized by the parameter α. The other one is the strength of the imposed shear flow determined by the shear rate $\dot{\gamma}$.

All coupling parameters are fixed at similar magnitude to allow an equal impact of the corresponding effects on the system behavior. We set $a_1 = b_1 = -1$, $a_2 = b_2 = c_2 = \nu_1 = 1$, and $b_3 + 2b_4 = 1$ (as noted in Sec. II the terms with the coefficients b_3 and b_4 coincide in two spatial dimensions, and the terms with the coefficients c_1 and ν_2 vanish in our geometry). Intermediate damping rates are used for the deformations and for the spinning motion by imposing $\kappa = 0.5$ and $\zeta = -0.1$, respectively. We obtained our results by directly numerically integrating Eqs. (3) and (8)–(10). After that we reparameterized them for illustrative purposes using Eqs. (5) and (17).

Our results are summarized in Fig. 1. We present in Fig. 1(a) a phase diagram in the parameter plane of the self-propulsion strength α and the shear rate $\dot{\gamma}$. Various qualitatively different types of dynamical states are found and explained in more detail below. They are indicated in the phase diagram by the different symbols. At the position included as $\dot{\gamma} = 0$, we describe in words the type of motion observed at zero shear rate for the different self-propulsion strengths α. Increasing the shear rate towards the right boundary of the phase diagram, we can see how the shear flow influences the dynamic behavior of the particle.

Each of the observed dynamic states is characterized separately in the rows of Figs. 1(b)–(f). The location in the phase diagram is indicated by the corresponding symbol below the panel number. We present typical real-space trajectories by the black dotted lines in the first column. Black arrows indicate the direction of migration. The trajectories are drawn from the laboratory frame. Therefore their appearance strongly depends on the initial y-coordinate: the advective flow velocity increases in y-direction due to the shear geometry and leads to a stretching of the trajectories in x-direction. In particular, the direction of motion also depends on the y-coordinate: the flow field points to the right for $y > 0$, whereas it points to the left for $y < 0$. To avoid confusion, we remark again that the dynamics that is described by Eqs. (8)–(10) itself is not affected in this way, because only the relative velocity $\mathbf{v}$ with respect to the shear flow is considered and only gradients of the flow velocity enter via the tensors $\mathbf{A}$ and $\mathbf{W}$.

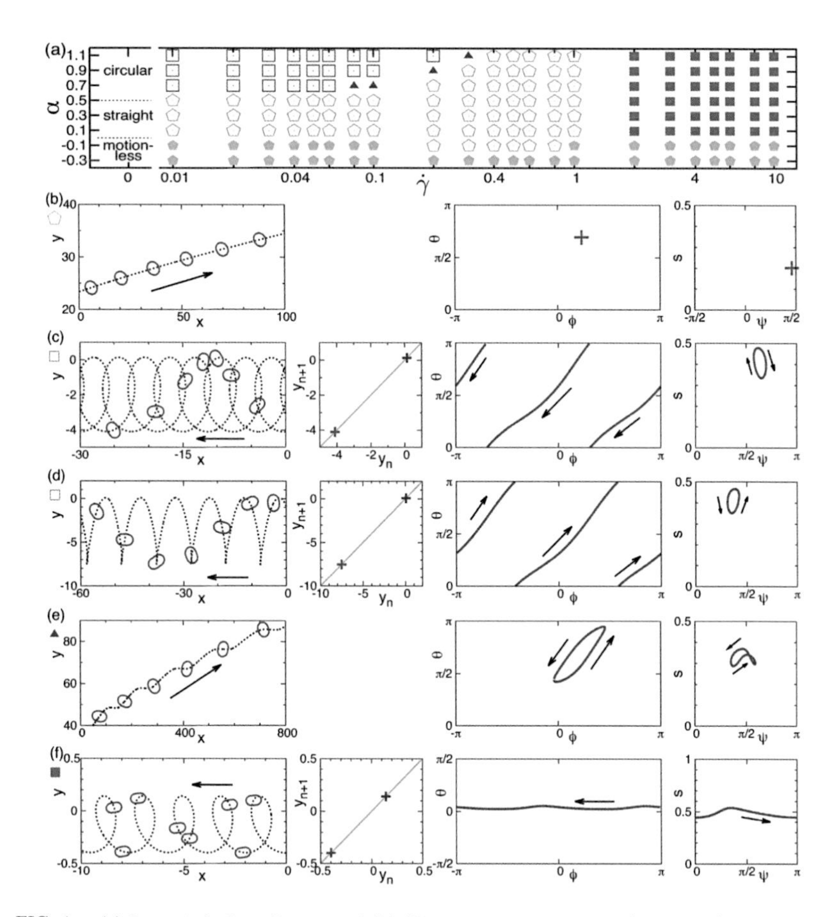

FIG. 1. (a) Dynamical phase diagram and (b)–(f) trajectories in real space (1st column), return maps (2nd column), attractors in θ-ϕ space (3rd column) as well as in s-ψ space (4th column) of the typical dynamical motions, obtained by solving Eqs. (3) and (8)–(10) numerically in two dimensions without active rotational motion ($\zeta = -0.1$); (b) *active straight motion* for $\alpha = 0.5$ and $\dot{\gamma} = 0.1$ indicated by the green open pentagons in panel (a); (c) and (d) *cycloidal I motions* of clockwise and counter-clockwise rotations of the particle deformations, respectively, for $\alpha = 0.9$ and $\dot{\gamma} = 0.1$ marked by the red open squares in panel (a); (e) *winding I motion* for $\alpha = 0.7$ and $\dot{\gamma} = 0.08$ indicated by the purple filled triangles in panel (a); (f) *cycloidal II motion* for $\alpha = 0.1$ and $\dot{\gamma} = 2$ marked by the gray filled squares in panel (a). Arrows in panels (b)–(f) show the directions of motion. Some snapshots of the particle, the size of which is adjusted for illustration, are superimposed to the trajectory in real space. Turquoise filled pentagons in panel (a) represent the *passive straight motion* with $v = 0$. The variable $\mathbf{\Omega}$ corresponding to active spinning equals $\mathbf{0}$ for all of these types of motion. A return map for the *active straight motion* and *winding I motion* in panels (b) and (e) does not exist because the y-component of the velocity does not change its sign in these motions.

In order to illustrate the current state of deformation and orientation along the trajectories that are drawn in the first column of Figs. 1(b)–(f), representative snapshots of the particle are superimposed in red. For the purpose of best visualization, the size of the particle was adjusted and the deformations are not drawn as pure ellipsoids. To further characterize the modes of migration, return maps of the corresponding motion in real space are included in the second column of Figs. 1(b)–(f). We extract the local maximum and minimum values of y along each real-space trajectory for several thousand up-down oscillations. The maximum and minimum values are labeled as y_n ($n = 1, 2, \dots$) and plotted as blue and red points, respectively, in a return map y_{n+1} vs. y_n. In other words, we calculated the return maps at the Poincaré sections where $v_y = 0$ for $dv_y/dt < 0$ and $dv_y/dt > 0$, respectively. The diagonal line in the return maps is included for illustration and does not represent any data points.

Finally, the attractors in phase space are drawn in the third and fourth columns. Black arrows indicate the direction of motion along the attractors. We show plots in θ-ϕ space (third column) and in s-ψ space (fourth column). As introduced in Eqs. (5) and (17), θ and ϕ describe the orientation of the long axis of the deformation tensor $\mathbf{S}$ and the orientation of the relative velocity vector $\mathbf{v}$, respectively. θ and ϕ are observed from the laboratory frame. This is different for s and ψ. First, s measures the magnitude of deformation as is obvious from Eq. (6). Second, ψ is defined as the relative angle between the long axis of deformation and the velocity orientation,

$$\psi = \theta - \phi \, . \tag{35}$$

Thus s and ψ are measured in the co-moving particle frame. Both attractors, in the θ-ϕ space and in the s-ψ space, are independent of the initial y-coordinate.

We now go through the different dynamic states depicted in Fig. 1. The most trivial state is represented by the turquoise filled pentagon symbols in Fig. 1(a). They indicate a *passive straight motion*. In the absence of any external flow field, a particle in this state is motionless and has a circular shape. When the shear flow is switched on, the particle is elongated due to the elongational contribution from the flow field. Nevertheless, its active self-propulsion velocity remains zero, $v = 0$, so that it is just passively advected with the flow field. We do not include plots of these trivial trajectories in Fig. 1.

Next, a particle that moves straight in a condition without shear flow continues to move straight in the presence of shear flow at low shear rates $\dot{\gamma}$. It features a time-independent steady state of deformation. Such a situation is marked by the green open pentagons in the phase diagram (Fig. 1(a)) and shown in Fig. 1(b) for $\alpha = 0.5$ and $\dot{\gamma} = 0.1$. The real-space trajectory is only bended a little because the particle is advected in x-direction with the fluid flow that increases in the y-direction due to the shear geometry. Since $v \neq 0$, we term this type of motion an *active straight motion* in this paper. We note from the line $\alpha = -0.1$ in the phase diagram (Fig. 1(a)) that with increasing shear rate $\dot{\gamma}$ a transition from passive to active straight motion can be induced. Interestingly, the phase behavior is reentrant, and we again observe passive straight motion at very high shear rates. The reason for this behavior are shear-rate dependent deformations $\mathbf{S}$ that are induced in Eq. (9) via the shear flow. They in turn couple to the relative velocity $\mathbf{v}$ in Eq. (8) and at intermediate shear rates induce active self-propulsion.

If a particle undergoes a circular motion when the shear is absent, it exhibits what we call a *cycloidal I motion* under a small nonzero shear rate as indicated by the red open squares in Fig. 1(a). In this state, a particle moves on a cycloidal trajectory with $v \neq 0$ and with its deformation axes rotating as depicted in Figs. 1(c) and (d), both for $\alpha = 0.9$ and $\dot{\gamma} = 0.1$. Both, clockwise (c) and counter-clockwise (d) rotations are possible. Whether clockwise or counter-clockwise rotation appears during the cycloidal I motion generally depends on the initial conditions. At higher shear rates close to the stability boundary of the cycloidal I motion, however, cycloidal I motion of counter-clockwise rotation becomes unstable first, before the one with clockwise rotation. This is because the rotational part of the shear flow is oriented in clockwise direction as well and breaks the rotational symmetry of space. However, the effect occurs within a thinner parameter region than the grid size in Fig. 1(a) resolves. Therefore we do not mark this region in the phase diagram (Fig. 1(a)).

So far, we have only discussed types of motion that result directly as a generalization of the types of motion found for vanishing flow field $\dot{\gamma} = 0$.[16,20] Quite contrarily, the following types of motion are qualitatively different and newly observed in the presence of the shear flow.

When the cycloidal I motion has become unstable at high shear rates, the particle exhibits a *winding I motion*. The corresponding narrow region in the phase diagram (Fig. 1(a)) is marked by the purple filled triangles. It is located between the cycloidal I motion and the active straight motion. For this winding I motion, the long axis of the particle does not make full rotations in the laboratory frame. It only oscillates in time around the velocity vector, as shown in Fig. 1(e) for $\alpha = 0.7$ and $\dot{\gamma} = 0.08$. In particular, the trajectories in θ-ϕ space exhibit a closed loop indicating an oscillation. In contrast to the active straight motion, both, the relative velocity and the deformations of the particle, are time-dependent.

Finally, at high shear rates, also the active straight motion becomes unstable, and a *cycloidal II motion* appears. It is indicated by the gray filled squares in Fig. 1(a) and further characterized in Fig. 1(f) for $\alpha = 0.1$ and $\dot{\gamma} = 2$. Again the trajectory in real space is of cycloidal shape. However, as can be seen from the trajectory in θ-ϕ space, the value of θ stays close to zero with only small oscillations around it. Thus, in contrast to the cycloidal I motion, the elongation axis of deformation remains approximately horizontal for all times.

V. FULL DYNAMICS

In Secs. III and IV, we considered simplified special cases of the dynamic equations (3) and (8)–(10) to identify the basic states of motion. First we neglected deformations in Sec. III, then we excluded active contributions from the rotational spinning motion in Sec. IV. Nevertheless, the dynamics in both cases was already quite complex. This complexity is increased even further when we now investigate the full active dynamics. For example qualitatively new quasi-periodic and chaotic states arise.

We use the same methods and the same parameter values as in Sec. IV to study the full set of dynamic equations (3) and (8)–(10). The only difference is that now active rotations of the particle, which we call active spinning motions, are taken into account. They are induced by setting ζ to a positive value, $\zeta = 1.5$, in Eq. (10). Since the rotational symmetry of space is broken by the shear flow given by Eq. (13) with $\dot{\gamma} > 0$, we distinguish between two cases. First, we consider clockwise active rotations of the particle, after that counter-clockwise spinning motions. The rotational part of the shear flow itself is oriented in the clockwise direction. Our results are again presented in terms of the quantities introduced in Eqs. (5), (17), and (35).

A. Clockwise active rotation

Without an externally imposed shear flow, i.e. for $\dot{\gamma} = 0$, the situation of active spinning has been recently investigated by some of the present authors.[19,20] For the parameters that we have chosen in this paper, two types of motion have been found in the absence of the shear flow: circular and quasi-periodic motions. We repeat these results on the left border of our phase diagram (Fig. 2(a)) in the column $\dot{\gamma} = 0$. There, with increasing self-propulsion strength α, the circular motion is reentrant.[19,20]

With the shear flow now turned on and an active spinning in clockwise direction, the circular motion changes to the corresponding cycloidal I motion that was already obtained in Sec. IV and characterized in Fig. 1(c) for $\zeta = -0.1$. It covers a major part of our phase diagram (Fig. 2(a)) and is indicated by the red open squares. A cycloidal trajectory naturally results, when advection due to the flow is superimposed to a circular motion.

Next, a *quasi-periodic motion* is marked by the black crosses in the phase diagram (Fig. 2(a)). It occurs at intermediate self-propulsion strengths α. Interestingly, this type of motion is suppressed with increasing shear-rate $\dot{\gamma}$. We further characterize it in Fig. 2(b) for $\alpha = 0.5$ and $\dot{\gamma} = 0.1$. Obviously, the motion is not simply periodic as evident from the real-space trajectory and from the trajectories indicated in the θ-ϕ and s-ψ phase spaces by the black dots. However, it is quasi-periodic and not chaotic, because the return maps give discrete closed loops as shown in the second column of Fig. 2(b). The difference between

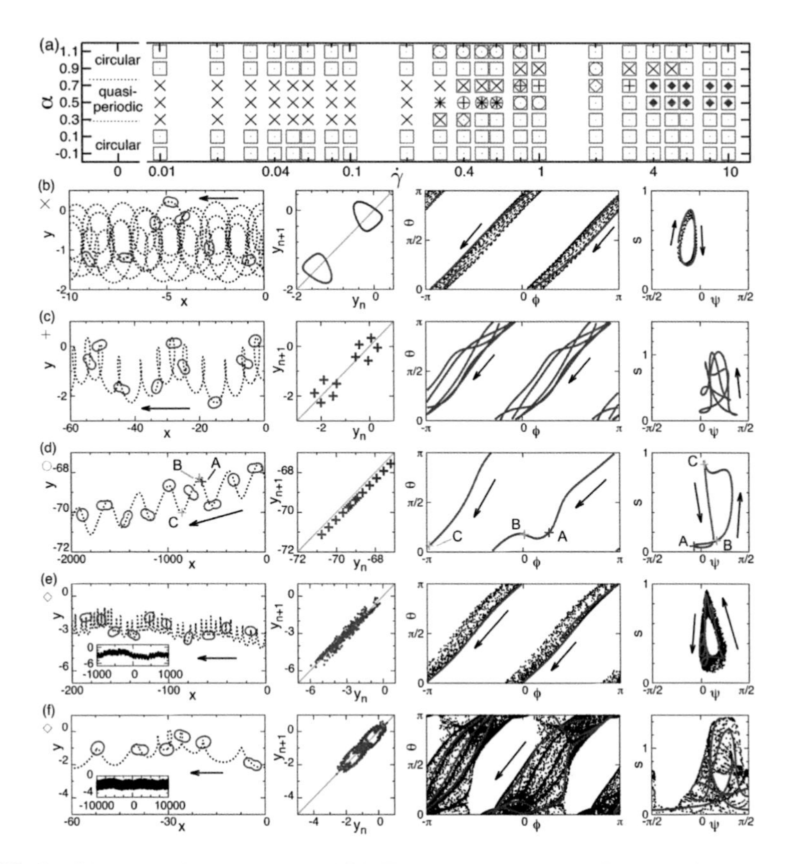

FIG. 2. (a) Dynamical phase diagram and (b)–(f) trajectories in real space (1st column), return maps (2nd column), attractors in θ-ϕ space (3rd column) as well as in s-ψ space (4th column) of the typical dynamical motions, obtained by solving Eqs. (3) and (8)–(10) numerically in two dimensions for clockwise active rotations ($\zeta = 1.5$); (b) *quasi-periodic motion* for $\alpha = 0.5$ and $\dot{\gamma} = 0.1$ indicated by the black crosses in panel (a); (c) *periodic motion* for $\alpha = 0.7$ and $\dot{\gamma} = 1$ marked by the black plus symbols in panel (a); (d) *winding II motion* for $\alpha = 0.5$ and $\dot{\gamma} = 0.8$ identified by the blue open circles in panel (a); (e) and (f) *chaotic motions* for $\alpha = 0.3$ and $\dot{\gamma} = 0.4$ as well as for $\alpha = 0.7$ and $\dot{\gamma} = 2$, respectively, marked by the purple open diamonds in panel (a). Arrows in panels (b)–(f) show the directions of motion. Some snapshots of the particle, the size of which is adjusted for illustration, are superimposed to the trajectory in real space. In panels (e) and (f), insets show the corresponding trajectories over longer time intervals. The trajectories in θ-ϕ and s-ψ phase space in panels (b), (e), and (f) are indicated by the black points. We also include short-time trajectories as red lines. Red open squares stand for the *cycloidal I motion* as already characterized in Fig. 1(c) for $\zeta = -0.1$. The blue filled diamonds in panel (a) represent an *undulated cycloidal I motion* as it is further illustrated by Fig. 3(c). Superimposed symbols in panel (a) indicate the observation of different trajectory types depending on the initial conditions.

the quasi-periodic motion in the absence of the shear flow and the one in the presence of the shear flow is simply that the particle stays within a finite area in the former case while in the latter case it escapes over time in the positive or negative x-direction.

When the shear rate $\dot{\gamma}$ is increased, several new types of motion are found that we have not observed before. Interestingly, all of them are sensitive to the initial conditions. We find a coexistence of at least two types of motion at every point of the phase diagram that we investigated for these new dynamic states, which leads to the superposition of the symbols in Fig. 2(a).

First, at higher shear rates, a *periodic motion* that cannot be observed at low shear rates is found at some positions in the phase diagram. This dynamic state is marked by the black plus symbols in Fig. 2(a) and further illustrated in Fig. 2(c) for $\alpha = 0.7$ and $\dot{\gamma} = 1$. The real-space trajectory appears as a commensurately modulated cycloid. We can distinguish this kind of motion from the quasi-periodic motion by the return map in the second column of Fig. 2(c), where thousands of measured trajectory extrema condense on ten discrete points in contrast to the closed loop object in Fig. 2(b). There are even some coexistence points of periodic and quasi-periodic motion in the phase diagram, induced by different initial conditions.

Next, in analogy to the winding I motion of Sec. IV, a *winding II motion* is identified at the positions of the blue open circles in the phase diagram (Fig. 2(a)). We characterize it in Fig. 2(d) for $\alpha = 0.5$ and $\dot{\gamma} = 0.8$. Since the particle in real space continuously descends in y-direction, the discrete points in the return map descend along the diagonal. The winding II motion can easily be distinguished from its counterpart, the winding I motion in Fig. 1(e), by the trajectory in θ-ϕ phase space. When observed from the laboratory frame, the particle features full rotations of its long axis of deformation in the winding II state, while only oscillations of this long axis occur in the winding I state. To facilitate the connection between the real- and phase-space trajectories in Fig. 2(d), we marked corresponding points by the capital letters "A," "B," and "C." We found a three-state coexistence region including the winding II motion, the periodic motion, and the quasi-periodic motion in the phase diagram (Fig. 2(a)) around the point $\alpha = 0.5$ and $\dot{\gamma} = 0.6$.

Most interestingly, we now also find *chaotic states* of the dynamic behavior of our single deformable active particle subjected to linear shear flow. In the phase diagram (Fig. 2(a)) they are marked by the purple open diamonds. Figures 2(e) and (f) show the characteristics of two chaotic dynamic states for $\alpha = 0.3$ and $\dot{\gamma} = 0.4$ as well as for $\alpha = 0.7$ and $\dot{\gamma} = 2$, respectively. The inset figures in the real-space trajectory plots in the first column of Figs. 2(e) and (f) depict the trajectory over longer time intervals. Closer inspection shows that the attractors in the θ-ϕ and s-ψ phase spaces of Fig. 2(e) are similar to the ones of the quasi-periodic motion in Fig. 2(b). However, the return maps are different enough to distinguish these two separate types of motion: the return map of the quasi-periodic motion forms a simple closed loop, while that of the chaotic motion in Fig. 2(e) is dispersed around the diagonal $y_{n+1} = y_n$. Likewise, we can distinguish the chaotic motion in Fig. 2(f) from the undulated cycloidal I motion discussed below in Fig. 3(c) via their return maps, although the attractors in θ-ϕ space and s-ψ space are similar.

At large shear rates $\dot{\gamma}$ and intermediate self-propulsion strengths α, another dynamic state was observed. It is marked by the blue filled diamonds in the phase diagram (Fig. 2(a)) and we call it an *undulated cycloidal I motion.* Since it also appears in the case of counter-clockwise active rotations of the particle, we discuss it below together with the dynamic states observed in that case.

B. Counter-clockwise active rotation

Finally, we analyze the case of counter-clockwise spinning motions of the active particle, i.e. active rotations in the direction opposite to that of the rotational part of the shear flow. Without the shear flow at $\dot{\gamma} = 0$, the rotational symmetry in space is not broken, and the dynamical states of clock- and counter-clockwise rotations are identical (except for the sense of rotation).

At low shear rates, the dynamics for both senses of rotation is still similar as can be inferred when comparing the corresponding phase diagrams Figs. 2(a) and 3(a) for low values of $\dot{\gamma}$. Again, a cycloidal I motion appears for both high and low self-propulsion strengths α. Likewise, a quasi-periodic motion emerges at intermediate self-propulsion strengths α. They are marked by the red open squares and black crosses in the phase diagram (Fig. 3(a)) and were discussed in Figs. 1(d) and 2(b), respectively. Increasing the shear rate $\dot{\gamma}$, the quasi-periodic motion becomes unstable in favor of the cycloidal I motion. Also the periodic motion, indicated by the black pluses and previously characterized in Fig. 2(c), as well as the winding II motion, marked by the blue open circle and previously depicted in Fig. 2(d), are recovered. Coexistence of different dynamic states again occurs and is shown by the superposition of different symbols in the phase diagram (Fig. 3(a)).

Interestingly, at large shear rates $\dot{\gamma} \gtrsim 1$, we observe an active-straight motion at all investigated self-propulsion strengths α. It is indicated in the phase diagram (Fig. 3(a)) by the green open pentagons and was previously discussed in Fig. 1(b). The origin of the emergence of the active straight motion at these shear rates can be easily understood: it appears when the rotation due to the active spinning motion in the counter-clockwise direction and the rotation due to the external flow in the clockwise direction balance each other.

At still larger shear rates $\dot{\gamma}$, this balance is no longer maintained and different types of motion appear. At high self-propulsion strength α, the particle next undergoes a *winding III motion* as denoted by the blue open downward triangles in the phase diagram (Fig. 3(a)).

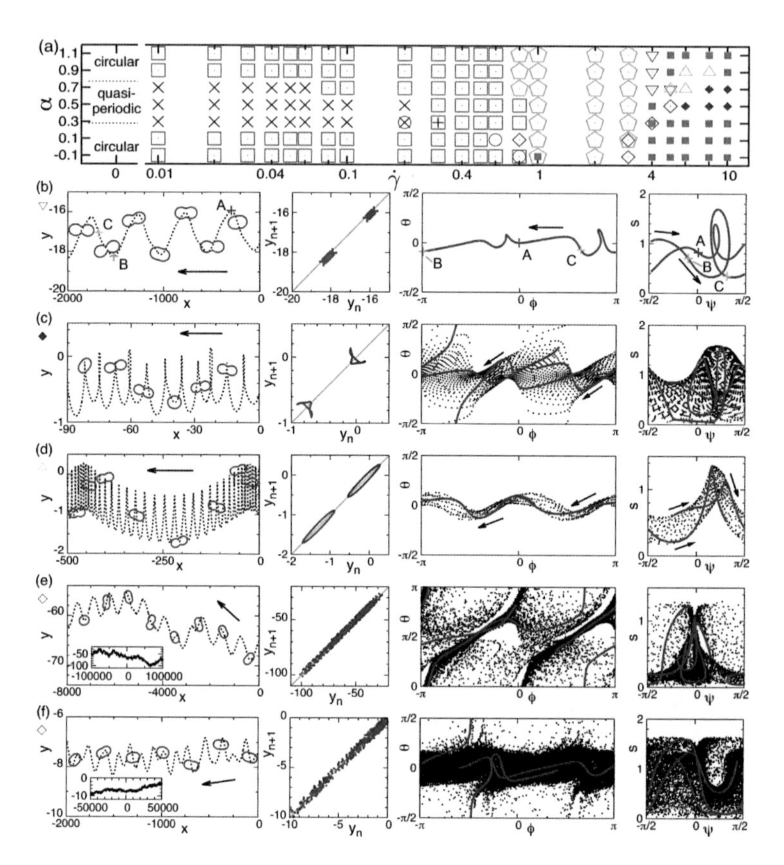

FIG. 3. (a) Dynamical phase diagram and (b)–(f) trajectories in real space (1st column), return maps (2nd column), attractors in θ-ϕ space (3rd column) as well as in s-ψ space (4th column) of the typical dynamical motions, obtained by solving Eqs. (3) and (8)–(10) numerically in two dimensions for counter-clockwise active rotations ($\zeta = 1.5$); (b) *winding III motion* for $\alpha = 0.9$ and $\dot{\gamma} = 4$ indicated by the blue open downward triangles in panel (a); (c) *undulated cycloidal I motion* for $\alpha = 0.5$ and $\dot{\gamma} = 6$ specified by the blue filled diamonds in panel (a); (d) *undulated cycloidal II motion* for $\alpha = 0.9$ and $\dot{\gamma} = 6$ marked by the blue open upward triangles in panel (a); (e) and (f) *chaotic motions* for $\alpha = 0.1$ and $\dot{\gamma} = 0.8$ as well as for $\alpha = 0.3$ and $\dot{\gamma} = 4$, respectively, indicated by the purple open diamonds in panel (a). Arrows in panels (b)–(f) show the directions of motion. Some snapshots of the particle, the size of which is adjusted for illustration, are superimposed to the trajectory in real space. In panels (e) and (f), insets show the corresponding trajectories over longer time intervals. The trajectories in θ-ϕ and s-ψ phase space in panels (c)–(f) are indicated by the black points. We also include short-time trajectories as red lines. The dynamic states corresponding to the other symbols in panel (a) that are not further characterized in panels (b)–(f) have already been explained in Figs. 1 and 2. Superimposed symbols in panel (a) indicate the observation of different trajectory types depending on the initial conditions.

This motion is characterized in Fig. 3(b) for $\alpha = 0.9$ and $\dot{\gamma} = 4$. In contrast to the winding I and winding II motions in Figs. 1(e) and 2(d), respectively, the angle θ as viewed from the laboratory frame always remains of small magnitude with values close to zero. This means that the particle always remains elongated along the horizontal direction. Only small oscillations of the long axis of deformation occur that are due to the competition between the active rotation in the counter-clockwise direction and the clockwise rotation induced by the shear flow. In Fig. 3(b), capital letters "A," "B," and "C" are again used to mark corresponding points along the trajectories in real space, in θ-ϕ phase space, and in s-ψ phase space.

In Sec. IV we have already found a dynamic mode that features an almost horizontal elongation of the particle at all times. It was the cycloidal II motion, obtained without active spinning, and depicted in Fig. 1(f). Indeed, we find this type of motion again when increasing the shear rate $\dot{\gamma}$ from the winding III motion at high self-propulsion strengths α. In addition to that, it is also the dominant dynamic mode at large shear rate $\dot{\gamma}$ but small self-propulsion strength α. We indicate it again by the gray filled squares in Fig. 3(a).

Apart from the cycloidal I and II motions, depicted previously in Figs. 1(c) and (d) as well as in Fig. 1(f), respectively, there are two other types of cycloidal motions. One of them is the undulated cycloidal I motion that has already been found for clockwise active spinning motion. It is marked by the blue filled diamonds in the phase diagrams (Figs. 2(a) and 3(a)). We now further characterize it in Fig. 3(c) for $\alpha = 0.5$ and $\dot{\gamma} = 6$. An undulation of the cycloidal amplitude is apparent from the real-space trajectory as well as from the return map. As can be seen in θ-ϕ phase space, the long axis of deformation makes full rotations in the laboratory frame.

The other further cycloidal type is the new *undulated cycloidal II motion* that we find only for counter-clockwise particle spinning and that we mark by the blue open upward triangles in Fig. 3(a). We illustrate this dynamic mode in Fig. 3(d) for $\alpha = 0.9$ and $\dot{\gamma} = 6$. In contrast to the undulated cycloidal I motion, there is no full rotation of the long axis of deformation in the laboratory frame as becomes obvious in θ-ϕ phase space.

Again we also observe chaotic motions, which are represented by the purple open diamonds in Fig. 3(a). Characteristics of these chaotic motions are displayed in Figs. 3(e) and (f) for $\alpha = 0.1$ and $\dot{\gamma} = 0.8$ as well as for $\alpha = 0.3$ and $\dot{\gamma} = 4$, respectively. A qualitative difference between the two depicted chaotic motions becomes obvious from the plots in phase space. While in the first case of Fig. 3(e) the long axis of deformation of the particle tends to rotate together with the velocity direction, it has a tendency to remain horizontal in the second case of Fig. 3(f). Both tendencies can be inferred from the dark bands in the θ-ϕ plots. Generally, we find that the trajectories in phase space in Figs. 3(e) and (f) are more delocalized than for the chaotic states of clockwise rotations that we have illustrated in Figs. 2(e) and (f).

VI. SUMMARY AND CONCLUSIONS

In this paper, we have investigated the dynamics of a deformable active particle in shear flow. For that purpose, we have considered a soft deformable particle with two types of activity: one is a spontaneous propulsion and the other one is a spontaneous spinning motion. The deformation of the particle is described by a symmetric traceless tensor variable, and its rotation by an anti-symmetric tensor variable. Further variables are the position of the center of mass and its relative velocity with respect to the flow. The externally imposed linear shear flow is included by taking into account its deformational and its rotational impact. Using symmetry arguments, we derive coupled dynamic equations for all of these variables. Our equations reduce to known models in the two limits of vanishing shear flow and vanishing particle deformability. On the one hand, in the limit of vanishing shear flow, we reproduce the previous results of Refs. 19,20. On the other hand, for vanishing particle deformability, we obtain an approximate analytical solution that is consistent with previous investigations.[42]

Various types of motion arise as numerical solutions of the full set of dynamical equations, including active straight motion, periodic motions, motions on regular and undulated cycloids, winding motions, as well as quasi-periodic and chaotic motions induced at high shear rates. In order to characterize and distinguish these dynamical states, we have analyzed and categorized them via their trajectories, corresponding return maps, as well as their attractors in phase space. Also the two situations of clockwise and counter-clockwise rotations with respect to the direction of the shear flow are distinguished and lead to partially different results, in particular at high shear rates.

Our predictions can be verified in experiments on self-propelled droplets exposed to shear flow. For instance, in some experiments[10,44] self-propelled droplets on liquid-air interfaces can be exposed to linear shear fields by putting the carrier liquid between two parallel confining walls that move alongside into opposite directions. This induces an approximately planar linear shear gradient at the surface of the carrier liquid, if the liquid container is sufficiently deep. Since the motion of the droplets is confined to the liquid-air interface, the geometry is quasi-two dimensional. Using this experimental set-up, it is in principle possible to verify the phenomena predicted by our analysis.

Future studies should address several extensions of our model: first of all, different prescribed flow fields can be explored using our equations. Most noticeable examples include a Poiseuille flow[43] or an imposed vorticity field. We expect again a manifold of different types of motion in these flow fields presuming that the different flow topologies will induce different types of motion. The next step is to extend our analysis to a finite concentration of particles and to include steric interactions between them. This is a complex problem which is already very difficult for rigid self-propelled particles.[38,54–61] Another step is to extend the current analysis of the model to three spatial dimensions. In a previous study, some of us demonstrated that – without shear flow – a particle can exhibit additional qualitatively different types of motion when comparing a three- to a two-dimensional set-up.[20] In the case without shear flow, these were additional helical and superhelical types of motion.[20] Thus we expect that further new types of dynamics can arise in three dimensions when the shear flow is included. Finally one could access the dynamics of propelled vesicles by using our analysis as a starting point. Here one should impose the constraints of constant volume and constant surface area of the deformable particle. The dynamics of passive vesicles in shear flow has been explored quite extensively in recent years[62,63] with various specific effects like tank-treading motion, lifting,[64,65] wrinkling,[66] tumbling and swinging.[67,68] It would indeed be interesting to generalize all these effects to self-propelled vesicles.

ACKNOWLEDGMENTS

This work was supported by the Japan Society for the Promotion of Science (JSPS) Core-to-Core Program "International research network for non-equilibrium dynamics of soft matter" and "Non-equilibrium dynamics of soft matter and information," and by a Grant-in-Aid for Scientific Research C (No. 23540449) from JSPS and a Grant-in-Aid for Scientific Research A (No. 24244063) from Mext. M.T. is supported by the Japan Society for the Promotion of Science Research Fellowship for Young Scientists. A.M.M. and H.L. gratefully acknowledge support from the Deutsche Forschungsgemeinschaft (DFG) through the German–Japanese project "Nichtgleichgewichtsphänomene in Weicher Materie/Soft Matter" No. LO 418/15. R.W. gratefully acknowledges financial support from a Postdoctoral Research Fellowship (Grant No. WI 4170/1-1) of the Deutsche Forschungsgemeinschaft.

REFERENCES

[1]S. J. Ebbens and J. R. Howse, Soft Matter **6**, 726 (2010).
[2]S. Ramaswamy, Annu. Rev. Condens. Matter Phys. **1**, 323 (2010).
[3]M. E. Cates, Rep. Prog. Phys. **75**, 042601 (2012).
[4]M. C. Marchetti, J. F. Joanny, S. Ramaswamy, T. B. Liverpool, J. Prost, M. Rao, and R. A. Simha, Rev. Mod. Phys. **85**, 1143 (2013).
[5]P. Romanczuk, M. Bär, W. Ebeling, B. Linder, and L. Schimansky-Geier, Eur. Phys. J. Spec. Top. **202**, 1 (2012).
[6]W. F. Paxton, K. C. Kistler, C. C. Olmeda, A. Sen, S. K. St. Angelo, Y. Cao, T. E. Mallouk, P. E. Lammert, and V. H. Crespi, J. Am. Chem. Soc. **126**, 13424 (2004).
[7]H. R. Jiang, N. Yoshinaga, and M. Sano, Phys. Rev. Lett. **105**, 268302 (2010).
[8]R. Kapral, J. Chem. Phys. **138**, 020901 (2013).

[9]H. Boukellal, O. Campas, J.-F. Joanny, J. Prost, and C. Sykes, Phys. Rev. E **69**, 061906 (2004).

[10]K. Nagai, Y. Sumino, H. Kitahata, and K. Yoshikawa, Phys. Rev. E **71**, 065301(R) (2005).

[11]T. Toyota, N. Maru, M. M. Hanczyc, T. Ikegami, and T. Sugawara, J. Am. Chem. Soc. **131**, 5012 (2009).

[12]K. Keren, Z. Pincus, G. M. Allen, E. L. Barnhart, G. Marriott, A. Mogilner, and J. A. Theriot, Nature (London) **453**, 475 (2008).

[13]L. Bosgraaf and P. J. M. Van Haastert, PLoS ONE **4**, e5253 (2009).

[14]L. Li, S. F. Nørrelykke, and E. C. Cox, PLoS ONE **3**, e2093 (2008).

[15]Y. T. Maeda, J. Inoue, M.Y. Matsuo, S. Iwaya, and M. Sano, PLoS ONE **3**, e3734 (2008).

[16]T. Ohta and T. Ohkuma, Phys. Rev. Lett. **102**, 154101 (2009).

[17]T. Hiraiwa, M. Y. Matsuo, T. Ohkuma, T. Ohta, and M. Sano, Europhys. Lett. **91** 20001 (2010).

[18]T. Hiraiwa, K. Shitara, and T. Ohta, Soft Matter **7**, 3083 (2011).

[19]M. Tarama and T. Ohta, J. Phys.: Condens. Matter **24**, 464129 (2012).

[20]M. Tarama and T. Ohta, Prog. Theor. Exp. Phys. (2013) 013A01.

[21]M. Tarama and T. Ohta, Phys. Rev. E **87**, 062912 (2013).

[22]H. Wada and R. R. Netz, Phys. Rev. E **80**, 021921, (2009).

[23]S. I. Nishimura, M. Ueda, and M. Sasai, PLoS Comput. Biol. **5**, e1000310 (2009).

[24]D. Shao, W.-J. Rappel, and H. Levine, Phys. Rev. Lett. **105**, 108104 (2010).

[25]K. Doubrovinski and K. Kruse, Phys. Rev. Lett. **107**, 258103 (2011).

[26]F. Ziebert, S. Swaminathan, and I. S. Aranson, J. R. Soc. Interface **9**, 1084 (2012).

[27]G. Miño, T. E. Mallouk, T. Darnige, M. Hoyos, J. Dauchet, J. Dunstan, R. Soto, Y. Wang, A. Rousselet, and E. Clement, Phys. Rev. Lett. **106**, 048102 (2011).

[28]S. Rafaï, L. Jibuti, and P. Peyla, Phys. Rev. Lett. **104**, 098102 (2010).

[29]A. Bagorda and C. A. Parents, J. Cell Sci. **121**, 2621 (2008).

[30]B. M. Friedrich and F. Jülicher, Proc. Natl. Acad. Sci. USA **104**, 13256 (2007).

[31]G. Jékely, Philos. Trans. R. Soc. London, Ser. B **364**, 2795 (2009).

[32]J. Tailleur and M. E. Cates, Europhys. Lett. **86**, 60002 (2009).

[33]M. Tarama and T. Ohta, Eur. Phys. J. B **83**, 391 (2011).

[34]R. Wittkowski and H. Löwen, Phys. Rev. E **85**, 021406 (2012).

[35]T. Hiraiwa, A. Baba, and T. Shibata, Eur. Phys. J. E **36**, 32 (2013).

[36]W. R. DiLuzio, L. Turner, M. Mayer, P. Garstecki, D. B. Weibel, H. C. Berg, and G. M. Whitesides, Nature (London) **435**, 1271 (2005).

[37]S. van Teeffelen and H. Löwen, Phys. Rev. E **78**, 020101(R) (2008).

[38]H. H. Wensink and H. Löwen, Phys. Rev. E **78**, 031409 (2008).

[39]K. Drescher, J. Dunkel, L. H. Cisneros, S. Ganguly, and R. E. Goldstein, Proc. Natl. Acad. Sci. USA **108**, 10940 (2011).

[40]J. O. Kessler, Nature (London) **313**, 218 (1985).
[41]Y.-G. Tao and R. Kapral, Soft Matter **6**, 756 (2010).
[42]B. ten Hagen, R. Wittkowski, and H. Löwen, Phys. Rev. E **84**, 031105 (2011).
[43]A. Zöttel and H. Stark, Phys. Rev. Lett. **108** 218104 (2012).
[44]F. Takabatake, N. Magome, M. Ichikawa, and K. Yoshikawa, J. Chem. Phys. **134**, 114704 (2011).
[45]H. Pleiner, M. Liu, and H. R. Brand, Rheol. Acta **41**, 375 (2002).
[46]H. Stark and T. C. Lubensky, Phys. Rev. E **67**, 061709 (2003).
[47]T. Ohta, T. Ohkuma, and K. Shitara, Phys. Rev. E **80**, 056203 (2009).
[48]K. Shitara, T. Hiraiwa, and T. Ohta, Phys. Rev. E **83**, 066208 (2011).
[49]M. Makino and M. Doi, J. Phys. Soc. Jpn. **73**, 2739 (2004).
[50]P. L. Maffettone and M. Minale, J. Non-Newtonian Fluid Mech. **78**, 227 (1998).
[51]B. ten Hagen, S. van Teeffelen, and H. Löwen, J. Phys.: Condens. Matter **23**, 194119 (2011).
[52]F. Kümmel, B. ten Hagen, R. Wittkowski, I. Buttinoni, R. Eichhorn, G. Volpe, H. Löwen, and C. Bechinger, Phys Rev. Lett. **110**, 198302 (2013).
[53]L. Baraban, D. Makarov, R. Streubel, I. Mönch, D. Grimm, S. Sanchez, and O. G. Schmidt, ACS Nano **6**, 3383 (2012).
[54]T. Vicsek, A. Czirók, E. Ben-Jacob, I. Cohen, and O. Shochet, Phys. Rev. Lett. **75**, 1226 (1995).
[55]J. Toner and Y. Tu, Phys. Rev. Lett. **75**, 4326 (1995).
[56]R. A. Simha and S. Ramaswamy, Phys. Rev. Lett. **89**, 058101 (2002).
[57]A. Sokolov, I. S. Aranson, J. O. Kessler, and R. E. Goldstein, Phys. Rev. Lett. **98** 158102 (2007).
[58]F. Ginelli, F. Peruani, M. Bär, and H. Chaté, Phys. Rev. Lett. **104**, 184502 (2010).
[59]Y. Itino, T. Ohkuma, and T. Ohta, J. Phys. Soc. Jpn. **80**, 033001 (2011).
[60]A. M. Menzel and T. Ohta, Europhys. Lett. **99**, 58001 (2012).
[61]H. H. Wensink, J. Dunkel, S. Heidenreich, K. Drescher, R. E. Goldstein, H. Löwen, and J. M. Yeomans, Proc. Natl. Acad. Sci. USA **109**, 14308 (2012).
[62]M. Kraus, W. Wintz, U. Seifert, and R. Lipowsky, Phys. Rev. Lett. **77**, 3685 (1996).
[63]H. Noguchi and G. Gompper, Phys. Rev. Lett. **93**, 258102 (2004).
[64]U. Seifert, Phys. Rev. Lett. **83**, 876 (1999).
[65]I. Cantat and C. Misbah, Phys. Rev. Lett. **83**, 880 (1999).
[66]R. Finken and U. Seifert, J. Phys.: Condens. Matter **18**, L185 (2006).
[67]H. Noguchi and G. Gompper, Phys. Rev. Lett. **98**, 128103 (2007).
[68]S. Kessler, R. Finken, and U. Seifert, J. Fluid Mech. **605**, 207 (2008).

Statement of the author: This theoretical work emerged from several research stays of the Japanese Ph.D. student Mitsusuke Tarama at our institute in Düsseldorf. He performed the numerical calculations under the supervision of Hartmut Löwen and Takao Ohta. The study combines previous models for deformable particles in a quiescent solvent [288, 294, 296] and for rigid particles in shear flow [306]. The latter model was developed in my master thesis. I was involved in many discussions with Mitsusuke Tarama, Andreas M. Menzel, Raphael Wittkowski, and Hartmut Löwen regarding the generalization and merging of the two previous approaches and helped to relate the new findings to the case of non-deformable active particles. Moreover, I contributed to writing and finalizing the manuscript.

CHAPTER

Brownian motion and the hydrodynamic friction tensor for colloidal particles of complex shape

The content of this chapter has been published in a similar form in *Physical Review E* **88**, 050301(R) (2013) by Daniela J. Kraft, Raphael Wittkowski, Borge ten Hagen, Kazem V. Edmond, David J. Pine, and Hartmut Löwen (see reference [216]).

Spherical colloidal particles have served as a model system for investigating Brownian motion since the pioneering studies of Einstein [1] and Perrin [2]. Such particles are characterized by a translational diffusion coefficient that is linked to the Stokes friction coefficient through the well-known Stokes-Einstein relation [1]. Particles encountered in nature and industry usually have complex nonspherical shapes, and describing their Brownian motion raises fundamental questions about how translational and rotational diffusion are coupled. However, in most studies, translational and rotational diffusion are considered separately, which is valid only for certain highly symmetrical particle shapes.

Recently, model colloids with well-characterized but complex shapes have become available [3, 4], which permits the quantitative study of the hydrodynamic coupling between translational and rotational diffusion for nontrivial particle shapes for the first time.

In general, the dynamics of a colloidal particle suspended in a liquid is described by a Langevin equation that equates the Stokes friction forces and torques with random thermal forces and torques on a particle. For an arbitrary colloidal particle suspended in a liquid, the friction forces and torques are described by a symmetric second-rank hydrodynamic friction tensor $\mathcal{H}$ [5, 6], which includes off-diagonal terms coupling the three translational and three rotational degrees of freedom. In all, $\mathcal{H}$ has 21 independent elements. For spherical particles, $\mathcal{H}$ is diagonal, with two distinct entries corresponding to the inverse translational and rotational friction coefficients [7]. For rodlike particles, both the translational and rotational entries involve two different coefficients, corresponding to parallel and perpendicular particle orientation, but $\mathcal{H}$ remains diagonal, meaning that translation and rotation remain decoupled [8].

For a general biaxial particle, $\mathcal{H}$ involves nonzero off-diagonal elements that couple translational and rotational motion. The corresponding Langevin equation involves intricate multiplicative noise terms due to this coupling, which makes a description of the Brownian dynamics much more difficult. Although a first theoretical treatment dates back to Perrin [9], it was not reconsidered until much later, and only by a few authors [5, 10] who never explicitly applied it to experiments for biaxial nonorthotropic particle shapes [11].

In this Rapid Communication, we report experimental measurements and theoretical calculations of the hydrodynamic friction tensor for various anisotropic colloidal particles, including a general irregular biaxial shape with three fused spheres of different diameters. The particle shape and size are determined by confocal as well as scanning electron microscopy (SEM). We track the Brownian trajectories of these anisotropic colloidal particles with full orientational resolution in real space by confocal microscopy. This three-dimensional (3D) real-space technique allows for tracking the motion of arbitrarily complex colloidal particles, even in crowded environments. Based on the generalized Stokes-Einstein relation, we then propose a theoretical framework to extract all independent hydrodynamic friction coefficients from the short-time limit of appropriate correlation functions. Our results are consistent with low-Reynolds-number hydrodynamic calculations of the friction tensor assuming stick boundary conditions of the solvent at the particle surface, where the experimentally determined particle shapes are taken as an input. Since the full orientational resolution of the individual particle trajectories reveals the couplings between different degrees of freedom of Brownian motion, the information obtained by our analysis is much more basic and detailed than averaged quantities derived from dynamic light scattering [12] or sedimentation [13] experiments of biaxial colloidal particles. Our method can be used to analyze the Brownian dynamics of any rigid irregularly shaped colloidal particles.

For our experiments, we have prepared anisotropic colloidal particles from fluorescently labeled, cross-linked PMMA [poly(methyl methacrylate)] spheres [14] using an emulsion-evaporation method [3]. The resulting cluster shapes are uniquely set by minimization of the second moment of the mass distribution [3] as confirmed by SEM. We have specifically chosen regular trimers and tetramers as well as an irregular trimer shown in Fig. 1 for their different symmetry properties. We idealized the particle cluster shapes as a composition of fused spheres and measured only the radii and relative distances between the spheres.

For the regular clusters, RITC-labeled (rhodamine-B-isothiocyanate-labeled) PMMA spheres 2.1 ± 0.1 μm (trimer) and 2.4 ± 0.1 μm (tetramer) in diameter are employed. For the irregular trimer, 2.1 ± 0.1 μm and 1.3 ± 0.1 μm RITC-dyed spheres are combined with 1.7 ± 0.1 μm spheres labeled with NBD-MAEM (4-methylaminoethyl methacrylate-7-nitrobenzo-2-oxa-1,3-diazol), which allows us to easily distinguish the different spheres with a confocal microscope. All sphere diameters are measured by static light scattering. The clusters are dispersed in a tetrabutylammonium bromide saturated ~78/22 (weight/weight) cyclohexyl bromide/*cis*-decalin (CHB/decalin) mixture, which nearly matches the particles' density and index of refraction, and has a dynamic (shear) viscosity $\eta = 2.22$ mPa s. The dispersion is then put into rectangular glass capillaries (Vitrotubes, 100 μm×5 mm×50 mm) and the ends are sealed with optical adhesive (Norland, NOA81).

The three-dimensional motion of the anisotropic particles is observed using a Leica TCS SP5 confocal microscope equipped with an argon laser ($\lambda_1 = 488$ nm and $\lambda_2 = 543$ nm) and an oil-immersion objective (Leica, 63×, 1.4 NA) [15]. The imaging speed is typically 70 stacks in z direction per approximately 0.8 s. All experiments are conducted at room temperature $T = 294$ K. The positions of the individual spheres of each cluster are tracked using IDL routines (see Ref. [16]). From these sphere positions we calculate the center-of-mass positions, orientations, and bond lengths of the clusters. For the regular trimer the bond length is 1.5 μm, for the regular tetramer the centers of any two spheres are separated by 2.3 μm, and for the irregular trimer the bond lengths are 2.2 μm between the big and the medium sphere as well as between the big and the small sphere, and 1.7 μm between the medium and the small sphere.

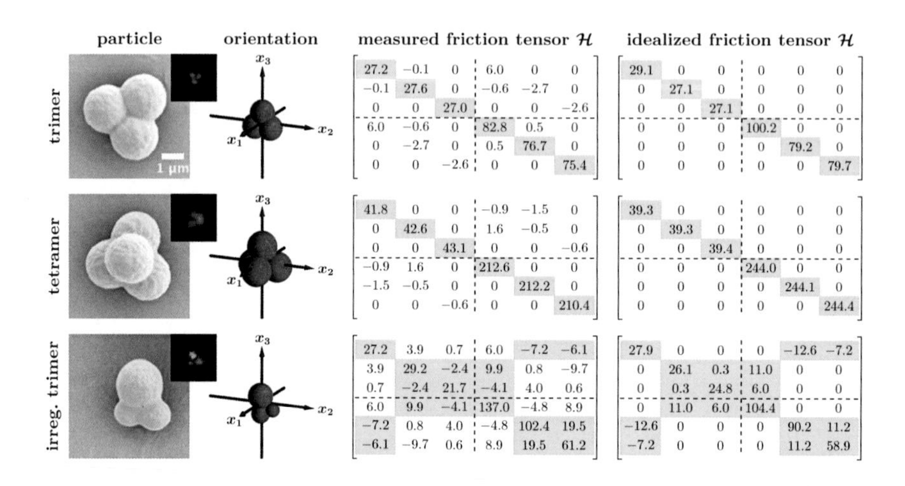

FIG. 1. SEM images (gray) and images obtained by confocal microscopy (colored insets) of three different colloidal particles (regular trimer, regular tetramer, and irregular trimer) and the corresponding hydrodynamic friction tensors $\mathcal{H}$ determined in experiments and predicted for idealized particle shapes by hydrodynamic calculations using HYDROSUB. The tensors shown are given in dimensions $[\mathcal{H}_{i,j=1,2,3}] = \mu\mathrm{m}$, $[\mathcal{H}^{i=1,2,3}_{j=4,5,6}] = [\mathcal{H}^{i=4,5,6}_{j=1,2,3}] = \mu\mathrm{m}^2$, and $[\mathcal{H}_{i,j=4,5,6}] = \mu\mathrm{m}^3$. The four 3×3 submatrices of $\mathcal{H}$ contain the translational friction coefficients (upper left), the rotational friction coefficients (lower right), and the translational-rotational friction coefficients for the coupling of translational and rotational motion (upper right and lower left). Since $\mathcal{H}$ depends on the center-of-mass position and orientation of the particles, the coordinate systems used are illustrated. For the diagonal elements of $\mathcal{H}$, the statistical error of the experimental data is 3% for the regular trimer and tetramer and 10% for the irregular trimer. The absolute statistical error of the off-diagonal elements is 1 and 5, respectively, in the given units.

Apart from temperature T and solvent viscosity η, the Brownian dynamics of a single rigid colloidal particle depends only on its shape and size, which enter in the 6×6-dimensional symmetric hydrodynamic friction tensor $\mathcal{H}$ [5, 6]. The latter relates the translational velocity $\vec{v}$ and the angular velocity $\vec{\omega}$ of the particle to the hydrodynamic drag force $\vec{F}$ and torque $\vec{T}$ that the particle experiences in the viscous solvent: $\vec{K} = -\eta\mathcal{H}\vec{\mathfrak{v}}$ with $\vec{K} = (\vec{F}, \vec{T})$ and $\vec{\mathfrak{v}} = (\vec{v}, \vec{\omega})$.

There are two possibilities for determining $\mathcal{H}$ for a given particle. It can either be obtained from its shape and size by a hydrodynamic calculation that involves solving the Stokes equation with stick boundary conditions for the solvent at the particle surface [6], or it can be extracted from appropriate equilibrium short-time correlation functions. We have used the software HYDROSUB [17] to follow the first route, where we used the experimentally determined particle shape, idealized by fused spheres, as input [18]. For a trimer and a tetramer of equal spheres as well as for an irregular trimer, results are shown in Fig. 1 [19]. For convenience, we have chosen the coordinate systems in such a way that the center of mass of a particle coincides with the origin of coordinates and the particle's planes of symmetry coincide with the coordinate planes, whenever this is possible. This choice of particle-fixed coordinate systems leads to a particularly simple structure of the hydrodynamic friction tensor with many vanishing nondiagonal elements [6]. The remaining nonvanishing elements are highlighted in Fig. 1.

The second route to access $\mathcal{H}$ is to measure the trajectory of the Brownian particle with full orientational resolution, i. e., the combined knowledge of the center-of-mass position $\vec{x}(t)$ and the three mutually perpendicular normalized orientation vectors $\hat{u}_i(t)$ with $i = 1, 2, 3$ in dependence of time t. The key idea is now to consider a set of appropriate dynamical cross-correlation functions

$$C_{ij}(t) = \langle X_i(t) X_j(t) \rangle \tag{1}$$

with $i, j \in \{1, \dots, 6\}$, where $\langle \cdot \rangle$ denotes a noise average and the six-dimensional positional-orientational displacement vector $\vec{X}(t) = (\Delta\vec{x}(t), \Delta\hat{u}(t))$ is defined by $\Delta\vec{x}(t) = \vec{x}(t) - \vec{x}(0)$ and $\Delta\hat{u}(t) = \frac{1}{2}\sum_{i=1}^{3} \hat{u}_i(0) \times \hat{u}_i(t)$, where the latter is the appropriate expression for orientational displacements. The short-time limit of this set of cross-correlation functions gives access to the hydrodynamic friction tensor $\mathcal{H}$ via

$$\mathcal{D} = \frac{1}{2} \lim_{t \to 0} \frac{\mathrm{d}C(t)}{\mathrm{d}t} , \qquad \mathcal{H} = \frac{k_{\mathrm{B}} T}{\eta} \mathcal{D}^{-1} , \tag{2}$$

where $\mathcal{D}$ denotes the (generalized) diffusion tensor and k_{B} Boltzmann's constant. A larger value for an element of $\mathcal{H}$ therefore means a higher hydrodynamic friction and thus a slower diffusion. From this second route, based on the experimentally determined trajectories, we obtain the results presented in Fig. 1. The experimental results for $\mathcal{H}$ are in good agreement with our hydrodynamic calculations; deviations are due to an idealization of the particle shape in the hydrodynamic calculations and due to the statistical error originating from the limited length of the measured trajectories.

In reality, the particles are not compositions of perfect spheres, but have rough surfaces and deformations near the overlap areas of the spheres. Additionally, the spheres that make up the clusters have a polydispersity of about 5%. While uncertainties in the size of the particles only lead to small deviations in the translational-translational elements $\mathcal{H}_{i,j=1,2,3} \propto l$ with the length scale l, these deviations are of greater relevance for the translational-rotational coupling elements $\mathcal{H}^{i=1,2,3}_{j=4,5,6} \propto l^2$ and lead to large deviations in the rotational-rotational elements $\mathcal{H}_{i,j=4,5,6} \propto l^3$. Nondiagonal tensor elements, which should vanish by

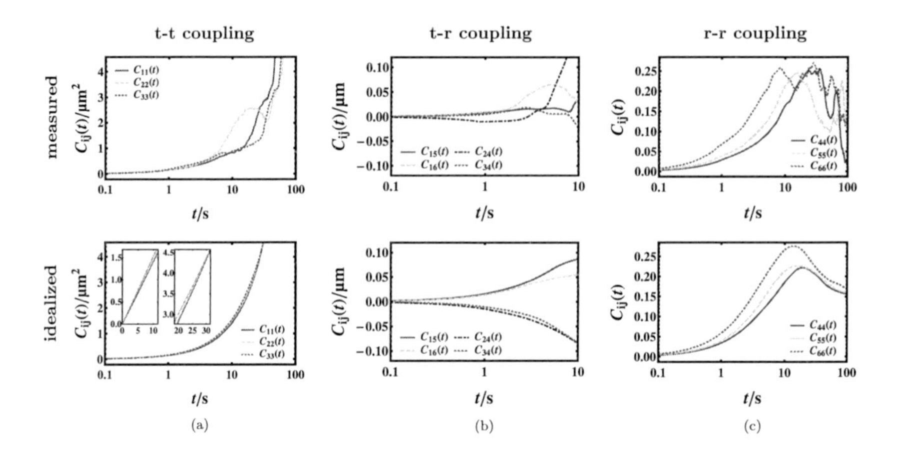

FIG. 2. Time evolution of ten representative correlation functions $C_{ij}(t)$ for the irregular trimer determined in experiments (upper row) and predicted from a simulation (lower row) based on the idealized hydrodynamic friction tensor (see Fig. 1). The insets in the lower left plot show the same quantities on linear scales.

symmetry considerations (not highlighted in Fig. 1), indeed have nonzero values in the experimentally determined friction tensor due to both irregularities in the actual particle shape and the statistical error because of the finite length of the trajectories.

We finally address the full time dependence of the basic cross-correlation functions $C_{ij}(t)$. Results for $C_{ij}(t)$ obtained from our measured trajectories of the irregular trimer particle are presented in Fig. 2. Figure 2(a) shows the translational mean-square displacements for the different Cartesian components, which increase linearly for short times with the slope governed by the anisotropic short-time diffusion coefficients. At long times, there is another linear function in time, which is the same for all coordinates [20], since it is governed by the orientationally averaged long-time diffusion coefficient [see insets in the lower plot in Fig. 2(a)]. Several cross-correlations between translational and rotational displacements are shown in Fig. 2(b). The absolute values of these nontrivial correlations are initially increasing with time, but decorrelate for longer times. Finally, the rotational-rotational correlations shown in Fig. 2(c) are clearly positive and build up continuously as functions of time until they decay again and approach the same constant value for very long times. Here, we restrict the presentation of correlation functions to this irregular particle, since it is the only particle with nonzero translational-rotational coupling elements and thus provides the most nontrivial dynamics.

An idealized time evolution of the correlation functions $C_{ij}(t)$ can be obtained from a given hydrodynamic friction tensor $\mathcal{H}$. With this tensor as an input, the Brownian motion of a colloidal particle can be simulated by solving its Langevin equation (see Ref. [10]) numerically using a stochastic integrator of strong order 1.5 [21]. Our results for the predicted correlation functions are presented in Fig. 2 as well. A comparison with the experimental results for $C_{ij}(t)$ reveals again good agreement with deviations resulting from choosing the idealized instead of the experimentally determined hydrodynamic friction tensor and from the statistical error. Note that the statistical error is obvious for the experimentally determined correlation functions, while it is extremely small for the simulated correlation functions, where trajectories with 10^6 time steps have been calculated.

In this work, we track the Brownian dynamics of individual colloidal particles with various anisotropic shapes and extract the hydrodynamic friction tensor from an analysis of appropriate short-time correlation functions. The framework of our analysis can in principle be applied to any rigid particle with an arbitrary shape and therefore to a broad range of relevant suspensions. Using confocal microscopy we obtain real-space 3D measurements of any complex particles, even in crowded environments.

Future work should address the effect of aligning external fields such as gravity or magnetic fields [22] that gain major importance in the context of directed self-assembly [23]. Moreover, nondilute colloidal suspensions would be interesting, where direct particle-particle interactions [24] and solvent-flow mediated hydrodynamic interactions [25] will lead to even more intricate translational-rotational couplings. This would provide a basis to understand the rheological properties of concentrated dispersions of irregularly shaped colloidal particles such as clay [26] and asphalt [27].

ACKNOWLEDGMENTS

We thank Tom Lubensky, HyunJoo Park, Gary Hunter, Andrew Hollingsworth, and Marco Heinen for helpful discussions. The theoretical work was supported by the German Research Foundation (DFG) within SFB TR6 (project D3) and by the European Research Council (ERC Advanced Grant INTERCOCOS, Project No. 267499). The experimental work was supported by the US National Science Foundation (Grants No. CBET-1236378 and No. DMR-1105455). D.J.K. gratefully acknowledges financial support through a Rubicon Grant (680-50-1019) by the Netherlands Organization for Scientific Research (NWO). R.W. gratefully acknowledges financial support from a Postdoctoral Research Fellowship (WI 4170/1-1) of the DFG.

---

[1] A. Einstein, Ann. Phys. (Berlin) **322**, 549 (1905).

[2] J. Perrin, Ann. Chim. Phys. **18**, 5 (1909).

[3] V. N. Manoharan, M. T. Elsesser, and D. J. Pine, Science **301**, 483 (2003).

[4] S. C. Glotzer and M. J. Solomon, Nat. Mater. **6**, 557 (2007); D. J. Kraft, W. S. Vlug, C. M. van Kats, A. van Blaaderen, A. Imhof, and W. K. Kegel, J. Am. Chem. Soc. **131**, 1182 (2009); D. J. Kraft, J. Groenewold, and W. K. Kegel, Soft Matter **5**, 3823 (2009).

[5] H. Brenner, J. Colloid. Interf. Sci. **23**, 407 (1967).

[6] J. Happel and H. Brenner, *Low Reynolds Number Hydrodynamics: With Special Applications to Particulate Media*, 2nd ed., Mechanics of Fluids and Transport Processes, Vol. 1 (Kluwer Academic, Dordrecht, 1991).

[7] G. Nägele, Phys. Rep. **272**, 215 (1996); V. Degiorgio, R. Piazza, and R. B. Jones, Phys. Rev. E **52**, 2707 (1995).

[8] J. K. G. Dhont, *An Introduction to Dynamics of Colloids*, 1st ed., Studies in Interface Science, Vol. 2 (Elsevier Science, Amsterdam, 1996); Y. Han, A. M. Alsayed, M. Nobili, J. Zhang, T. C. Lubensky, and A. G. Yodh, Science **314**, 626 (2006).

[9] F. Perrin, J. Phys. Radium **5**, 497 (1934); **7**, 1 (1936).

[10] M. X. Fernandes and J. Garcia de la Torre, Biophys. J. **83**, 3039 (2002); M. Makino and M. Doi, J. Phys. Soc. Jpn. **73**, 2739 (2004); R. Wittkowski and H. Löwen, Phys. Rev. E **85**, 021406 (2012).

[11] Notice that there is related earlier work that considers only the translational part of the friction tensor [28] or just the viscosity [29].

[12] M. Hoffmann, C. S. Wagner, L. Harnau, and A. Wittemann, ACS Nano **3**, 3326 (2009).

[13] K. Zahn, R. Lenke, and G. Maret, J. Phys. II **4**, 555 (1994).

[14] M. T. Elsesser, A. D. Hollingsworth, K. V. Edmond, and D. J. Pine, Langmuir **27**, 917 (2011).

[15] See Supplemental Material at `http://link.aps.org/supplemental/10.1103/PhysRevE.88.050301` for movies on a 2D projection of the 3D motion of the anisotropic particles.

[16] G. L. Hunter, K. V. Edmond, M. T. Elsesser, and E. R. Weeks, Opt. Express **19**, 17189 (2011).

[17] J. Garcia de la Torre and V. A. Bloomfield, Q. Rev. Biophys. **14**, 81 (1981); B. Carrasco and J. Garcia de la Torre, Biophys. J. **76**, 3044 (1999); J. Garcia de la Torre and B. Carrasco, Biopolymers **63**, 163 (2002).

[18] G. C. Abade, B. Cichocki, M. L. Ekiel-Jeżewska, G. Nägele, and E. Wajnryb, J. Chem. Phys. **132**, 014503 (2010).

[19] Clearly, our method is also applicable to dimers, where the tensor $\mathcal{H}$ is diagonal. We obtain agreement with the theoretical calculations within a few percent.

[20] B. Cichocki, M. L. Ekiel-Jeżewska, and E. Wajnryb, J. Chem. Phys. **136**, 071102 (2012).

[21] P. E. Kloeden and E. Platen, *Numerical Solution of Stochastic Differential Equations*, 1st ed., Applications of Mathematics: Stochastic Modelling and Applied Probability, Vol. 23 (Springer, Berlin, 2006).

[22] D. van der Beek, P. Davidson, H. H. Wensink, G. J. Vroege, and H. N. W. Lekkerkerker, Phys. Rev. E **77**, 031708 (2008).

[23] M. Grzelczak, J. Vermant, E. M. Furst, and L. M. Liz-Marzán, ACS Nano **4**, 3591 (2010).

[24] A. Yethiraj, Soft Matter **3**, 1099 (2007).

[25] T. G. M. van de Ven, *Colloidal Hydrodynamics* (Academic, London, 1989).

[26] W. K. Kegel and H. N. W. Lekkerkerker, Nat. Mater. **10**, 5 (2011).

[27] I. N. Evdokimov, N. Y. Eliseev, and D. Y. Eliseev, J. Petrol. Sci. Eng. **30**, 199 (2001).

[28] J. B. Hubbard and J. F. Douglas, Phys. Rev. E **47**, R2983 (1993); H. Boukari, R. Nossal, D. L. Sackett, and P. Schuck, Phys. Rev. Lett. **93**, 098106 (2004).

[29] J. Bicerano, J. F. Douglas, and D. A. Brune, J. Macromol. Sci. Polymer Rev. **39**, 561 (1999).

Statement of the author: This combined theoretical and experimental study on the hydrodynamic friction tensor of anisotropic particles evolved from a collaboration with the experimental physicists Daniela J. Kraft, Kazem V. Edmond, and David J. Pine from New York University. While Daniela J. Kraft was mainly responsible for the experiments, Raphael Wittkowski performed the simulations. I was involved in numerous discussions with Raphael Wittkowski and Hartmut Löwen in Düsseldorf and helped to finalize the manuscript. The method of determining the diffusion and coupling coefficients from short-time correlation functions was developed in the context of our study on L-shaped microswimmers (see reference [218]). Here, we also put forward the idea of using the software of the HYDRO suite to predict the hydrodynamic friction tensor theoretically.

CHAPTER 6

Transport powered by bacterial turbulence

The content of this chapter has been published in a similar form in *Physical Review Letters* **112**, 158101 (2014) by Andreas Kaiser, Anton Peshkov, Andrey Sokolov, Borge ten Hagen, Hartmut Löwen, and Igor S. Aranson (see reference [217]).

Suspensions of bacteria or synthetic microswimmers show fascinating collective behavior emerging from their self-propulsion [1–4], which results in many novel active states such as swarming [5–8] and active "turbulence" [9–14]. In contrast to hydrodynamic turbulence, the apparent turbulent (or swirling) state occurs at exceedingly low Reynolds numbers but at relatively large bacterial concentrations. Here we address the question of whether one can systematically extract energy out of the seemingly turbulent state established by swimming bacteria and how the bacterial turbulence may power microengines and transport mesoscopic carriers through the suspension. A related question is what processes on a scale of an individual swimmer are responsible for the energy rectification from this "active heat bath."

In our experiments we analyze the motion of a microwedgelike carrier ("bulldozer") submersed in a suspension of swimming bacteria, *Bacillus subtilis*. Experimental studies are combined with particle-resolved computer simulations. A broad span of bacterial densities is examined, ranging from the dilute regime over the turbulent one to the jammed state. Because of the activity of the suspension, the bulldozerlike particle is set into a rectified motion along its wedge cusp [15, 16], in contrast to tracers with symmetric shape [17]. Its averaged propagation speed becomes maximal within the turbulentlike regime of collective swimming. Our simulations and experiments indicate that the directed motion is caused by polar ordered bacteria trapped inside the carrier in a region near the cusp which is shielded from outer turbulent fluctuations. The orientation of trapped bacteria yields a double-peaked distribution centered in the direction of the average carrier motion. Consequently, the bacterial turbulence powers efficiently the transport of carriers through the suspension.

Converting bacterial self-propulsion into mechanical energy has been considered previously for shuttles and cogwheels [18–23]. Most of the studies were restricted to low swimmer concentrations where swirling is absent. While fundamental microscopic mechanisms of energy transduction and interaction with solid walls on a scale of a single bacterium are fairly well understood, see, e.g., [24–26], the role of collective motion has not been elucidated so far. Reference [27] demonstrated that while the energy generated by individual bacteria dissipates on the microscale, the increase in swimming velocity due to collective motion is significant. Here we put forward an idea of how the collective bacterial swimming strongly amplifies the energy transduction. Complementary, in high-Reynolds-number turbulent flows, the motion of suspended inertial particles is not directed [28], lacking a conversion of turbulent fluctuations into useful mechanical work.

Experiment.— Mesoscopic wedgelike carriers were fabricated by photolithography [20, 29]. To control the orientation of the carriers with an external magnetic field, we mixed a liquid photoresist SU-8 with micron-size magnetic particles before spin coating. The arm length of the wedgelike carriers is $L = 262\,\mu\text{m}$ (see Fig. 1), and the wedge angle is 90°. Experiments were conducted on a suspension of *Bacillus subtilis*, a flagellated rod-shaped swimming bacterium $\ell \sim 5\,\mu\text{m}$ long and $0.7\,\mu\text{m}$ wide. The suspension of bacteria was grown for 8–12 h

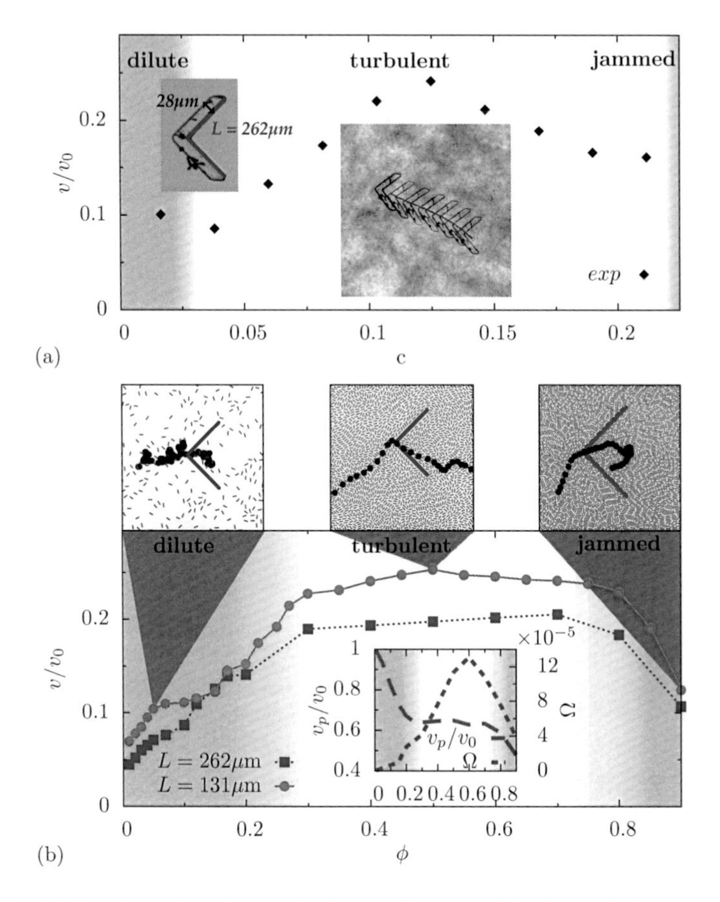

FIG. 1. (a) Experimental carrier speed v/v_0 as a function of the bacterial three-dimensional volume fraction c. The insets show the temporal progress of the carrier positions (right) and a snapshot of the carrier (left) indicating the characteristic spatial dimensions and the schematic representation (blue line). See also movie 1 in the Supplemental Material [32]. (b) Numerically obtained transport speed for varying swimmer concentration ϕ and carriers of two different contour lengths: $L = 262\,\mu\text{m}$ (squares) corresponding to the experimental conditions and half the length $L = 131\,\mu\text{m}$ (circles). The magnitude of vorticity Ω and the averaged bacterial swimming speed v_p/v_0 for various concentrations are shown in the inset. See also movies 2 and 3 in the Supplemental Material [32].

in Terrific Broth growth medium (Sigma Aldrich). To monitor the concentration of bacteria during the growth phase, we continuously measured the optical scattering of the medium using an infrared proximity sensor. At the end of the exponential growth phase the bacteria were washed and centrifuged to achieve the desired concentration. Then a small drop of concentrated bacterial suspension was placed between four movable fibers and stretched up to the thickness of $\sim 100\,\mu$m (see Ref. [30]). Both surfaces of the free-standing liquid film were exposed to air, significantly increasing the oxygen diffusion rate into the bulk of the film. We measured the bacteria concentration (or, equivalently, three-dimensional volume fraction) by means of optical coherence tomography (see Ref. [31]) before and after the experiments in order to monitor the effect of film evaporation.

Two pairs of orthogonal Helmholtz coils were used to create a uniform magnetic field in the bulk of the liquid film. The carrier was carefully inserted in the film by a digital micropipet. In the course of our experiment, the orientation of the magnetic carrier was reversed every 20 s to prevent migration of the carrier out of the field of view. We also confirm that the average speed of the wedge does not depend on the direction of the motion. The influence of gravity was negligible in our experiment. The motion of the wedge was captured by a digital high-resolution microscope camera [Fig. 1(a) and [32]] for the duration of 2–4 min. Both displacement and orientation of the carrier were tracked by a custom-designed software based on MATLAB toolboxes.

Simulation.— We model the bacteria by rodlike objects with repulsive interactions and an effective self-propulsion using parameters matching the experimental conditions. The excluded volume interaction between the rods is described by n "Yukawa" segments positioned equidistantly with the distance $d = 0.85\,\mu$m along a stiff axis of length ℓ; i.e., a repulsive Yukawa potential is imposed between the segments of different rods [33]. In order to properly take into account collisions between the bacteria, an important modification is introduced in the model compared to that of Ref. [11]. Experiments [26, 30, 34] demonstrated that two bacteria swim away from each other after the collision. This effect results in a suppression of clustering for small and moderate bacteria concentrations. However, in previous simulations [11], the bacteria had a propensity to swim parallel after the collision and to form dense clusters with a smecticlike alignment. In order to describe the experimentally observed swim-off effect and the resulting suppression of clustering, we incorporate an asymmetric effective bacterial shape by enlarging the interaction prefactor of the first segments of each rod with respect to the other segments by a factor of 3 (see [32]). The resulting total interaction potential between a swimmer pair α, β is then given by $U_{\alpha\beta} = \sum_{i=1}^{n}\sum_{j=1}^{n} U_i U_j \exp[-r_{ij}^{\alpha\beta}/\lambda]/r_{ij}^{\alpha\beta}$ with $U_1^2/U_j^2 = 3$ $(j = 2\ldots n)$, where λ is a screening length obtained from the experimental effective rod aspect ratio $\ell/\lambda = 5$, and $r_{ij}^{\alpha\beta} = |\mathbf{r}_i^\alpha - \mathbf{r}_j^\beta|$ is the distance between segment i of rod α and segment j of rod β $(\alpha \neq \beta)$. The carrier is implemented correspondingly by tiling the wedgelike contour with length $L = 262\,\mu$m [see

Fig. 1(a)] with Yukawa segments. The ratio L/ℓ is matched to the experimental situation. For comparison we also perform simulations for smaller arm lengths. The self-propulsion is taken into account via a formal effective force F_0 acting along the rod axis $\hat{\mathbf{u}} = (\cos\varphi, \sin\varphi)$. By imposing a large interaction strength $U_j^2 = 2.5F_0\ell$, we ensure that the bacteria and the wedge do not overlap. The model neglects long-range hydrodynamic interactions between the swimmers. These interactions do not change the overall morphology of the bacterial flow. The most noticeable effect of the hydrodynamic interactions is a 7- to 10-fold increase of the collective flow velocity compared to the speed of individual bacteria [30, 35]. Certainly, this phenomenon cannot be attained by our model.

Since bacterial swimming occurs at exceedingly low Reynolds numbers, the overdamped equations of motion for the positions and orientations of the rods are

$$\mathbf{f}_{\mathcal{T}} \cdot \partial_t \mathbf{r}_\alpha(t) = -\nabla_{\mathbf{r}_\alpha} U + F_0 \hat{\mathbf{u}}_\alpha(t)\,, \tag{1}$$

$$\mathbf{f}_{\mathcal{R}} \cdot \partial_t \hat{\mathbf{u}}_\alpha(t) = -\nabla_{\hat{\mathbf{u}}_\alpha} U\,. \tag{2}$$

Here, $U = (1/2)\sum_{\alpha,\beta(\alpha\neq\beta)} U_{\alpha\beta} + \sum_\alpha U_{\alpha<}$ is the total potential energy, where $U_{\alpha<}$ denotes the interaction energy of rod α with the carrier. (In general, a subscript $<$ refers to a quantity associated with the wedgelike carrier.) The one-body translational and rotational friction tensors for the rods $\mathbf{f}_{\mathcal{T}}$ and $\mathbf{f}_{\mathcal{R}}$ can be decomposed into parallel $f_\parallel$, perpendicular $f_\perp$, and rotational f_R components which depend solely on the aspect ratio ℓ/λ and are taken from Ref. [36]. The resulting self-propulsion speed of a single rod $v_0 = F_0/f_\parallel$ is matched to the experimental value 15 μm/s [30] leading to the time unit $\tau = \ell/v_0$. Since at a relatively large bacteria concentration thermal fluctuations and tumbling are not important, we neglect all stochastic noise terms (our experimental studies in Ref. [9] showed that tumbling of *Bacillus subtilis* becomes significant only for very low oxygen concentrations). Also, hydrodynamic interaction between the bacteria and the air-water interface is neglected. According to the experiment, the motion of the carrier is mostly translational and induced by the carrier-bacteria interactions. The hydrodynamic friction tensor $\mathbf{f}_<$ of the wedge is calculated for the specific geometry with the dimensions shown in the sketch (left inset) in Fig. 1(a). For this purpose, the shape of the carrier is approximated by a large number of beads that are rigidly connected. The corresponding calculations based on the Stokes equation for the flow field around a particle at a low Reynolds number are performed with the software package HYDRO++ [37, 38].

The resulting equation of motion for the carrier is

$$\mathbf{f}_< \cdot \partial_t \mathbf{r}_<(t) = -\nabla_{\mathbf{r}_<} \sum_\alpha U_{\alpha<}(t)\,. \tag{3}$$

We simulated $N \sim 10^4$ rods and a single carrier in a square simulation domain with the area $A = (3L/\sqrt{2})^2$ and periodic boundary conditions in both directions. The dimensionless

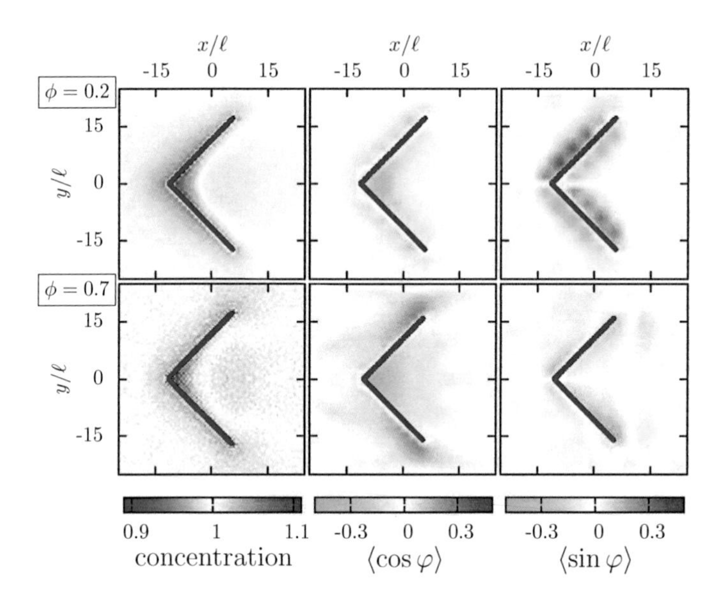

FIG. 2. Intensity plots for different bacteria concentrations, $\phi = 0.2$ (top row) and $\phi = 0.7$ (bottom row): local bacterial concentration around the carrier, normalized by the total concentration (left) as well as averaged bacteria orientations $\langle \cos \varphi \rangle$ (middle) and $\langle \sin \varphi \rangle$ (right).

packing fraction $\phi = N\lambda\ell/A$ corresponds to the bacterial volume fraction c in the experiments.

Results and discussion.— The shape reflection symmetry of the wedge around its apex will exclude any averaged directed motion perpendicular to the apex while there is no such symmetry in the apex direction. Hence, due to rectification of random fluctuations, the carrier will proceed on average along its cusp. The transport efficiency of the carrier can then be characterized by its average migration speed v in this direction. We have examined the carrier motion in a wide range of bacterial bulk concentrations including a dilute regime, where bacterial swimming is almost uncorrelated, as well as an intermediate turbulent and a final jammed regime. These regimes can be characterized by suitable order parameters. For that purpose, we define the mean magnitude of vorticity $\Omega = (1/2)\langle |[\nabla \times \mathbf{V}(\mathbf{r},t)] \cdot \hat{\mathbf{e}}_z|^2 \rangle$ for a bacterial velocity field $\mathbf{V}(\mathbf{r},t)$ coarse grained over three bacterial lengths, which is a convenient indicator for bacterial turbulence [11, 39]. The average swimming speed v_p of the bacteria obtained by averaging the displacements after a time $t = 10^{-3}\tau$ indicates jamming at high concentrations. Simulation results for these two order parameters are presented for a

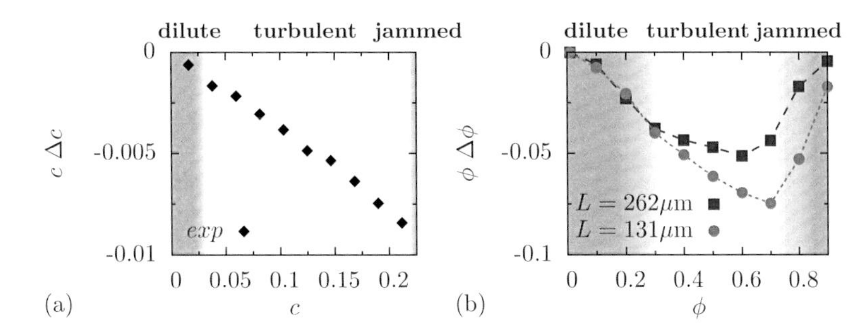

FIG. 3. Concentration difference between the wake of the carrier and its front $\Delta c = c_w - c_f$ obtained from (a) experiments and (b) simulations.

bacterial suspension in the absence of the carrier [see inset in Fig. 1(b)]. The results indicate three different states: "dilute" ($\phi \lesssim 0.25$), "turbulent" ($0.25 \lesssim \phi \lesssim 0.75$), and "jammed" ($0.75 \lesssim \phi$). The same sequence of states is found in the experiments [40].

Figures 1(a) and 1(b) show that the transport efficiency v/v_0 of the carrier peaks in the turbulent regime where it attains a significant fraction of the net bacterial velocity v_0. Experimentally, this fraction is found to be about 0.25, which is confirmed by the simulations. Snapshots from the experiments and simulations (see insets in Fig. 1 and [32]) show a directed motion along the wedge apex though there are considerable fluctuations which we discuss later. For a very dilute regime, there are only a few bacteria pushing the carrier such that v tends to zero (note that in the experiment no motion of the carrier was observed in a very dilute regime since the resulting bacterial forces are not sufficient to overcome the friction of the carrier with the surface). We have also performed simulations for different carrier lengths and opening angles to determine the geometry leading to optimal transport. While we have chosen the optimal apex angle for our experiments, a slightly higher transport efficiency can be achieved with a smaller carrier length; see [32].

In the following we discuss the underlying reason for the optimal carrier transport in the turbulent regime. First, the bacteria inside the wedge close to the cusp are on average orientationally ordered along the wedge orientation (x direction). The orientational ordering is revealed by the intensity plots for $\langle\cos\varphi\rangle$, $\langle\sin\varphi\rangle$, which were used as appropriate orientational order parameters (see the green "hot spot" in Fig. 2). The hot spot sets the carrier into motion along the x direction. Similar to a moving bulldozer piling up sand, the carrier motion causes an accumulation of particles in the front and a depleted wake, while not destroying the ordering of particles inside the wedge (see the intensity plot in Fig. 2 as well as Fig. 3 for experimental and simulation data for the averaged concentration difference between inside and outside bacteria). This concentration difference decreases the transport

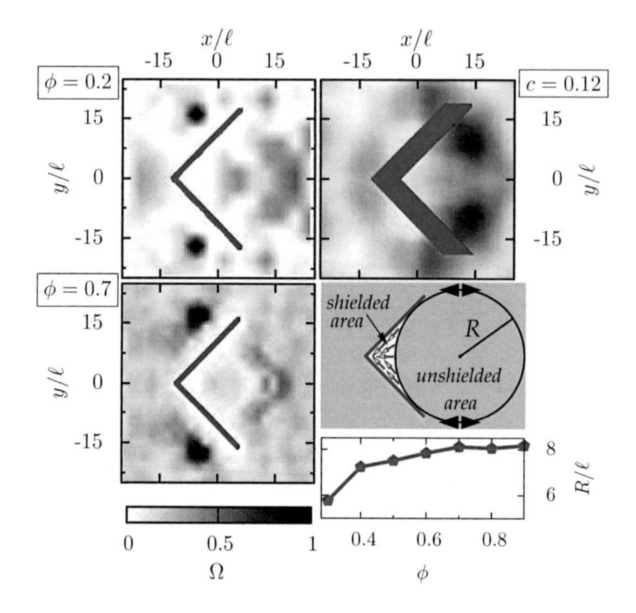

FIG. 4. Normalized local magnitude of vorticity obtained from simulations (left column) and experiment (top right). Bottom right: illustration of swirl shielding in the carrier cusp—bacteria in the shielded area (light colored) are indicated by arrows and the unshielded area is marked by dark color. The typical swirl radius R for different bacteria concentrations is obtained as the first minimum of the equal-time spatial velocity autocorrelation function [39].

speed. But the driving effect increases with the increase in bacterial concentration. When the turbulence sets in, there is a shielding of turbulent fluctuations near the walls of the carrier (see the intensity plots of the local magnitude of vorticity in Fig. 4). The shielding is, however, more pronounced inside than outside the wedge. Intuitively this implies that the hot spot is shielded from swirls which would sweep away the driving bacteria. (Concomitantly, the outside swirls shown in Fig. 4 are induced by the bulldozer motion but do not cause the motion.) The intuitive concept of swirl shielding is sketched in Fig. 4 where also a typical swirl size as a function of the density is shown. A typical swirl of this size can never reach the hot spot area as schematically shown by the shielded area in Fig. 4. Swimmers within this area are trapped, leading to large transport velocity correlation times for the carrier; see [32]. For simple geometric reasons, there is no such swirl shielded zone for the outside bacteria as a swirl can sweep them away. At very high concentrations ($\phi \approx 0.8$), the bacteria are jammed (see Fig. 1), which is manifested in a reduced carrier mobility. Thus,

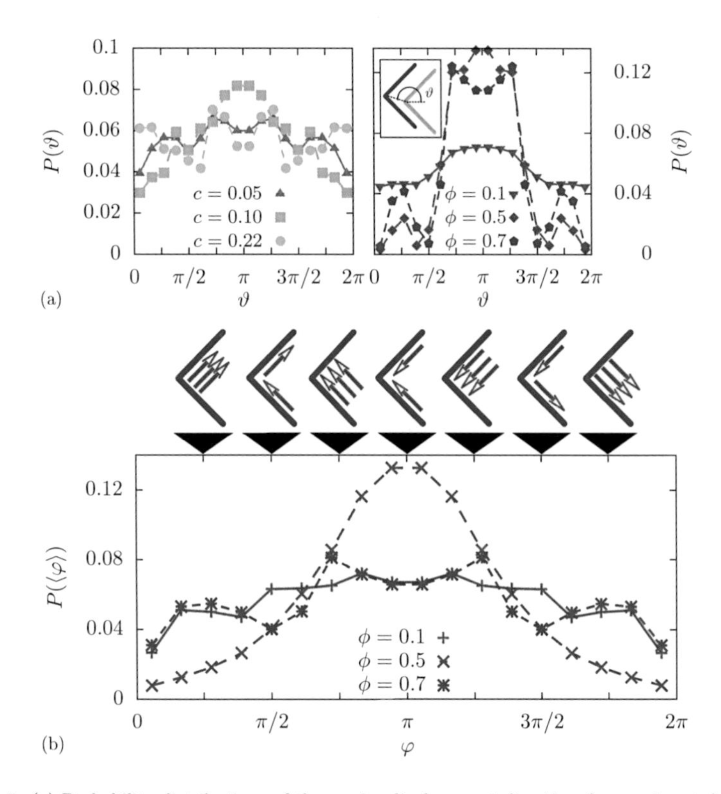

FIG. 5. (a) Probability distributions of the carrier displacement direction ϑ: experimental results (left) measured after a time of 0.03 s and numerical results (right) after a time of $10^{-3}\tau$ for various bacterial concentrations. The definition of the displacement angle ϑ is shown in the inset. (b) Orientational ordering of the bacteria within the swirl shielded area. Characteristic configurations are illustrated by sketches.

the polar order of bacteria (see intensity plots in Fig. 2) inside the wedge and its shielding from the swirls are the two basic ingredients to understand the optimal transport in the turbulent state. Optimal transport is achieved when the carrier aperture width is comparable to a typical swirl size; see [32].

We plot the full distribution of the carrier velocity direction $P(\vartheta)$ in Fig. 5(a). For small bacterial densities ϕ the distribution is random, while for intermediate concentrations this distribution exhibits a single peak centered around the x axis. For even higher concentrations it becomes double peaked, corresponding to a motion perpendicular to the single wedge walls ($\pi \pm \pi/4$). This is correlated with the orientational order distribution of the inside bacteria in the shielded area, $P(\langle\varphi\rangle)$, implying that there is a flipping in the orientation of the inside bacteria, see Fig. 5(b), for high bacterial concentrations. The kinklike change in the direction of motion perpendicular to the wedge walls is also observed experimentally [see the trajectory in Fig. 1 and peaks in Fig. 5(a)].

Conclusion.— We have shown that mechanical energy of bacterial turbulent movements can be extracted to power the directed motion of a wedgelike carrier. Both polar ordering and swirl shielding inside the wedge yield an optimal transport velocity which becomes even bidirectional at high concentrations. This effect can be exploited to power and steer carriers and motors by bacterial turbulence or the collective motion of synthetic swimmers.

Work by A. K. and B. t. H. was supported by the ERC Advanced Grant INTERCOCOS (Grant No. 267499) and H. L. was supported by the SPP 1726 of the DFG. Work by A. S., A. P., and I. S. A. was supported by the U.S. Department of Energy (DOE), Office of Science, Basic Energy Sciences (BES), Work by A. S., A. P., and I. S. A. was supported by the U.S. Department of Energy (DOE), Office of Science, Basic Energy Sciences (BES), Materials Science and Engineering Division.

---

* kaiser@thphy.uni-duesseldorf.de

† aronson@anl.gov

[1] M.C. Marchetti, J.F. Joanny, S. Ramaswamy, T.B. Liverpool, J. Prost, M. Rao, and R. A. Simha, Rev. Mod. Phys. **85**, 1143 (2013).

[2] M.E. Cates, Rep. Prog. Phys. **75**, 042601 (2012).

[3] P. Romanczuk, M. Bär, W. Ebeling, B. Linder, and L. Schimansky-Geier, Eur. Phys. J. Spec. Top. **202**, 1 (2012).

[4] I.S. Aranson, Phys. Usp. **56**, 79 (2013).

[5] V. Narayan, S. Ramaswamy, and N. Menon, Science **317**, 105 (2007).

[6] Y. Yang, V. Marceau, and G. Gompper, Phys. Rev. E **82**, 031904 (2010).

[7] X. Chen, X. Dong, A. Be'er, H.L. Swinney, and H.P. Zhang, Phys. Rev. Lett. **108**, 148101 (2012).

[8] F. Ginelli, F. Peruani, M. Bär, and H. Chaté, Phys. Rev. Lett. **104**, 184502 (2010).

[9] A. Sokolov and I.S. Aranson, Phys. Rev. Lett. **109**, 248109 (2012).

[10] D. Saintillan and M.J. Shelley, Phys. Fluids **20**, 123304 (2008).

[11] H.H. Wensink, J. Dunkel, S. Heidenreich, K. Drescher, R.E. Goldstein, H. Löwen, and J.M. Yeomans, Proc. Natl. Acad. Sci. U.S.A. **109**, 14308 (2012).

[12] K.-A. Liu and L. I, Phys. Rev. E **88**, 033004 (2013).

[13] Y. Yang, F. Qiu, and G. Gompper, Phys. Rev. E **89**, 012720 (2014).

[14] S. Zhou, A. Sokolov, O. D. Lavrentovich, and I. S. Aranson, Proc. Natl. Acad. Sci. U.S.A. **111**, 1265 (2014).

[15] P. Galajda, J. Keymer, P. Chaikin, and R. Austin, J. Bacteriol. **189**, 8704 (2007).

[16] M. B. Wan, C. J. Olson Reichhardt, Z. Nussinov, and C. Reichhardt, Phys. Rev. Lett. **101**, 018102 (2008).

[17] G. Miño, T. E. Mallouk, T. Darnige, M. Hoyos, J. Dauchet, J. Dunstan, R. Soto, Y. Wang, A. Rousselet, and E. Clement, Phys. Rev. Lett. **106**, 048102 (2011).

[18] L. Angelani, R. DiLeonardo, and G. Ruocco, Phys. Rev. Lett. **102**, 048104 (2009).

[19] L. Angelani and R. DiLeonardo, New J. Phys. **12**, 113017 (2010).

[20] A. Sokolov, M. M. Apodaca, B. A. Grzyboski, and I. S. Aranson, Proc. Natl. Acad. Sci. U.S.A. **107**, 969 (2010).

[21] D. Wong, E. E. Beattie, E. B. Steager, and V. Kumar, Appl. Phys. Lett. **103**, 153707 (2013).

[22] H. H. Wensink, V. Kantsler, R. E. Goldstein, and J. Dunkel, Phys. Rev. E **89**, 010302 (2014).

[23] R. DiLeonardo, L. Angelani, D. Dell'Arciprete, G. Ruocco, V. Iebba, S. Schippa, M. P. Conte, F. Mecarini, F. De Angelis, and E. Di Fabrizio, Proc. Natl. Acad. Sci. U.S.A. **107**, 9541 (2010).

[24] E. Lauga, W. R. DiLuzio, G. M. Whitesides, and H. A. Stone, Biophys. J. **90**, 400 (2006).

[25] G. Li and J. X. Tang, Phys. Rev. Lett. **103**, 078101 (2009).

[26] K. Drescher, J. Dunkel, L. H. Cisneros, S. Ganguly, and R. E. Goldstein, Proc. Natl. Acad. Sci. U.S.A. **108**, 10940 (2011).

[27] T. Ishikawa, N. Yoshida, H. Ueno, M. Wiedeman, Y. Imai, and T. Yamaguchi, Phys. Rev. Lett. **107**, 028102 (2011).

[28] E. Calzavarini, M. Cencini, D. Lohse, and F. Toschi, Phys. Rev. Lett. **101**, 084504 (2008).

[29] J. Gachelin, G. Miño, H. Berthet, A. Lindner, A. Rousselet, and É. Clément, Phys. Rev. Lett. **110**, 268103 (2013).

[30] A. Sokolov, I. S. Aranson, J. O. Kessler, and R. E. Goldstein, Phys. Rev. Lett. **98**, 158102 (2007).

[31] A. Sokolov, R. E. Goldstein, F. I. Feldchtein, and I. S. Aranson, Phys. Rev. E **80**, 031903 (2009).

[32] See Supplemental Material at `http://link.aps.org/supplemental/10.1103/PhysRevLett.112.158101` for movies and additional data.

[33] T. Kirchhoff, H. Löwen, and R. Klein, Phys. Rev. E **53**, 5011 (1996).

[34] I. S. Aranson, A. Sokolov, J. O. Kessler, and R. E. Goldstein, Phys. Rev. E **75**, 040901 (2007).

[35] C. Dombrowski, L. Cisneros, S. Chatkaew, R. E. Goldstein, and J. O. Kessler, Phys. Rev. Lett. **93**, 098103 (2004).

[36] M. M. Tirado, C. L. Martinez, and J. Garcia de la Torre, J. Chem. Phys. **81**, 2047 (1984).

[37] J. Garcia de la Torre, S. Navarro, M. C. Lopez Martinez, F. G. Diaz, and J. J. Lopez Cascales, Biophys. J. **67**, 530 (1994).

[38] B. Carrasco and J. Garcia de la Torre, J. Chem. Phys. **111**, 4817 (1999).

[39] H. H. Wensink and H. Löwen, J. Phys. Condens. Matter **24**, 464130 (2012).

[40] A. Sokolov and I. S. Aranson, Phys. Rev. Lett. **103**, 148101 (2009).

Statement of the author: In this chapter, the findings originating from a collaboration between Düsseldorf and the experimentalists Anton Peshkov, Andrey Sokolov, and Igor S. Aranson from the Materials Science Division at the Argonne National Laboratory, Illinois are presented. Anton Peshkov performed the experiments and Andrey Sokolov was mainly involved in analyzing the experimental data. The computer simulations modeling the wedgelike carrier in a bacterial bath were performed by Andreas Kaiser. I calculated the hydrodynamic friction tensor for the specific shape of the carrier based on a bead model. In order to interpret the simulation results, we had many fruitful discussions between Hartmut Löwen, Andreas Kaiser, and me. Igor S. Aranson, who initiated the collaboration together with Hartmut Löwen, provided helpful input to understand the theoretical results in the context of the experimental observations.

Circular motion of asymmetric self-propelling particles

The content of this chapter has been published in a similar form in *Physical Review Letters* **110**, 198302 (2013) by Felix Kümmel, Borge ten Hagen, Raphael Wittkowski, Ivo Buttinoni, Ralf Eichhorn, Giovanni Volpe, Hartmut Löwen, and Clemens Bechinger (see reference [218]) and in *Physical Review Letters* **113**, 029802 (2014) by Felix Kümmel, Borge ten Hagen, Raphael Wittkowski, Daisuke Takagi, Ivo Buttinoni, Ralf Eichhorn, Giovanni Volpe, Hartmut Löwen, and Clemens Bechinger (see reference [447]), respectively.

Circular Motion of Asymmetric Self-Propelling Particles

Felix Kümmel,[1] Borge ten Hagen,[2] Raphael Wittkowski,[3] Ivo Buttinoni,[1]
Ralf Eichhorn,[4] Giovanni Volpe,[1,*] Hartmut Löwen,[2] and Clemens Bechinger[1,5]

[1] *2. Physikalisches Institut, Universität Stuttgart, D-70569 Stuttgart, Germany*
[2] *Institut für Theoretische Physik II: Weiche Materie, Heinrich-Heine-Universität Düsseldorf, D-40225 Düsseldorf, Germany*
[3] *SUPA, School of Physics and Astronomy, University of Edinburgh, Edinburgh, EH9 3JZ, United Kingdom*
[4] *Nordita, Royal Institute of Technology, and Stockholm University, SE-10691 Stockholm, Sweden*
[5] *Max-Planck-Institut für Intelligente Systeme, D-70569 Stuttgart, Germany*

Micron-sized self-propelled (active) particles can be considered as model systems for characterizing more complex biological organisms like swimming bacteria or motile cells. We produce asymmetric microswimmers by soft lithography and study their circular motion on a substrate and near channel boundaries. Our experimental observations are in full agreement with a theory of Brownian dynamics for asymmetric self-propelled particles, which couples their translational and orientational motion.

PACS numbers: 82.70.Dd, 05.40.Jc

Micron-sized particles undergoing active Brownian motion [1] currently receive considerable attention from experimentalists and theoreticians because their locomotion behavior resembles the trajectories of motile microorganisms [2–5]. Therefore, such systems allow interesting insights into how active matter [6] organizes into complex dynamical structures. During the last decade, different experimental realizations of microswimmers have been investigated, where, e.g., artificial flagella [7] or thermophoretic [8] and diffusiophoretic [9] driving forces lead to active motion of micron-sized objects. So far, most studies have concentrated on spherical or rodlike microswimmers whose dynamics is well described by a persistent random walk with a transition from a short-time ballistic to a long-time diffusive behavior [10]. Such simple rotationally symmetric shapes, however, usually provide only a crude approximation for self-propelling microorganisms, which are often asymmetric around their propulsion axis. Then, generically, a torque is induced that significantly perturbs the swimming path and results in a characteristic circular motion.

In this Letter, we experimentally and theoretically study the motion of asymmetric self-propelled particles in a viscous liquid. We observe a pronounced circular motion whose curvature radius is independent of the propulsion strength but only depends on the shape of the swimmer. Based on the shape-dependent particle mobility matrix, we propose two coupled Langevin equations for the translational and rotational motion of the particles under an intrinsic force, which dictates the swimming velocity. The anisotropic particle shape then generates an additional velocity-dependent torque, in agreement with our measurements. Furthermore, we also investigate the motion of asymmetric particles in lateral confinement. In agreement with theoretical predictions we find either a stable sliding along the wall or a reflection, depending on the contact angle.

Asymmetric L-shaped swimmers with arm lengths of 9 and 6 μm were fabricated from photoresist SU-8 by photolithography [11]. In short, a 2.5 μm thick layer of SU-8 is spin coated onto a silicon wafer, soft baked for 80 s at 95 °C and then exposed to ultraviolet light through a photomask. After a postexposure bake at 95 °C for 140 s the entire wafer with the attached particles is coated with a 20 nm thick Au layer by thermal evaporation. When the wafer is tilted to approximately 90° relative to the evaporation source, the Au is selectively deposited at the front side of the short arms as schematically shown in Figs. 1(a) and 1(b). Finally, the coated particles are released from the wafer by an ultrasonic bath treatment. A small amount of L-shaped particles is suspended in a homogeneous mixture of water and 2,6-lutidine at critical concentration (28.6 mass percent of lutidine), which is kept several degrees below its lower critical point ($T_C = 34.1$ °C) [12]. To confine the particle's motion to two dimensions, the suspension is contained in a sealed sample cell with 7 μm height. The particles are localized above the lower wall at an average height of about 100 nm due to the presence of electrostatic and gravitational forces. Under these conditions, they cannot rotate between the two configurations shown in Figs. 1(a) and 1(b), which will be

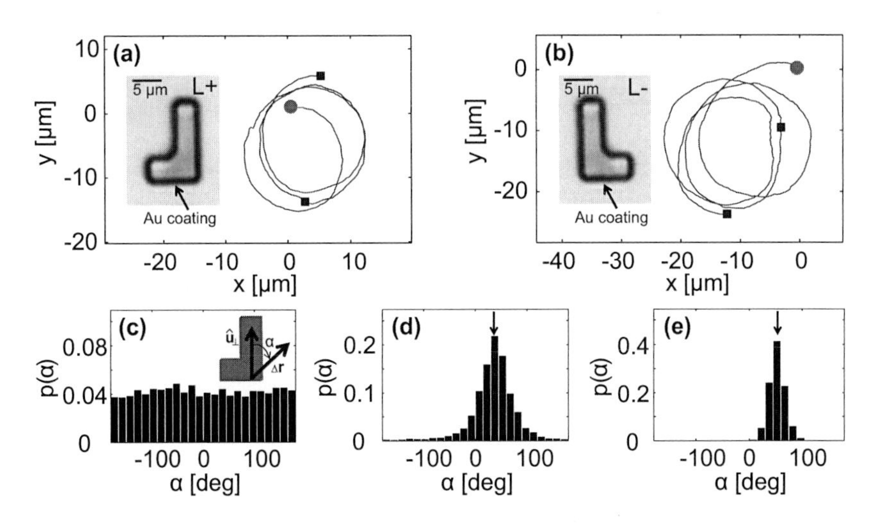

FIG. 1. (a),(b) Trajectories of an (a) $L+$ and (b) $L-$ swimmer for an illumination intensity of 7.5 μW/μm^2. Bullets and square symbols correspond to initial particle positions and those after 1 min each, respectively. The insets show microscope images of two different swimmers with the Au coating (not visible in the bright-field image) indicated by an arrow. (c),(d),(e) Probability distributions $p(\alpha)$ of the angle α [see inset in (c)] between the normal vector $\hat{\mathbf{u}}_\perp$ of the metal coating and the displacement vector $\Delta\mathbf{r}$ of an $L+$ particle in time intervals of 12 s each for illumination intensities (c) $I = 0$ μW/μm^2, (d) 5 μW/μm^2, and (e) 7.5 μW/μm^2.

denoted as $L+$ (left) and $L-$ (right) in the following. When the sample cell is illuminated by light ($\lambda = 532$ nm) with intensities ranging on the order of several μW/μm^2, the metal cap becomes slightly heated above the critical point and thus induces a local demixing of the solvent [13, 14]. This leads to a self-phoretic particle motion similar to what has been observed in other systems [15–17].

Figures 1(a) and 1(b) show trajectories of $L+$ and $L-$ swimmers obtained by digital video microscopy for an illumination intensity of 7.5 μW/μm^2, which corresponds to a mean propulsion speed of 1.25 μm/s. In contrast to spherical swimmers, here a pronounced circular motion with clockwise ($L+$) and counterclockwise ($L-$) direction of rotation is observed. For the characterization of trajectories we determined the center-of-mass position $\mathbf{r}(t) = (x(t), y(t))$ and the normalized orientation vector $\hat{\mathbf{u}}_\perp$ of the particles [see inset of Fig. 1(c)]. From this, we derived the angle α between the displacement vector $\Delta\mathbf{r}$ and the particle's body orientation $\hat{\mathbf{u}}_\perp$. Figures 1(c)–1(e) show how the normalized probability distribution

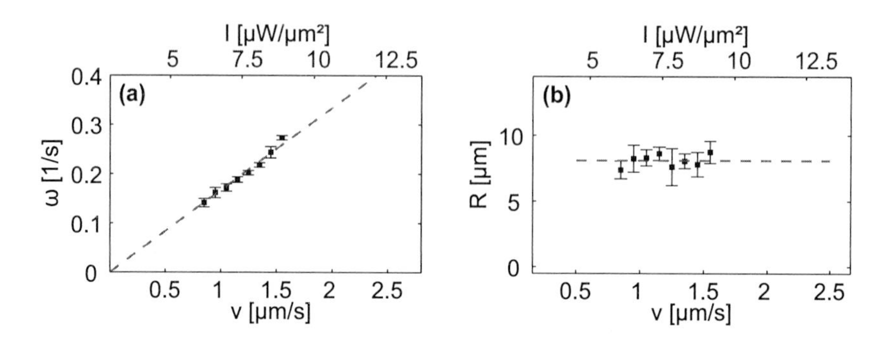

FIG. 2. (a) Angular velocity ω and (b) radius R of the circular motion of an $L+$ swimmer plotted as functions of the linear velocity $v = |\mathbf{v}|$ and the illumination intensity $I \sim v$. The dashed lines correspond to a linear fit with nonzero and zero slope, respectively.

$p(\alpha)$ changes with increasing illumination intensity I. In the case of pure Brownian motion [see Fig. 1(c)] $p(\alpha) \approx$ const since the orientational and translational degrees of freedom are decoupled when only random forces are acting on the particle. In the presence of a propulsion force which is constant in the body frame of the particle, however, the translational and rotational motion of an asymmetric particle are coupled. This leads to a peaked behavior of $p(\alpha)$ as shown in Figs. 1(d) and 1(e). The peak's halfwidth decreases with increasing illumination intensity since the contribution of the Brownian motion is more and more dominated by the propulsive part. In addition, the peaks are shifted to positive (negative) values for a particle swimming in a (counter)clockwise direction. The position of the peak is given by $\alpha = \pi\Delta t/\tau$, where τ is the intensity-dependent cycle duration of the circle swimmer [cf., Fig. 2(b)] and Δt is the considered time interval. This estimate [see arrows in Figs. 1(d) ($\tau = 60$ s) and 1(e) ($\tau = 40$ s)] is in good agreement with the experimental data. The shift of the maximum of $p(\alpha)$ documents a torque responsible for the observed circular motion of such asymmetric swimmers. In contrast to an externally applied constant torque [18], here it is due to viscous forces acting on the self-propelling particle. This is supported by the experimental observation that the particle's angular velocity $\omega(t) = d\alpha/dt$ increases linearly with its total translational velocity $v(t)$ [see Fig. 2(a)]. As a direct result of the linear relationship between ω and v, the radius R of the circular trajectories becomes independent of the propulsion speed, which is set by the illumination intensity [see Fig. 2(b)].

For a theoretical description of the motion of asymmetric swimmers, we consider an effective propulsion force $\mathbf{F}$ [19], which is constant in a body-fixed coordinate system that rotates with the active particle. With the unit vectors $\hat{\mathbf{u}}_\perp = (-\sin\phi, \cos\phi)$ and $\hat{\mathbf{u}}_\parallel = (\cos\phi, \sin\phi)$ [see Figs. 1(c) and 3(a)], where—in the case of L-shaped particles—ϕ is the angle between the short arm and the x axis, the propulsion force $\mathbf{F}$ is given by $\mathbf{F} = F\hat{\mathbf{u}}_{\text{int}}$ with $\hat{\mathbf{u}}_{\text{int}} = (c\hat{\mathbf{u}}_{\|} + \hat{\mathbf{u}}_{\perp})/\sqrt{1+c^2}$ with the constant c depending on how the force is aligned relative to the particle shape. If the propulsion force is aligned along the long axis $\hat{\mathbf{u}}_{\perp}$, one obtains $c = 0$, i.e., $\hat{\mathbf{u}}_{\text{int}} = \hat{\mathbf{u}}_{\perp}$. In the case of an asymmetric particle, the propulsion force leads also to a velocity-dependent torque relative to the particle's center of mass. For $c = 0$ this torque is given by $M = lF$ with l the effective lever arm [see Fig. 3(a)]. Our theoretical model is valid for arbitrary particle shapes and values of c and l. However, for the sake of clarity, we set $c = 0$ as this applies for the L-shaped particles considered here. Accordingly, we obtain the following coupled Langevin equations, which describe the motion of an asymmetric microswimmer

$$\begin{aligned} \dot{\mathbf{r}} &= \beta F\big(\mathrm{D}_{\mathrm{T}}\hat{\mathbf{u}}_{\perp} + l\mathbf{D}_C\big) + \boldsymbol{\zeta}_{\mathbf{r}}\,, \\ \dot{\phi} &= \beta F\big(lD_R + \mathbf{D}_C\cdot\hat{\mathbf{u}}_{\perp}\big) + \zeta_{\phi}\,. \end{aligned} \tag{1}$$

Here, $\beta = 1/(k_BT)$ is the inverse effective thermal energy of the system. These Langevin equations contain the translational short-time diffusion tensor $\mathrm{D}_{\mathrm{T}}(\phi) = D_{\|}\hat{\mathbf{u}}_{\|}\otimes\hat{\mathbf{u}}_{\|} + D_{\|}^{\perp}(\hat{\mathbf{u}}_{\|}\otimes\hat{\mathbf{u}}_{\perp} + \hat{\mathbf{u}}_{\perp}\otimes\hat{\mathbf{u}}_{\|}) + D_{\perp}\hat{\mathbf{u}}_{\perp}\otimes\hat{\mathbf{u}}_{\perp}$ with the dyadic product $\otimes$ and the translation-rotation coupling vector $\mathbf{D}_C(\phi) = D_C^{\|}\hat{\mathbf{u}}_{\|} + D_C^{\perp}\hat{\mathbf{u}}_{\perp}$ [20]. The translational diffusion coefficients $D_{\|}$, $D_{\|}^{\perp}$, and $D_{\perp}$, the coupling coefficients $D_C^{\|}$ and $D_C^{\perp}$, and the rotational diffusion constant D_R are determined by the specific shape of the particle. Finally, $\boldsymbol{\zeta}_{\mathbf{r}}(t)$ and $\zeta_{\phi}(t)$ are Gaussian noise terms of zero mean and variances $\langle\boldsymbol{\zeta}_{\mathbf{r}}(t_1)\otimes\boldsymbol{\zeta}_{\mathbf{r}}(t_2)\rangle = 2\mathrm{D}_{\mathrm{T}}\delta(t_1-t_2)$, $\langle\boldsymbol{\zeta}_{\mathbf{r}}(t_1)\zeta_{\phi}(t_2)\rangle = 2\mathbf{D}_C\delta(t_1-t_2)$, and $\langle\zeta_{\phi}(t_1)\zeta_{\phi}(t_2)\rangle = 2D_R\delta(t_1-t_2)$ [23].

In the case of vanishing noise, Eq. (1) immediately leads to a circular trajectory with radius

$$R = \sqrt{\frac{(D_{\|}^{\perp} + lD_C^{\|})^2 + (D_{\perp} + lD_C^{\perp})^2}{(D_C^{\perp} + lD_R)^2}}\,. \tag{2}$$

In agreement with the experimental observation [see Fig. 2(b)] the radius does not depend on the particle velocity set by the propulsion force. Rather, the value of R is determined only by the particle's geometry, which defines its diffusional properties. Moreover, the translational and angular particle velocities are $v = \beta F\sqrt{(D_{\|}^{\perp} + lD_C^{\|})^2 + (D_{\perp} + lD_C^{\perp})^2}$ and $\omega = \beta F(D_C^{\perp} + lD_R)$. Both quantities are proportional to the internal force F and ensure $R = v/|\omega|$ in perfect agreement with the experimental results shown in Fig. 2(a).

For a quantitative comparison with the experimental data, most importantly, the diffusion and coupling coefficients for the particles under study have to be determined. They constitute the components of the generalized diffusion matrix and are, in principle, obtained from solving the Stokes equation that describes the low Reynolds number flow field around a particle close to the substrate [24]. This procedure can be approximated by using a bead model [25], where the L-shaped particle is assembled from a large number of rigidly connected small spheres. Exploiting the linearity of the Stokes equation, the hydrodynamic

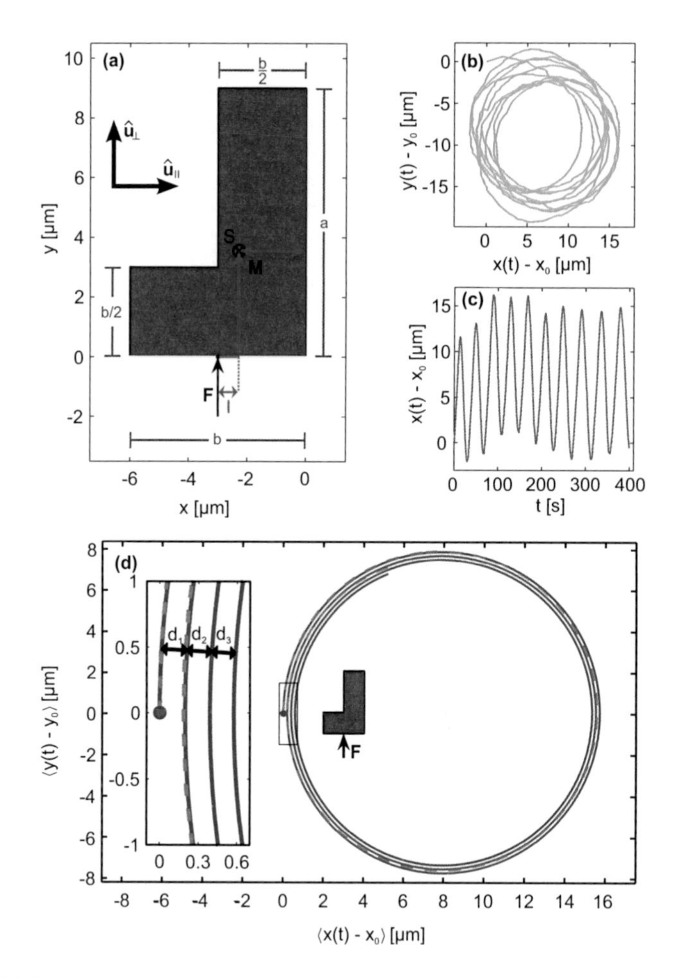

FIG. 3. (a) Geometrical sketch of an ideal $L+$ swimmer as considered in our model. The dimensions are $a = 9$ μm, $b = 6$ μm, $x_S = 2.29$ μm, and $y_S = 3.55$ μm (for homogeneous mass density and an additional 20 nm thick Au layer). The internal force **F** induces a torque **M** on the center-of-mass S depending on the lever arm l. (b),(c) Visualization of the experimental trajectory (for an illumination intensity of $I = 7.5\ \mu\mathrm{W}/\mu\mathrm{m}^2$) that is used for the quantitative analysis of the fluctuation-averaged trajectory in (d). The dashed curve in (d) is the experimental one, and the solid curve shows the theoretical prediction with the starting point indicated by a red bullet. Inset: closeup of the framed area in the plot.

TABLE I. Diffusion coefficients for the L-shaped particle in Fig. 3(a) on a substrate: translational diffusion along the long ($D_\perp$) and the short ($D_\parallel$) axis of the L-shaped particle as well as rotational diffusion constant D_R.

	Experiment	Theory
$D_\perp$ [10^{-3} µm^2 s^{-1}]	8.1 ± 0.6	8.3
$D_\parallel$ [10^{-3} µm^2 s^{-1}]	7.2 ± 0.4	7.5
D_R [10^{-4} s^{-1}]	6.2 ± 0.8	6.1

interactions between any pair of those beads can be superimposed to calculate the generalized mobility tensor of the L-shaped particle and from that its diffusion and coupling coefficients; details of the calculation are outlined in Ref. [25]. This method is well established for arbitrarily shaped particles in bulk solution [25, 26]. We take into account the presence of the substrate by using the Stokeslet close to a no-slip boundary [27] to model the hydrodynamic interactions between the component beads in the bead model. For the L-shaped particles considered here, we find that the value of $D_\perp$ exceeds the terms $D_\parallel^\perp$, $lD_C^\parallel$, and $lD_C^\perp$ in the numerator of Eq. (2) by more than one order of magnitude (given that l is in the range of 1 µm). On the other hand, the value of $D_C^\perp$ is negligible compared to lD_R. This finally yields

$$R = |D_\perp/(lD_R)| \tag{3}$$

as an approximate expression for the trajectory radius and, correspondingly,

$$\omega = \beta D_R l F \tag{4}$$

for the angular velocity.

We determined the diffusion coefficients $D_\perp$, $D_\parallel$, and D_R experimentally under equilibrium conditions (i.e., in the absence of propulsion) from the short-time correlations of the translational and orientational components of the particle's trajectories [28, 29] (see Table I). The experimental values are in good agreement with the theoretical predictions.

Inserting the experimentally determined values for the diffusion coefficients and the mean trajectory radius $R = 7.91$ µm into Eq. (3), we obtain the effective lever arm $l = -1.65$ µm. This value is about a factor of 2 larger compared to an ideally shaped L particle [see Fig. 3(a)] with its propulsion force perfectly centered at the middle of the Au layer. This deviation suggests that the force is shifted by 0.94 µm in lateral direction, which is most likely caused by small inhomogeneities of the Au layer due to shadowing effects during the grazing incidence metal evaporation. Accordingly, from Eq. (4) we obtain the intensity-dependent propulsion force $F/I = 0.83 \times 10^{-13}$ N µm^2/µW.

To compare the trajectories obtained from the Langevin equations (1) with experimental data, we divided the measured trajectories into smaller segments and superimposed them such that the initial slopes and positions of the segments overlap. After averaging the data we obtained the noise-averaged mean swimming path, which is predicted to be a logarithmic spiral (*spira mirabilis*) [30] that is given in polar coordinates by

$$r(\phi) = \beta F \sqrt{\frac{D_\perp^2}{D_R^2 + \omega^2}} \exp\left(- \frac{D_R}{\omega}(\phi - \phi_0) \right). \tag{5}$$

Qualitatively, such spirals can be understood as follows: in the absence of thermal noise, the average swimming path corresponds to a circle with radius R given by Eq. (3). In the presence of thermal noise, however, single trajectory segments become increasingly different as time proceeds. This leads to decreasing distances d_i between adjacent turns of the mean swimming path [$d_i/d_{i+1} = \exp(2\pi D_R/|\omega|)$, see Fig. 3(d)] and, finally, to the convergence in a single point for $t \to \infty$. Because of the alignment of the initial slope, this point is shifted relative to the starting point depending on the alignment angle and the circulation direction of the particle.

The solid curve in Fig. 3(d) is the theoretical prediction [see Eq. (5)] with the measured values of $D_\perp$, D_R, and ω. On the other hand, the dashed curve in Fig. 3(d) visualizes the noise-averaged trajectory determined directly from the experimental data [see Figs. 3(b) and 3(c)]. The agreement of the two curves constitutes a direct verification of our theoretical model on a fundamental level.

Finally, we also address the motion of asymmetric swimmers under confinement, e.g., their interaction with a straight wall. This is shown in Fig. 4(a) exemplarily for an $L+$ swimmer which approaches the wall at an angle θ. Because of the internal torque associated with the active particle motion, it becomes stabilized at the wall and smoothly glides to the right along the interface. In contrast, for a much larger initial contact angle the internal torque rotates the front part of the particle away from the obstacle, the motion is unstable, and the swimmer is reflected by the wall [see Fig. 4(b)] [31]. Figure 4(c) shows the observed dependence of the motional behavior as a function of the approaching angle.

The experimental findings are in line with an instability analysis based on a torque balance condition of an L-shaped particle at wall contact as a function of its contact angle θ. For $\theta_{\text{crit}} < \theta < \pi$ [see Figs. 4(b) and 4(e)] with a critical angle θ_{crit}, the particle is reflected, while for $0 < \theta < \theta_{\text{crit}}$ [see Figs. 4(a) and 4(d)] stable sliding with an angle θ_{sl} occurs. Both, θ_{sl} and θ_{crit} are given as stable and unstable solutions, respectively, of the torque balance condition

$$|l| = [(a - y_s)\cos\theta - x_s \sin\theta]\sin\theta\,. \tag{6}$$

For $l = -0.71\ \mu\text{m}$ corresponding to an ideal L-shaped particle with the propulsion force centered in the middle of the Au layer, we obtain $\theta_{\text{sl}} = 8.0°$ and $\theta_{\text{crit}} = 59.2°$, which is in

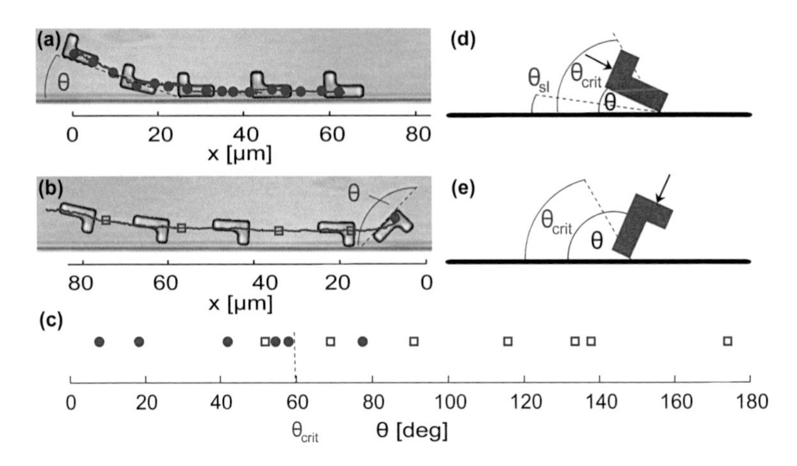

FIG. 4. (a),(b) Trajectories of an $L+$ swimmer approaching a straight wall at different angles (symbols correspond to positions after 1 min each). (c) Experimentally determined particle motion for different contact angles θ. Bullets and open squares correspond to particle sliding and reflection. (d),(e) Visualization of the predicted types of motion for an $L+$ swimmer with arrows indicating the direction of the propulsion force: (d) stable sliding and (e) reflection. The angles are defined in the text.

good agreement with the measured value of about $\theta_{\text{crit}} = 60°$ [see Fig. 4(c)]. The observed scatter in the experimental data around the critical angle is due to thermal fluctuations that wash out the sharp transition between the sliding and the reflection regime.

In conclusion, we have demonstrated that due to viscous forces of the surrounding liquid, asymmetric microswimmers are subjected to a velocity-dependent torque. This leads to a circular motion, which is observed in experiments in agreement with a theoretical model based on two coupled Langevin equations. In a channel geometry, this torque leads either to a reflection or a stable sliding motion along the wall. An interesting question for the future addresses how asymmetric swimmers move through patterned media. In the presence of a drift force, one may expect Shapiro steps in the particle current similar to what has also been found in colloidal systems driven by a circular drive [33]. Another appealing outlook addresses the motion of chiral swimmers in the presence of external fields such as gravity [34]. In the case of asymmetric particles, this leads to an orientational alignment during their sedimentation, which may result in a preferential motion relative to gravity similar to the gravitactic behavior of asymmetric cells as, e.g., *Chlamydomonas* [35, 36].

We thank M. Aristov for assistance in particle preparation and M. Heinen for helpful discussions. This work was supported by the DFG within SPP 1296 and SFB TR6-C3 as well as by the Marie Curie-Initial Training Network Comploids funded by the European Union Seventh Framework Program (FP7). R. W. gratefully acknowledges financial support from a Postdoctoral Research Fellowship (WI 4170/1-1) of the DFG.

---

* Present address: Department of Physics, Bilkent University, Cankaya, Ankara 06800, Turkey.

[1] P. Romanczuk, M. Bär, W. Ebeling, B. Lindner, and L. Schimansky-Geier, Eur. Phys. J. Special Topics **202**, 1 (2012).

[2] S. J. Ebbens and J. R. Howse, Soft Matter **6**, 726 (2010).

[3] T. Mirkovic, N. S. Zacharia, G. D. Scholes, and G. A. Ozin, ACS Nano **4**, 1782 (2010).

[4] R. Pontier-Bres, F. Prodon, P. Munro, P. Rampal, E. Lemichez, J. F. Peyron, and D. Czerucka, PLoS ONE **7**, e33796 (2012).

[5] M. E. Cates, Rep. Prog. Phys. **75**, 042601 (2012).

[6] M. C. Marchetti, J. F. Joanny, S. Ramaswamy, T. B. Liverpool, J. Prost, M. Rao, and R. A. Simha, Rev. Mod. Phys. **85**, 1143 (2013).

[7] R. Dreyfus, J. Baudry, M. L. Roper, M. Fermigier, H. A. Stone, and J. Bibette, Nature (London) **437**, 862 (2005).

[8] H. R. Jiang, N. Yoshinaga, and M. Sano, Phys. Rev. Lett. **105**, 268302 (2010).

[9] Y. Hong, N. M. K. Blackman, N. D. Kopp, A. Sen, and D. Velegol, Phys. Rev. Lett. **99**, 178103 (2007).

[10] J. R. Howse, R. A. L. Jones, A. J. Ryan, T. Gough, R. Vafabakhsh, and R. Golestanian, Phys. Rev. Lett. **99**, 048102 (2007).

[11] S. Badaire, C. Cottin-Bizonne, W. Joseph, A. Yang, and A. D. Stroock, J. Am. Chem. Soc. **129**, 40 (2007).

[12] C. A. Grattoni, R. A. Dawe, C. Y. Seah, and J. D. Gray, J. Chem. Eng. Data **38**, 516 (1993).

[13] I. Buttinoni, G. Volpe, F. Kümmel, G. Volpe, and C. Bechinger, J. Phys. Condens. Matter **24**, 284129 (2012).

[14] G. Volpe, I. Buttinoni, D. Vogt, H. Kümmerer, and C. Bechinger, Soft Matter **7**, 8810 (2011).

[15] J. Palacci, C. Cottin-Bizonne, C. Ybert, and L. Bocquet, Phys. Rev. Lett. **105**, 088304 (2010).

[16] W. F. Paxton, A. Sen, and T. E. Mallouk, Chem. Eur. J. **11**, 6462 (2005).

[17] J. Vicario, R. Eelkema, W. R. Browne, A. Meetsma, R. M. La Crois, and B. L. Feringa, Chem. Commun. (Cambridge) **31**, 3936 (2005).

[18] M. Mijalkov and G. Volpe, Soft Matter **9**, 6376 (2013).

[19] B. ten Hagen, S. van Teeffelen, and H. Löwen, J. Phys. Condens. Matter **23**, 194119 (2011).

[20] Alternatively, it is also possible to use the resistance matrix formalism [21, 22].

[21] E. Lauga and T. R. Powers, Rep. Prog. Phys. **72**, 096601 (2009).

[22] E. M. Purcell, Am. J. Phys. **45**, 3 (1977).

[23] Because of the multiplicative noise, an additional drift term has to be taken into account, when Eqs. (1) are solved numerically.

[24] J. Happel and H. Brenner, *Low Reynolds Number Hydrodynamics: With Special Applications to Particulate Media*, Mechanics of Fluids and Transport Processes Vol. 1 (Kluwer Academic Publishers, Dordrecht, 1991), 2nd ed.

[25] B. Carrasco and J. Garcia de la Torre, J. Chem. Phys. **111**, 4817 (1999).

[26] J. Garcia de la Torre, S. Navarro, M. C. Lopez Martinez, F. G. Diaz, and J. J. Lopez Cascales, Biophys. J. **67**, 530 (1994).

[27] J. R. Blake, Proc. Cambridge Philos. Soc. **70**, 303 (1971).

[28] Y. Han, A. M. Alsayed, M. Nobili, J. Zhang, T. C. Lubensky, and A. G. Yodh, Science **314**, 626 (2006).

[29] Y. Han, A. Alsayed, M. Nobili, and A. G. Yodh, Phys. Rev. E **80**, 011403 (2009).

[30] S. van Teeffelen and H. Löwen, Phys. Rev. E **78**, 020101 (2008).

[31] Note that this effect occurring for hard colloidal swimmers is different from the situation of *Escherichia coli* bacteria confined in the proximity of a surface [32]. In the latter case, the circular motion is generated by the rotation of the body and the flagella, coupled with the hydrodynamic interactions with the boundary.

[32] E. Lauga, W. R. DiLuzio, G. M. Whitesides, and H. A. Stone, Biophys. J. **90**, 400 (2006).

[33] C. Reichhardt, C. J. Olson, and M. B. Hastings, Phys. Rev. Lett. **89**, 024101 (2002).

[34] R. Wittkowski and H. Löwen, Phys. Rev. E **85**, 021406 (2012).

[35] A. M. Roberts, J. Exp. Biol. **213**, 4158 (2010).

[36] D. P. Häder, Adv. Space Res. **24**, 843 (1999).

Statement of the author: This is the first article originating from the long-lasting and extremely fruitful collaboration with the group of Clemens Bechinger in Stuttgart. It constitutes one of the centerpieces of my doctorate.

I developed the theoretical model and derived the analytical expressions with helpful input from Raphael Wittkowski and Hartmut Löwen. Felix Kümmel performed the bulk of the experiments. He was introduced to some of the experimental techniques by Ivo Buttinoni. The idea of studying L-shaped particles in experiments originates from Giovanni Volpe. I was largely involved in analyzing the experimental data and wrote substantial parts of the paper. Hartmut Löwen and Clemens Bechinger helped to motivate the present study in a wider context and established the contact with Ralf Eichhorn. He was able to improve the numerical calculation of the diffusion coefficients, which I had originally computed using the software HYDRO++, by taking the substrate into account additionally.

I presented the results of this article at many international conferences such as the DPG Spring Meeting 2013 in Regensburg, the International Soft Matter Conference 2013 in Rome, the East Asia Joint Seminars on Statistical Physics 2013 in Kyoto, Japan, and at the State Key Laboratory of Nonlinear Mechanics in Beijing, China.

Kümmel *et al.* Reply

Felix Kümmel,[1] Borge ten Hagen,[2] Raphael Wittkowski,[3] Daisuke Takagi,[4] Ivo Buttinoni,[1]
Ralf Eichhorn,[5] Giovanni Volpe,[1,*] Hartmut Löwen,[2] and Clemens Bechinger[1,6]

[1] *2. Physikalisches Institut, Universität Stuttgart, D-70569 Stuttgart, Germany*

[2] *Institut für Theoretische Physik II: Weiche Materie, Heinrich-Heine-Universität Düsseldorf, D-40225 Düsseldorf, Germany*

[3] *SUPA, School of Physics and Astronomy, University of Edinburgh, Edinburgh, EH9 3JZ, United Kingdom*

[4] *Department of Mathematics, University of Hawaii at Manoa, Honolulu, Hawaii 96822, USA*

[5] *Nordita, Royal Institute of Technology, and Stockholm University, SE-10691 Stockholm, Sweden*

[6] *Max-Planck-Institut für Intelligente Systeme, D-70569 Stuttgart, Germany*

PACS numbers: 82.70.Dd, 05.40.Jc

In a Comment [1] on our Letter on self-propelled asymmetric particles [2], Felderhof claims that our theory based on Langevin equations would be conceptually wrong. In this Reply we show that our theory is appropriate, consistent, and physically justified.

The motion of a self-propelled particle (SPP) is force- and torque-free if external forces and torques are absent. Nevertheless, as stated in our Letter [2], *effective* forces and torques [3–7] can be used together with the grand resistance matrix (GRM) [8] to describe the self-propulsion of force- and torque-free swimmers [9]. To prove this, we perform a hydrodynamic calculation based on slender-body theory for Stokes flow [10, 11]. This approach has been applied successfully to model, e.g., flagellar locomotion [12, 13] and avoids a general Faxén theorem for asymmetric particles. A key assumption of slender-body theory is that the width 2ϵ of the arms of the L-shaped particle is much smaller than the total arc length $L = a + b$, where a and b are the arm lengths.

The centerline position of the slender particle is $\mathbf{x}(s) = \mathbf{r} - \mathbf{r}_S + s\hat{\mathbf{u}}_{\parallel}$ for $-b \leq s \leq 0$ and $\mathbf{x}(s) = \mathbf{r} - \mathbf{r}_S + s\hat{\mathbf{u}}_{\perp}$ for $0 < s \leq a$. Here, $\mathbf{r}$ is the center-of-mass position of the particle in the laboratory frame of reference and $\mathbf{r}_S = (a^2\hat{\mathbf{u}}_{\perp} - b^2\hat{\mathbf{u}}_{\parallel})/(2L)$ is a vector in the particle's frame—defined by the unit vectors $\hat{\mathbf{u}}_{\parallel}$, $\hat{\mathbf{u}}_{\perp}$—such that $\mathbf{r} - \mathbf{r}_S$ is the point where the two arms meet at right angles. The fluid velocity on the particle surface is approximated by $\dot{\mathbf{x}} + \mathbf{v}_{\mathrm{sl}}$ with a prescribed slip velocity $\mathbf{v}_{\mathrm{sl}}(s)$. According to the leading-order slender-body approximation [10], the fluid velocity is related to the local force per unit length $\mathbf{f}(s)$ on the particle surface by $\dot{\mathbf{x}} + \mathbf{v}_{\mathrm{sl}} = c(\mathbf{I} + \mathbf{x}' \otimes \mathbf{x}')\mathbf{f}$ with $c = \log(L/\epsilon)/(4\pi\eta)$, the solvent viscosity η, the identity matrix $\mathbf{I}$, $\mathbf{x}' = \partial\mathbf{x}/\partial s$, and the dyadic product $\otimes$. The force density $\mathbf{f}$ satisfies the integral constraints of vanishing net force, $\int_{-b}^{a} \mathbf{f} ds = \mathbf{0}$, and vanishing net torque relative to the center of mass, $\hat{\mathbf{e}}_z \cdot \int_{-b}^{a}(-\mathbf{r}_S + s\mathbf{x}') \times \mathbf{f} ds = \int_{-b}^{0} s\hat{\mathbf{u}}_{\perp} \cdot \mathbf{f} ds - \int_{0}^{a} s\hat{\mathbf{u}}_{\parallel} \cdot \mathbf{f} ds = 0$, with $\hat{\mathbf{e}}_z = (0,0,1)^T$.

First, we consider a *passive* particle driven by an external force $\mathbf{F}_{\mathrm{ext}}$, which is constant in the particle's frame, and torque M_{ext}. For this case, we assume no-slip conditions for the fluid on the entire particle surface. Then the integral constraints with net force $\mathbf{F}_{\mathrm{ext}}$ and torque M_{ext} give

$$\eta\boldsymbol{\mathcal{H}} \left(\hat{\mathbf{u}}_{\parallel} \cdot \dot{\mathbf{r}}, \hat{\mathbf{u}}_{\perp} \cdot \dot{\mathbf{r}}, \dot{\phi}\right)^T = \left(\hat{\mathbf{u}}_{\parallel} \cdot \mathbf{F}_{\mathrm{ext}}, \hat{\mathbf{u}}_{\perp} \cdot \mathbf{F}_{\mathrm{ext}}, M_{\mathrm{ext}}\right)^T, \tag{1}$$

where

$$\boldsymbol{\mathcal{H}} = \frac{1}{2c\eta} \begin{pmatrix} 2a+b & 0 & -a^2b/(2L) \\ 0 & a+2b & -ab^2/(2L) \\ -a^2b/(2L) & -ab^2/(2L) & A \end{pmatrix} \tag{2}$$

with $A = [(8L^2 - 3ab)(a^3 + b^3) - 6L(a^4 + b^4)]/(12L^2)$ is the GRM that depends on the particle shape [8, 14].

In the *self-propelled* case, motivated by the slip flow generated near the Au coating in the experiments, we set $\mathbf{v}_{\mathrm{sl}} = -V_{\mathrm{sl}}\hat{\mathbf{u}}_{\perp}$ along the arm of length b and no slip ($\mathbf{v}_{\mathrm{sl}} = \mathbf{0}$) along the other arm. This results in

$$\eta\boldsymbol{\mathcal{H}}\left(\hat{\mathbf{u}}_{\parallel}\cdot\dot{\mathbf{r}},\hat{\mathbf{u}}_{\perp}\cdot\dot{\mathbf{r}},\dot{\phi}\right)^{T}=\left(0,bV_{\mathrm{sl}}/c,-ab^{2}V_{\mathrm{sl}}/(2cL)\right)^{T}. \tag{3}$$

We emphasize that the tensor $\boldsymbol{\mathcal{H}}$ in Eq. (3) is identical to the GRM in Eq. (1). Formally, both equations are *exactly* the same if $\hat{\mathbf{u}}_{\parallel}\cdot\mathbf{F}_{\mathrm{ext}}=0$, $\hat{\mathbf{u}}_{\perp}\cdot\mathbf{F}_{\mathrm{ext}}=bV_{\mathrm{sl}}/c$, and $M_{\mathrm{ext}}=-ab^{2}V_{\mathrm{sl}}/(2cL)$. This shows that the motion of a SPP with $\mathbf{v}_{\mathrm{sl}}=-V_{\mathrm{sl}}\hat{\mathbf{u}}_{\perp}$ along the arm of length b is identical to the motion of a passive particle driven by a net external force $\mathbf{F}_{\mathrm{ext}}=F\hat{\mathbf{u}}_{\perp}$ and torque $M_{\mathrm{ext}}=lF$ with the effective self-propulsion force $F=bV_{\mathrm{sl}}/c$ and effective lever arm $l=-ab/(2L)$. By transforming Eq. (3) from the particle's frame to the laboratory frame and introducing the generalized diffusion tensor $\boldsymbol{\mathcal{D}}=\boldsymbol{\mathcal{H}}^{-1}/(\beta\eta)$ [11], where β is the inverse effective thermal energy, one directly obtains the noise-free version of the equations of motion (EOMs) (1) in our Letter [2].

Clearly, for the same particle velocity, the flow and pressure fields generated by the SPP and the externally driven particle are different. However, the EOMs are the same. Therefore, we can formally use external forces and torques that move with the SPP to model its self-propelled motion. In that sense, the concept of *effective* forces and torques is justified, the application of the GRM is appropriate, and the EOMs in our Letter correctly describe the dynamics of the SPP.

---

* Present address: Department of Physics, Bilkent University, Cankaya, Ankara 06800, Turkey.

[1] B. U. Felderhof, Phys. Rev. Lett. **113**, 029801 (2014).

[2] F. Kümmel, B. ten Hagen, R. Wittkowski, I. Buttinoni, R. Eichhorn, G. Volpe, H. Löwen, and C. Bechinger, Phys. Rev. Lett. **110**, 198302 (2013).

[3] B. M. Friedrich and F. Jülicher, New J. Phys. **10**, 123025 (2008).

[4] G. Jékely, J. Colombelli, H. Hausen, K. Guy, E. Stelzer, F. Nédélec, and D. Arendt, Nature (London) **456**, 395 (2008).

[5] P. K. Radtke and L. Schimansky-Geier, Phys. Rev. E **85**, 051110 (2012).

[6] A. Nourhani, P.E. Lammert, A. Borhan, and V.H. Crespi, Phys. Rev. E **87**, 050301(R) (2013).

[7] N. A. Marine, P. M. Wheat, J. Ault, and J. D. Posner, Phys. Rev. E **87**, 052305 (2013).

[8] D. J. Kraft, R. Wittkowski, B. ten Hagen, K. V. Edmond, D. J. Pine, and H. Löwen, Phys. Rev. E **88**, 050301(R) (2013).

[9] Following common nomenclature, we do not distinguish between the terms "self-propulsion" and "swimming" with regard to the rigidity of the particle.

[10] G. K. Batchelor, J. Fluid Mech. **44**, 419 (1970).

[11] B. ten Hagen, R. Wittkowski, D. Takagi, F. Kümmel, C. Bechinger, and H. Löwen, arXiv:1410.6707.

[12] J. Lighthill, SIAM Rev. **18**, 161 (1976).

[13] E. Lauga and T. R. Powers, Rep. Prog. Phys. **72**, 096601 (2009).

[14] J. Happel and H. Brenner, *Low Reynolds Number Hydrodynamics: With Special Applications to Particulate Media*, Mechanics of Fluids and Transport Processes Vol. 1 (Kluwer Academic Publishers, Dordrecht, 1991), 2nd ed.

Statement of the author: The second part of this chapter contains the Reply to a Comment by B. U. Felderhof [410] on our Letter "Circular motion of asymmetric self-propelling particles" published in *Physical Review Letters* [218].

To unambiguously justify the concept of effective forces and torques, we thoroughly discussed various approaches, initially mainly between Hartmut Löwen, Raphael Wittkowski, and me. The ansatz based on slender-body theory originates from a discussion between Daisuke Takagi and Hartmut Löwen at University of Hawaii at Manoa. After that Daisuke Takagi put forward this idea by applying it to the L-shaped particle and therefore became coauthor of the Reply. I wrote parts of the paper and was responsible for producing the final version of the manuscript. Felix Kümmel, Ivo Buttinoni, Ralf Eichhorn, Giovanni Volpe, and Clemens Bechinger were coauthors of the original publication and gave their feedback on the Reply.

CHAPTER 8

Can the self-propulsion of anisotropic microswimmers be described by using forces and torques?

The content of this chapter has been published in a similar form in *Journal of Physics: Condensed Matter* **27**, 194110 (2015) by Borge ten Hagen, Raphael Wittkowski, Daisuke Takagi, Felix Kümmel, Clemens Bechinger, and Hartmut Löwen (see reference [219] for the preprint).

Can the self-propulsion of anisotropic microswimmers be described by using forces and torques?

Borge ten Hagen[1], Raphael Wittkowski[2], Daisuke Takagi[3], Felix Kümmel[4], Clemens Bechinger[4,5], and Hartmut Löwen[1]

[1] Institut für Theoretische Physik II: Weiche Materie, Heinrich-Heine-Universität Düsseldorf, D-40225 Düsseldorf, Germany
[2] SUPA, School of Physics and Astronomy, University of Edinburgh, Edinburgh EH9 3FD, UK
[3] Department of Mathematics, University of Hawaii at Manoa, Honolulu, Hawaii 96822, USA
[4] 2. Physikalisches Institut, Universität Stuttgart, D-70569 Stuttgart, Germany
[5] Max-Planck-Institut für Intelligente Systeme, D-70569 Stuttgart, Germany

E-mail: bhagen@thphy.uni-duesseldorf.de

Abstract. The self-propulsion of artificial and biological microswimmers (or active colloidal particles) has often been modelled by using a force and a torque entering into the overdamped equations for the Brownian motion of passive particles. This seemingly contradicts the fact that a swimmer is force-free and torque-free, i.e. that the net force and torque on the particle vanish. Using different models for mechanical and diffusiophoretic self-propulsion, we demonstrate here that the equations of motion of microswimmers can be mapped onto those of passive particles with the shape-dependent grand resistance matrix and formally external *effective* forces and torques. This is consistent with experimental findings on the circular motion of artificial asymmetric microswimmers driven by self-diffusiophoresis. The concept of effective self-propulsion forces and torques significantly facilitates the understanding of the swimming paths, e.g. for a microswimmer under gravity. However, this concept has its limitations when the self-propulsion mechanism of a swimmer is disturbed either by another particle in its close vicinity or by interactions with obstacles, such as a wall.

1. Introduction

The basic interest in liquid matter research has shifted from passive simple liquids [1] to complex liquids [2] over the past decades. While the individual building blocks of the liquid are passive thermal particles in this case, a more recent topic concerns *living liquids* which consist of self-propelling biological constituents, such as bacteria in a low-Reynolds-number fluid [3–6]. Being perpetually driven by the consumption of energy needed for the self-propulsion, even a single swimmer represents already a complicated non-equilibrium situation. In this respect, artificial self-phoretically-driven colloidal particles serve as very helpful model systems to mimic the self-propulsion of biological microswimmers and enable particle-resolved insights into the one-particle and collective phenomena [7–14].

Regarding the modelling of microswimmer propulsion, there are basically two levels of description. The first coarse-grained approach does not resolve the details of the propagation but models the resulting propulsion velocity $\boldsymbol{v}$ by the action of a formally external *effective* self-propulsion force $\boldsymbol{F}$ moving with the particle. This force then enters into the completely overdamped equations of motion for passive particles. In a demonstrative way, $\boldsymbol{F}$ can be interpreted as the constraining force that is required to prevent the microswimmer from moving, which can be achieved, e.g. by optical tweezers [15]. In its simplest form, i.e. for a sphere of hydrodynamic radius R, the self-propulsion force $\boldsymbol{F}$ is given by

$$\gamma \boldsymbol{v} = \boldsymbol{F} \tag{1}$$

with $\gamma = 6\pi\eta R > 0$ denoting a (scalar) Stokes friction coefficient for stick fluid boundary conditions on the surface of the particle and η denoting the dynamic (shear) viscosity of the fluid. There is plenty of literature by now in which this kind of force modelling was employed either in the context of single-particle motion [16–19] or to study various collective phenomena of self-propelled particles such as swarming [20–22], clustering [13, 23–29], and ratchet effects [30, 31]. The concept has also been used for several applications like particle separation [32, 33] or trapping [34–36] of self-propelled particles. Typically, the swimming direction is along a particle-fixed axis (usually a symmetry axis of the particle) denoted by a unit vector $\hat{\boldsymbol{u}}$. The direction of the self-propulsion force thus rotates with the particle orientation. Therefore it is called an 'internal' force. The particle orientation $\hat{\boldsymbol{u}}$ obeys an overdamped equation of motion with rotational friction. This makes the problem non-Hamiltonian and non-trivial, in particular in the presence of noise [37, 38]. However, throughout this article thermal fluctuations are neglected as they are not relevant for the present study.

From (1) it is evident how to generalize this equation towards an asymmetric microswimmer with an arbitrary non-spherical shape by using the 6×6-dimensional grand resistance matrix $\boldsymbol{\mathcal{H}}$ [39] (also called 'hydrodynamic friction tensor' [40]) and a six-dimensional generalized velocity $\boldsymbol{V} = (\boldsymbol{v}, \boldsymbol{\omega})$ composed of the translational velocity $\boldsymbol{v}$ and the angular velocity $\boldsymbol{\omega}$ of the particle. Correspondingly, the right-hand side of (1) is replaced by the six-dimensional generalized force $\boldsymbol{K} = (\boldsymbol{F}, \boldsymbol{M})$ composed of an internal force $\boldsymbol{F}$ and an internal torque $\boldsymbol{M}$, which are fixed in the particle frame, such that‡

$$\eta \boldsymbol{\mathcal{H}} \boldsymbol{V} = \boldsymbol{K}. \tag{2}$$

Equation (2) comprises many different situations including circle swimming in two [12, 16] and helical swimming in three dimensions [18], which are induced by the simultaneous presence of a constant force and a torque in the particle frame. Many papers have adopted this formalism in order to describe the motion of microswimmers [15,41–56]. On this coarse-grained level of modelling there is no explicit solvent velocity field, which is just disregarded.

The second level involves a more detailed description of the propulsion mechanism of an individual swimmer. At this level, several modes of propagation have to be distinguished. We shall discriminate in the following between mechanical self-propulsion and self-diffusiophoretic motion. Some type of mechanical self-propulsion is usually realized in biological microswimmers such as bacteria [57–59], algae [60, 61], or spermatozoa [62–65]. In particular, flagellar locomotion [66–68] has been studied intensely. But also different propulsion mechanisms have been analysed theoretically [69, 70]. In general, the active motion is induced by mutually moving objects such as rotating screws [71] or non-reciprocally translated spheres as embodied in the paradigmatic three-sphere swimmer of Najafi and Golestanian [72]. Diffusiophoretic self-propulsion is typically described by imposing a slip solvent velocity on the particle surface [73, 74], which then puts the swimmer into motion [75–78]. A similar way of modelling is also well-established for squirmers [79], where a tangential surface velocity is prescribed [80–83]. In both cases, the swimming mode itself is intrinsic and therefore occurs in the absence of any external force or torque, which is typically expressed by the fact that a swimmer is force-free and torque-free [84]. Clearly, on this level of description the solvent velocity field enters explicitly.

At first sight, the description on the first level using internal forces and torques seems to be in contradiction to the fact that a swimmer is force-free and torque-free. It has therefore been criticized [85]. However, in fact there is no such contradiction if the self-propulsion force and torque in the equations of motion are interpreted as formal or effective quantities [86]. For spherical particles, the magnitude of the effective force just fixes the propulsion speed via the formal relation (1). We emphasize that this concept is not capable of revealing any insights about the detailed self-propulsion mechanism itself. In particular, it may not be confused with the disputable assumption of an osmotic driving force [87, 88] on a microscopic level. How the effective force has to be interpreted in the context of a mechanical self-propulsion mechanism was made explicit in [89] for the motion of a three-sphere swimmer with one big and two small spheres on a linear array. The force contribution acting on the big sphere due to two point forces located at the positions of the small spheres is not given as a simple sum

‡ In two spatial dimensions the grand resistance matrix $\boldsymbol{\mathcal{H}}$ is only 3×3-dimensional and the generalized velocity $\boldsymbol{V}$ as well as the generalized force $\boldsymbol{K}$ are three-dimensional.

of these two forces. Instead, they have to be appropriately rescaled first, in order to provide the effective force which obeys (1).

The issue becomes less evident for anisotropic swimmers which perform in general a circular (in two dimensions) or helical (in three dimensions) motion and whose dynamics is governed by a translational-rotational coupling [12, 18]. The basic question covered in this article is whether a similar formal rescaling can be done in order to map the equations of motion of an asymmetric microswimmer onto those of a corresponding passive particle described by (2). Or in other words: can the grand resistance matrix $\boldsymbol{\mathcal{H}}$ with formal forces and torques $\boldsymbol{K}$ be used to describe the self-propulsion of microswimmers? From linear response theory it is clear that the leading part of the equations is linear such that there is a matrix relating the generalized velocities $\boldsymbol{V}$ to the generalized forces $\boldsymbol{K}$. But is this matrix identical to the grand resistance matrix of the corresponding passive particles? This has recently been put into question by Felderhof [85]. This important issue can be answered either by experiments or by a detailed theoretical analysis. In [12] it was shown that the experimental data for the planar motion of asymmetric microswimmers can indeed be described using the grand resistance matrix occurring in (2).

In this article we follow the alternative theoretical route: we model the swimmer motion explicitly according to the more detailed second level and check whether the resulting equation can be mapped onto the coarse-grained equation (2) using an effective force and torque. We do this explicitly for an asymmetric two-dimensional swimmer with an L-shape and discriminate between mechanical self-propulsion and self-diffusiophoretic propulsion in an unbounded fluid. In the first case, we consider an extension of the Golestanian swimmer [72, 89] towards an L-shaped particle. In the second case, we study slender biaxial particles consisting of two arms of different lengths that meet at a certain internal angle. To account for the diffusiophoretic self-propulsion mechanism, we consider an imposed slip velocity on one of the arms. The sense of rotation of the particles is found to depend on the internal angle. A previous but less general account for a rectangular internal angle was published elsewhere [14, 86]. For both the mechanical and the diffusiophoretic self-propulsion, we explicitly show that the matrix needed to describe the motion of the microswimmer is identical to the grand resistance matrix of the particle. This opens the way to think in terms of effective forces and torques, which is advantageous in order to get insights into the variety of swimming paths in more complicated situations. The benefit becomes most obvious when both external body forces and torques and internal effective self-propulsion forces and torques are present, such as in gravitaxis [14, 90, 91].

The paper is organized as follows: the mechanical self-propulsion mechanism and a generalization of the three-sphere swimmer of Golestanian towards an L-shaped microswimmer are discussed in section 2. Afterwards, in section 3, the self-propulsion of biaxial diffusiophoretic microswimmers is studied using slender body theory. Based on these considerations for two classes of microswimmers, section 4 contains general considerations for asymmetric particles with arbitrary shape and arbitrary self-propulsion. In section 5, we discuss limitations of the concept of effective forces and torques before we finally conclude in section 6.

2. Mechanical self-propulsion

Microswimmers with mechanical self-propulsion change their shapes in a non-reciprocal way [84] in order to propel themselves forward. These shape changes can be realized by deformations or—if the microswimmer consists of several individual parts—by translating or rotating different parts relative to each other through internal forces and torques $\boldsymbol{k}^{(\alpha)}$ that are applied on the individual parts of the microswimmer. The noise-free equations of motion of such a mechanical microswimmer with n parts can be formulated through a modified force and torque balance (see, e.g. (42) and (45) in [92])

$$\boldsymbol{K}_{\mathrm{St}} + \boldsymbol{K} = \boldsymbol{0}\,, \qquad \boldsymbol{K} = \sum_{\alpha=1}^{n} \boldsymbol{\mathcal{C}}^{(\alpha)}\boldsymbol{k}^{(\alpha)} \tag{3}$$

that ensures that the microswimmer is force-free and torque-free and that includes the Stokes friction force and torque $\boldsymbol{K}_{\mathrm{St}} = -\eta\boldsymbol{\mathcal{H}}\boldsymbol{V}$, which is proportional to the shape-dependent§ grand resistance matrix $\boldsymbol{\mathcal{H}}$ of the particle, and the *rescaled* internal forces and torques $\boldsymbol{\mathcal{C}}^{(\alpha)}\boldsymbol{k}^{(\alpha)}$ with rescaling matrices $\boldsymbol{\mathcal{C}}^{(\alpha)}$. For a specific particle shape, the rescaling matrices $\boldsymbol{\mathcal{C}}^{(\alpha)}$, which take into account the hydrodynamic interactions between the individual parts of the microswimmer, can in principle be calculated using the Green's function approach [92,93]. It is important to note that due to this rescaling the equations of motion of the microswimmer are given through a *modified* and not through a simple direct force and torque balance. The sum of the rescaled internal forces and torques is the formally external *effective* force and torque vector $\boldsymbol{K}$.

The simplest mechanical microswimmer is the linear three-sphere swimmer of Najafi and Golestanian [72], which consists of three spheres one behind the other that are translated relative to each other (see (5) in [89] for the equation of motion). In the following, we generalize their theoretical approach to an L-shaped microswimmer in two spatial dimensions in order to provide a minimal microscopic model for an asymmetric mechanical microswimmer.

Our L-shaped Golestanian-like microswimmer consists of two very small spheres and a much larger L-shaped particle (see figure 1). The small spheres have distances $d_1(t)$ and $d_2(t)$ from the short arm of the L-shaped particle, which vary as a non-reciprocal function of time (see [72] for details) as a consequence of internal forces $\boldsymbol{f}^{(\alpha)}$ with $\alpha \in \{1,2\}$ that act on the two spheres. Note that the forces $\boldsymbol{f}^{(\alpha)}$ are two-dimensional here as the motion is restricted to the x-y plane. We denote the length of the long arm of the L-shaped particle by a and its orientation by the unit vector $\hat{\boldsymbol{u}}_{\perp}$. The length of the short arm of the particle is b and the width of the arms is $b/2$.

§ Note that the grand resistance matrix depends only on the shape but not on the activity of the particle. It is therefore the same for active and passive particles as long as the activity is not accompanied by a change of the shape.

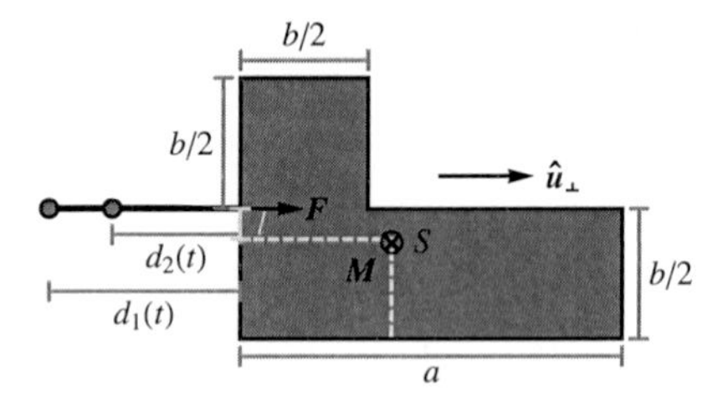

Figure 1. Schematic of a force-free and torque-free L-shaped Golestanian-like microswimmer of size a and orientation $\hat{\boldsymbol{u}}_{\perp}$. To put the microswimmer in motion, the distances $d_1(t)$ and $d_2(t)$ of the two small spheres from the L-shaped particle vary as a non-reciprocal function of time. The effective lever arm l relates the effective force $\boldsymbol{F}$ to the effective torque $M = ||\boldsymbol{M}||$ on the centre of mass S.

The small spheres are only translated relative to the larger L-shaped particle. There are no internal torques that rotate the spheres and the L-shaped particle relative to each other. Nevertheless, the resulting effective (here three-dimensional) generalized self-propulsion force $\boldsymbol{K} = (\boldsymbol{F}, M) = \boldsymbol{\mathcal{C}}^{(1)}\boldsymbol{f}^{(1)} + \boldsymbol{\mathcal{C}}^{(2)}\boldsymbol{f}^{(2)}$ includes not only an effective self-propulsion force $\boldsymbol{F}$, but also an effective self-propulsion torque M that acts on the centre of mass S of the particle. Since the effective force $\boldsymbol{F}$ acts perpendicularly on the middle of the short arm of the L-shaped particle, the effective torque is given by $\boldsymbol{M} = lF\hat{\boldsymbol{e}}_z$ with $\hat{\boldsymbol{e}}_z = (0, 0, 1)$ and $F = ||\boldsymbol{F}||$. The effective self-propulsion torque $\boldsymbol{M}$ can thus be expressed by the effective lever arm l, which denotes the distance in the direction of the short arm of the particle between the point on which the effective force acts and the point of reference for the calculation of the diffusion coefficients, which is the centre of mass in our case.

In two spatial dimensions, the grand resistance matrix of the L-shaped microswimmer, which in the limit of very small spheres is equal to the grand resistance matrix of the (passive) L-shaped part of the microswimmer, is 3×3-dimensional; the generalized velocity is the three-dimensional vector $\boldsymbol{V} = (\dot{x}, \dot{y}, \dot{\phi})$ with ϕ indicating the particle orientation, and the internal forces $\boldsymbol{f}^{(\alpha)}$ are two-dimensional (there are no internal torques). Lastly, the rescaling matrices $\boldsymbol{\mathcal{C}}^{(\alpha)}$ are 3×2-dimensional. With this notation, the force and torque balance (3), which comprises the conditions of zero external force and zero external torque, reduces to

$$\eta\boldsymbol{\mathcal{H}}\boldsymbol{V} = \sum_{\alpha=1}^{2}\boldsymbol{\mathcal{C}}^{(\alpha)}\boldsymbol{f}^{(\alpha)} \tag{4}$$

with the generalized effective force

$$\boldsymbol{K} = \begin{pmatrix}\boldsymbol{F}\\ M\end{pmatrix} = \sum_{\alpha=1}^{2}\boldsymbol{\mathcal{C}}^{(\alpha)}\boldsymbol{f}^{(\alpha)}\,. \tag{5}$$

From (4) it is obvious that the equations of motion for the mechanically driven microswimmer studied here can be written in terms of the grand resistance matrix

$\mathcal{H}$ and effective forces and torques as defined in (5). The specific shape of the particle enters into the equations of motion via the grand resistance matrix $\mathcal{H}$ and the rescaling matrices $\mathcal{C}^{(\alpha)}$. In principle, these quantities can be calculated for arbitrary particle shapes. However, in the context of experiments with active particles, the knowledge of the exact relations for the effective force and torque is usually not required as these quantities can directly be obtained from the measured trajectories [14].

3. Diffusiophoretic self-propulsion

3.1. Slender rigid particles

In contrast to mechanical microswimmers, self-diffusiophoretic microswimmers do not change their shapes in order to propel themselves forward. They instead move as a consequence of a local concentration gradient, which they generate themselves. The self-propulsion of such self-diffusiophoretic microswimmers can be modelled by prescribing a non-vanishing slip velocity on (parts of) the surface of the particle [73,74]. Interestingly, such a hydrodynamic modelling leads to equations of motion that are on a more coarse-grained model the same as the equations of motion of the corresponding passive particles complemented by effective forces and torques to model the self-propulsion. In the following, we prove this for biaxial particles with two arms of different lengths, which meet at a certain internal angle, through a hydrodynamic calculation based on slender-body theory for Stokes flow [94,95].

Applications of slender-body theory include Purcell's three-link swimmer [84,96] as well as the modelling of flagellar locomotion in general [66,68]. In this section, we apply slender-body theory to a rigid slender particle with two arms of different lengths as sketched in figure 2. The two straight arms of lengths a and b enclose the internal angle δ with $0 < \delta < \pi$. The special case $\delta = \pi/2$ corresponds to an L-shaped particle similar to what has been studied, using a different approach, in section 2. The results of the slender-body approach for this special case have already briefly been presented in [86]. Here, we provide a more general and detailed derivation, which shows that the concept of effective forces and torques follows intrinsically from a self-propulsion mechanism that involves a prescribed slip velocity. As required by slender-body theory, the sum $L = a + b$ of the lengths of the two arms of the particle is assumed to be significantly larger than the width 2ϵ of the arms. Although this is not always the case in related experiments, the physical structure of the equations of motion governing the dynamics of asymmetric microswimmers is clearly elucidated by the slenderness approximation $\epsilon \ll L$.

In order to describe the translational and rotational motion of the biaxial microswimmer, we define the centreline position

$$\boldsymbol{x}(s) = \boldsymbol{r} - \boldsymbol{r}_{\mathrm{CM}} + s\boldsymbol{x}' \qquad \text{for } -b \leq s \leq a\,. \tag{6}$$

Here, s is the arc length coordinate, $\boldsymbol{r}$ is the particle's centre-of-mass position, the vector $\boldsymbol{r}_{\mathrm{CM}} = (a^2\hat{\boldsymbol{u}}_{\mathrm{a}} - b^2\hat{\boldsymbol{u}}_{\parallel})/(2L)$ is directed from the meeting point of the arms to the

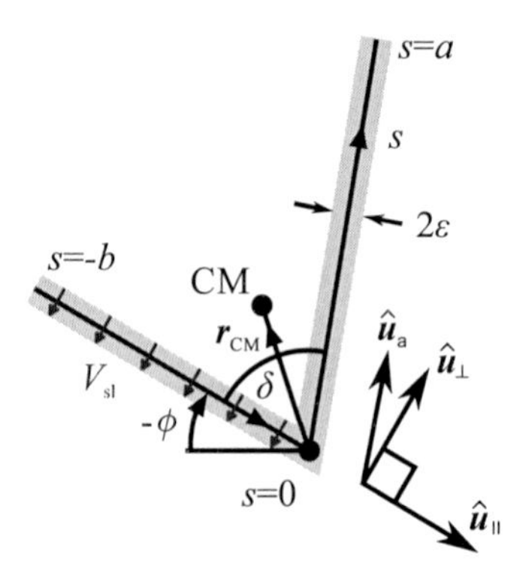

Figure 2. Sketch of a rigid biaxial particle with two straight arms of lengths a and b, the internal angle δ, the centre of mass CM, and a prescribed fluid slip velocity V_{sl} on the arm of length b. The particle-fixed rectangular coordinate system is indicated by the unit vectors $\hat{\boldsymbol{u}}_\parallel = (\cos(\phi), \sin(\phi))$ and $\hat{\boldsymbol{u}}_\perp = (-\sin(\phi), \cos(\phi))$, where $\hat{\boldsymbol{u}}_\parallel$ is parallel to the arm of length b, whereas the orientation of the arm of length a is denoted by the unit vector $\hat{\boldsymbol{u}}_{\mathrm{a}} = (-\cos(\delta - \phi), \sin(\delta - \phi))$.

centre of mass CM (see figure 2), and $\boldsymbol{x}' = \partial \boldsymbol{x}/\partial s$ is the unit tangent, which is $\hat{\boldsymbol{u}}_\parallel = (\cos(\phi), \sin(\phi))$ along the arm of length b $(-b \leq s \leq 0)$ and $\hat{\boldsymbol{u}}_{\mathrm{a}} = (-\cos(\delta-\phi), \sin(\delta-\phi))$ along the arm of length a $(0 \leq s \leq a)$. For slender arms with width $2\epsilon \ll L$, the fluid velocity at any point on the particle surface can be approximated by $\dot{\boldsymbol{x}}(s) + \boldsymbol{v}_{\mathrm{sl}}(s)$, where $\dot{\boldsymbol{x}}(s) = \partial \boldsymbol{x}/\partial t$ is the time derivative of $\boldsymbol{x}$ and $\boldsymbol{v}_{\mathrm{sl}}$ is the local slip velocity averaged over the perimeter of the surface adjacent to $\boldsymbol{x}(s)$. We consider a constant slip velocity $\boldsymbol{v}_{\mathrm{sl}} = -V_{\mathrm{sl}}\hat{\boldsymbol{u}}_\perp$ pointing in the direction $-\hat{\boldsymbol{u}}_\perp$ normal to the centreline along the arm of length b and a vanishing slip velocity along the other arm. The fluid velocity on the particle surface is related to the local force per unit length $\boldsymbol{f}(s)$ by the leading-order slender-body approximation

$$\dot{\boldsymbol{x}} + \boldsymbol{v}_{\mathrm{sl}} = c\,(\boldsymbol{\mathcal{I}} + \boldsymbol{x}' \otimes \boldsymbol{x}')\boldsymbol{f}\,, \tag{7}$$

where $c = \log(L/\epsilon)/(4\pi\eta)$ is a constant depending on the viscosity η of the solvent, $\boldsymbol{\mathcal{I}}$ is the identity matrix, and $\otimes$ denotes the dyadic product. This approximation is valid as long as δ is not too small so that lubrication effects in the corner between the two arms are negligible. The total force on the particle is given by

$$\boldsymbol{F}_{\mathrm{ext}} = \int_{-b}^{a} \boldsymbol{f}\,\mathrm{d}s \tag{8}$$

and the total torque on the particle is

$$M_{\mathrm{ext}} = \hat{\boldsymbol{e}}_z \cdot \int_{-b}^{a} (-\tilde{\boldsymbol{r}}_{\mathrm{CM}} + s\tilde{\boldsymbol{x}}') \times \tilde{\boldsymbol{f}}\,\mathrm{d}s\,. \tag{9}$$

To define the vector product $\times$ in (9), we technically add a third (zero) component to the vectors $\boldsymbol{r}_{\mathrm{CM}}$, $\boldsymbol{x}'$, and $\boldsymbol{f}$, which is indicated by the tilde. For force-free and torque-free microswimmers, $\boldsymbol{F}_{\mathrm{ext}}$ and M_{ext} are zero, whereas they are non-zero for externally driven particles.

The integral constraints (8) and (9) lead to a system of three dynamic equations for $\hat{\boldsymbol{u}}_{\parallel}\cdot\boldsymbol{r}$, $\hat{\boldsymbol{u}}_{\mathrm{a}}\cdot\boldsymbol{r}$, and ϕ. We begin by differentiating (6) with respect to time and multiplying with $\hat{\boldsymbol{u}}_{\parallel}$ and $\hat{\boldsymbol{u}}_{\mathrm{a}}$, respectively. This leads to

$$\hat{\boldsymbol{u}}_{\parallel}\cdot\dot{\boldsymbol{r}} + \frac{a^2\sin(\delta)}{2L}\dot{\phi} = \begin{cases} 2c\hat{\boldsymbol{u}}_{\parallel}\cdot\boldsymbol{f}\,, & \text{if } -b \leq s \leq 0\,, \\ c(\hat{\boldsymbol{u}}_{\parallel}\cdot\boldsymbol{f} - \cos(\delta)\hat{\boldsymbol{u}}_{\mathrm{a}}\cdot\boldsymbol{f}) + s\sin(\delta)\dot{\phi}\,, & \text{if } 0 < s \leq a \end{cases} \tag{10}$$

and

$$\hat{\boldsymbol{u}}_{\mathrm{a}}\cdot\dot{\boldsymbol{r}} + \frac{b^2\sin(\delta)}{2L}\dot{\phi} = \begin{cases} c(\hat{\boldsymbol{u}}_{\mathrm{a}}\cdot\boldsymbol{f} - \cos(\delta)\hat{\boldsymbol{u}}_{\parallel}\cdot\boldsymbol{f}) - s\sin(\delta)\dot{\phi} + V_{\mathrm{sl}}\sin(\delta)\,, & \text{if } -b \leq s \leq 0\,, \\ 2c\hat{\boldsymbol{u}}_{\mathrm{a}}\cdot\boldsymbol{f}\,, & \text{if } 0 < s \leq a\,. \end{cases} \tag{11}$$

In order to eliminate the unknown $\boldsymbol{f}$ from (10) and (11), the equations are integrated over s, separately from $-b$ to 0 and from 0 to a. After that, (8) and (9) are used to complete the system of differential equations. To apply the torque condition (9), equations (10) and (11) are first multiplied by s before integrating over the two sections. Thus, one obtains eight equations altogether. Taking suitable linear combinations of these equations finally leads to the dynamic equations

$$\eta\boldsymbol{\mathcal{H}}\begin{pmatrix} \hat{\boldsymbol{u}}_{\parallel}\cdot\dot{\mathbf{r}} \\ \hat{\boldsymbol{u}}_{\mathrm{a}}\cdot\dot{\mathbf{r}} \\ \dot{\phi} \end{pmatrix} = \begin{pmatrix} \hat{\boldsymbol{u}}_{\parallel}\cdot\boldsymbol{F}_{\mathrm{ext}} \\ \hat{\boldsymbol{u}}_{\mathrm{a}}\cdot\boldsymbol{F}_{\mathrm{ext}} \\ M_{\mathrm{ext}} \end{pmatrix} + V_{\mathrm{sl}}\begin{pmatrix} 0 \\ b\sin(\delta)/c \\ ab(a\cos(\delta)-b)/(2cL) \end{pmatrix} \tag{12}$$

with the grand resistance matrix

$$\boldsymbol{\mathcal{H}} = \frac{1}{2c\eta}\begin{pmatrix} 2a+b & a\cos(\delta) & \frac{ab(b\cos(\delta)-a)\sin(\delta)}{2L} \\ b\cos(\delta) & a+2b & \frac{ab(a\cos(\delta)-b)\sin(\delta)}{2L} \\ \frac{-a^2b\sin(\delta)}{2L} & \frac{-ab^2\sin(\delta)}{2L} & A \end{pmatrix} \tag{13}$$

and the constant

$$A = (a^3+b^3)\left(\frac{2}{3} - \frac{ab\sin^2(\delta)}{4L^2}\right) - \frac{a^4+b^4+2a^2b^2\cos(\delta)}{2L}\,. \tag{14}$$

These equations of motion describe the microswimmer in the non-rectangular particle's frame of reference defined by the unit vectors $\hat{\boldsymbol{u}}_{\parallel}$ and $\hat{\boldsymbol{u}}_{\mathrm{a}}$ and can be interpreted as follows. For force-free and torque-free biaxial swimmers driven by a fluid slip velocity V_{sl} along one of the arms (see figure 2) the first vector on the right-hand side of (12) vanishes. For passive particles driven by an external force $\boldsymbol{F}_{\mathrm{ext}}$ and torque M_{ext} the second vector on the right-hand side of (12) vanishes. With regard to the equations of motion the two cases are equivalent if $\hat{\boldsymbol{u}}_{\parallel}\cdot\boldsymbol{F}_{\mathrm{ext}} = 0$, $\hat{\boldsymbol{u}}_{\mathrm{a}}\cdot\boldsymbol{F}_{\mathrm{ext}} = V_{\mathrm{sl}}b\sin(\delta)/c$, and $M_{\mathrm{ext}} = V_{\mathrm{sl}}ab(a\cos(\delta)-b)/(2cL)$, although the respective hydrodynamic flow fields are clearly different [97,98]. This means that the motion of the self-propelled particle with slip velocity V_{sl} can be described using an effective force and an effective torque in combination with the same grand resistance matrix $\boldsymbol{\mathcal{H}}$ as for a corresponding passive particle.

As an additional result, we remark that the sense of rotation of the force-free and torque-free self-propelled particle switches when $b = a\cos(\delta)$. For a specific particle shape with the critical angle $\delta_{\text{crit}} = \arccos(b/a)$, the particle translates without rotation. For $0 < \delta < \delta_{\text{crit}}$, i.e. if the orthogonal projection of the arm with length a onto the arm with length b is longer than the latter arm, the particle rotates counter-clockwise, whereas for $\delta_{\text{crit}} < \delta < \pi$ the particle rotates clockwise.

We now consider the special case of an L-shaped particle, i.e. $\delta = \pi/2$. In this special case the unit vectors $\hat{\boldsymbol{u}}_{\perp}$ and $\hat{\boldsymbol{u}}_{\text{a}}$ are equal and the grand resistance matrix $\boldsymbol{\mathcal{H}}$ in (12) becomes [86]

$$\boldsymbol{\mathcal{H}}_{\text{L}} = \frac{1}{2c\eta}\begin{pmatrix} 2a+b & 0 & -a^2b/(2L) \\ 0 & a+2b & -ab^2/(2L) \\ -a^2b/(2L) & -ab^2/(2L) & A_{\text{L}} \end{pmatrix} \tag{15}$$

with the constant

$$A_{\text{L}} = (a^3+b^3)\left(\frac{2}{3} - \frac{ab}{4L^2}\right) - \frac{a^4+b^4}{2L}\,. \tag{16}$$

Thus, the equations of motion of a force-free and torque-free L-shaped diffusiophoretic microswimmer read

$$\eta\boldsymbol{\mathcal{H}}_{\text{L}}\begin{pmatrix} \hat{\boldsymbol{u}}_{\parallel}\cdot\dot{\mathbf{r}} \\ \hat{\boldsymbol{u}}_{\perp}\cdot\dot{\mathbf{r}} \\ \dot{\phi} \end{pmatrix} = V_{\text{sl}}\begin{pmatrix} 0 \\ b/c \\ -ab^2/(2cL) \end{pmatrix} \tag{17}$$

in the particle frame. The grand resistance matrix $\boldsymbol{\mathcal{H}}_{\text{L}}$ in (17) can be expressed in terms of the generalized diffusion tensor

$$\boldsymbol{\mathcal{D}}_0 = \frac{1}{\beta\eta}\boldsymbol{\mathcal{H}}_{\text{L}}^{-1} = \begin{pmatrix} D_{\parallel} & D_{\parallel}^{\perp} & D_{\text{C}}^{\parallel} \\ D_{\parallel}^{\perp} & D_{\perp} & D_{\text{C}}^{\perp} \\ D_{\text{C}}^{\parallel} & D_{\text{C}}^{\perp} & D_{\text{R}} \end{pmatrix}. \tag{18}$$

In order to transform the equations of motion (17) from the particle's frame of reference to the laboratory frame, we use the rotation matrix

$$\boldsymbol{\mathcal{M}}(\phi) = \begin{pmatrix} \cos\phi & -\sin\phi & 0 \\ \sin\phi & \cos\phi & 0 \\ 0 & 0 & 1 \end{pmatrix}. \tag{19}$$

With this definition, the generalized diffusion tensor $\boldsymbol{\mathcal{D}}(\phi)$ in the laboratory frame of reference is given by

$$\boldsymbol{\mathcal{D}}(\phi) = \boldsymbol{\mathcal{M}}(\phi)\boldsymbol{\mathcal{D}}_0\boldsymbol{\mathcal{M}}^{-1}(\phi)\,. \tag{20}$$

Using the orientation vectors $\hat{\boldsymbol{u}}_{\parallel}$ and $\hat{\boldsymbol{u}}_{\perp}$, it can also be written as

$$\boldsymbol{\mathcal{D}}(\phi) = \begin{pmatrix} \boldsymbol{\mathcal{D}}_{\text{T}}(\phi) & \boldsymbol{D}_{\text{C}}(\phi) \\ \boldsymbol{D}_{\text{C}}^{\intercal}(\phi) & D_{\text{R}} \end{pmatrix} \tag{21}$$

with the translational short-time diffusion tensor $\boldsymbol{\mathcal{D}}_{\text{T}}(\phi) = D_{\parallel}\hat{\boldsymbol{u}}_{\parallel}\otimes\hat{\boldsymbol{u}}_{\parallel} + D_{\parallel}^{\perp}(\hat{\boldsymbol{u}}_{\parallel}\otimes\hat{\boldsymbol{u}}_{\perp} + \hat{\boldsymbol{u}}_{\perp}\otimes\hat{\boldsymbol{u}}_{\parallel}) + D_{\perp}\hat{\boldsymbol{u}}_{\perp}\otimes\hat{\boldsymbol{u}}_{\perp}$, the translational-rotational coupling vector $\boldsymbol{D}_{\text{C}}(\phi) =$

$D_{\mathrm{C}}^{\parallel}\hat{\boldsymbol{u}}_{\parallel} + D_{\mathrm{C}}^{\perp}\hat{\boldsymbol{u}}_{\perp}$ and its transpose $\boldsymbol{D}_{\mathrm{C}}^{\intercal}(\phi)$, and the rotational diffusion coefficient D_{R}. Thus, one finally obtains the equations of motion for an L-shaped self-propelled particle

$$\begin{aligned} \dot{\boldsymbol{r}} &= \beta F[\boldsymbol{\mathcal{D}}_{\mathrm{T}}(\phi)\hat{\boldsymbol{u}}_{\perp} + l\boldsymbol{D}_{\mathrm{C}}(\phi)] \,, \\ \dot{\phi} &= \beta F[l D_{\mathrm{R}} + \boldsymbol{D}_{\mathrm{C}}(\phi)\cdot\hat{\boldsymbol{u}}_{\perp}] \,, \end{aligned} \tag{22}$$

where $F = bV_{\mathrm{sl}}/c$ is the effective self-propulsion force and $l = -ab/(2L)$ is an effective lever arm [12].

The above hydrodynamic calculation shows that also for a self-diffusiophoretic microswimmer the concept of effective forces and torques can be applied so that its equations of motion are obtained directly and much more easily from the equations of motion of the corresponding passive particle by introducing effective forces and torques that model the self-propulsion.

3.2. General rigid particles

On top of the previous derivation for an L-shaped particle based on slender body theory, we now describe how the effective forces and torques can in principle be calculated for an arbitrarily shaped particle propelled by self-diffusiophoresis. The procedure is based on the Lorentz reciprocal theorem for low-Reynolds-number hydrodynamics [39]. It was similarly applied in [99] in order to relate the translational and rotational velocities of a swimming microorganism to its surface distortions [100,101]. Such surface deformations and a streaming of the cell surface have been suggested as the swimming mechanism of cyanobacteria [102], for example. The modelling is similar to diffusiophoresis, where usually a slip velocity at the particle surface is assumed in the theoretical description [73,74].

In order to calculate the effective forces and torques for an arbitrarily shaped rigid particle with a surface S and vectors $\boldsymbol{s}$ that define the points on S, we first study how the slip velocity $\boldsymbol{v}_{\mathrm{sl}}(\boldsymbol{s})$ on the surface affects the translational and rotational velocities $\boldsymbol{v}$ and $\boldsymbol{\omega}$ of the particle. On the one hand, we consider the velocity field $\boldsymbol{u}$ and the stress field $\boldsymbol{\sigma}$ around a rigid self-propelled particle that translates at velocity $\boldsymbol{v}$ due to some slip velocity $\boldsymbol{v}_{\mathrm{sl,t}}(\boldsymbol{s})$ on the particle surface S. On the other hand, we assume that $\tilde{\boldsymbol{u}}$ and $\tilde{\boldsymbol{\sigma}}$ are the velocity and stress fields around a corresponding passive particle that moves with velocity $\tilde{\boldsymbol{v}}$ due to an external force $\boldsymbol{F}_{\mathrm{ext}}$. The Lorentz reciprocal theorem states that

$$\int_S \boldsymbol{n}\cdot\tilde{\boldsymbol{\sigma}}\cdot\boldsymbol{u}\,\mathrm{d}S = \int_S \boldsymbol{n}\cdot\boldsymbol{\sigma}\cdot\tilde{\boldsymbol{u}}\,\mathrm{d}S \,, \tag{23}$$

where $\boldsymbol{n}$ is the unit outward normal to the surface S and $\boldsymbol{n}\cdot\boldsymbol{\sigma}$ is the stress exerted by the fluid on the surface. Note that the right-hand side of (23) vanishes because $\tilde{\boldsymbol{u}} = \tilde{\boldsymbol{v}}$ is constant on the surface of the passive particle in the second case and $\int_S \boldsymbol{n}\cdot\boldsymbol{\sigma}\,\mathrm{d}S = \boldsymbol{0}$ because the self-propelled particle in the first case is force-free. Setting $\boldsymbol{u} = \boldsymbol{v} + \boldsymbol{v}_{\mathrm{sl,t}}$ on the surface of the self-propelled particle, we obtain

$$\int_S \boldsymbol{n}\cdot\tilde{\boldsymbol{\sigma}}\cdot(\boldsymbol{v} + \boldsymbol{v}_{\mathrm{sl,t}})\,\mathrm{d}S = 0 \,. \tag{24}$$

This indicates that $\boldsymbol{v}$ is directly related to $\boldsymbol{v}_{\mathrm{sl,t}}$. What is required is the stress $\boldsymbol{n}\cdot\tilde{\boldsymbol{\sigma}}$ on the passive particle, which can in principle be calculated by solving the Stokes equations for the flow around the particle. Equation (24) can also be written as

$$\boldsymbol{F}_{\mathrm{ext}}\cdot\boldsymbol{v}+\int_S \boldsymbol{n}\cdot\tilde{\boldsymbol{\sigma}}\cdot\boldsymbol{v}_{\mathrm{sl,t}}\,\mathrm{d}S=0\,. \tag{25}$$

Using a similar application of the Lorentz reciprocal theorem for a rigid self-propelled particle that rotates at angular velocity $\boldsymbol{\omega}$ due to some slip velocity $\boldsymbol{v}_{\mathrm{sl,r}}(\boldsymbol{s})$ on the particle surface S and a corresponding passive particle that rotates with angular velocity $\tilde{\boldsymbol{\omega}}$ due to an external torque $\boldsymbol{M}_{\mathrm{ext}}$, one can derive the relation

$$\boldsymbol{M}_{\mathrm{ext}}\cdot\boldsymbol{\omega}+\int_S \boldsymbol{n}\cdot\hat{\boldsymbol{\sigma}}\cdot\boldsymbol{v}_{\mathrm{sl,r}}\,\mathrm{d}S=0\,, \tag{26}$$

where $\boldsymbol{n}\cdot\hat{\boldsymbol{\sigma}}$ is the stress exerted by the fluid on the particle surface. By decomposing the slip velocity $\boldsymbol{v}_{\mathrm{sl}}$ into the parts that translate and rotate the self-propelled particle, $\boldsymbol{v}_{\mathrm{sl}}=\boldsymbol{v}_{\mathrm{sl,t}}+\boldsymbol{v}_{\mathrm{sl,r}}$, one could in principle predict the velocities $\boldsymbol{v}$ and $\boldsymbol{\omega}$. Although (25) is only a one-component equation, all components of $\boldsymbol{v}$ are accessible because different realizations of the external force $\boldsymbol{F}_{\mathrm{ext}}$ and the corresponding stress fields $\tilde{\boldsymbol{\sigma}}$ can be inserted. Analogously, $\boldsymbol{\omega}$ can be obtained from (26). Once the translational and rotational velocities are known, the effective force $\boldsymbol{F}$ can be obtained by integrating the stress $\tilde{\boldsymbol{\sigma}}\cdot\boldsymbol{n}$ for the special case $\tilde{\boldsymbol{v}}=\boldsymbol{v}$ over the particle surface S. Correspondingly, the effective torque $\boldsymbol{M}$ is the integral of $(\boldsymbol{s}-\boldsymbol{r}_0)\times(\hat{\boldsymbol{\sigma}}\cdot\boldsymbol{n})$ for the special case $\tilde{\boldsymbol{\omega}}=\boldsymbol{\omega}$ over the particle surface S, where $\boldsymbol{r}_0$ is the reference point for the torque (e.g. the centre of mass of the particle). Following this procedure, the effective forces and torques for an arbitrarily shaped self-propelled particle can be calculated in principle.

4. Considerations for general self-propulsion

Based on the previous explicit investigations of active particles with either mechanical or diffusiophoretic self-propulsion, here we briefly present some general considerations about effective forces and torques on a rigid self-propelled particle. For these general considerations, the particle shape and the origin of the self-propulsion do not matter.

Assume that a rigid self-propelled particle with surface S moves with translational velocity $\boldsymbol{v}$ and angular velocity $\boldsymbol{\omega}$. The only relevant condition is that the net force and the net torque vanish as required for swimmers [84]. On the other hand, we consider a passive rigid particle with the same surface S but without self-propulsion. By applying a suitably chosen external force $\boldsymbol{F}_{\mathrm{ext}}$ and torque $\boldsymbol{M}_{\mathrm{ext}}$ one can drive the passive particle so that it moves with exactly the same translational and rotational velocities as the self-propelled particle. This is possible because $\boldsymbol{F}_{\mathrm{ext}}$ and $\boldsymbol{M}_{\mathrm{ext}}$ together have six independent components, which is sufficient for controlling the six degrees of freedom of the rigid particle. Thus, the motion of a force-free and torque-free self-propelled particle is equivalent to the motion of a passive particle with the same shape that is driven by an appropriate effective force $\boldsymbol{F}=\boldsymbol{F}_{\mathrm{ext}}$ and torque $\boldsymbol{M}=\boldsymbol{M}_{\mathrm{ext}}$. For run-and-tumble particles [103–105] the orientations of the effective force and torque are constant in the particle frame, whereas for active Brownian particles [18, 56, 104] also their magnitudes are constant in the particle frame.

Due to the linearity of Stokes flow, the generalized effective force and torque vector $\boldsymbol{K} = (\boldsymbol{F}, \boldsymbol{M})$ can be written in the form $\eta\boldsymbol{\mathcal{H}}\boldsymbol{V}$ with the generalized velocity vector $\boldsymbol{V} = (\boldsymbol{v}, \boldsymbol{\omega})$ and a 6×6-dimensional grand resistance matrix $\boldsymbol{\mathcal{H}}$, which depends on the shape of the surface S. Thus, the validity of the concept of effective forces and torques is not affected by the specific type of self-propulsion of active particles. However, in order to relate these effective quantities directly to the physical process responsible for the self-propulsion, the details of the respective propulsion mechanism have to be known. The detailed calculations in sections 2 and 3 illustrate how the concept can be explicitly applied to specific situations.

5. Modifications and limitations of the concept of effective forces and torques

In the following, we address the questions when the concept of effective forces and torques has to be modified in order to still provide a valid theoretical model and when it finally reaches its limitations.

While we focused on particles with constant propulsion in this article, the concept can also be transferred to single active Brownian particles with time-dependent self-propulsion [55] in an unbounded fluid. This is particularly relevant in the context of run-and-tumble particles [60, 103] or when the swimming stroke itself results in variations of the propulsion speed.

The concept is also very useful if self-propelled particles in additional external fields such as gravity [14] are considered. All external body forces that do not affect the self-propulsion itself can easily be included in the generalized force vector $\boldsymbol{K}$. However, if the intrinsic propulsion mechanism of an active particle is disturbed by the external field, the concept of effective forces and torques reaches its limitations. As an example, we refer to bimetallic nanorods driven by electrophoresis [106, 107] in an external electric field.

The situation becomes more complicated when the solvent flow field which is generated by the self-propelled particles governs the particle dynamics and has to be taken into account. The far-field behaviour of this solvent flow field is different from that for a passive particle exposed to a body force. While the latter corresponds to a force monopole, the former is a force dipole such that pusher- and puller-like swimmers can be distinguished. While the details of the solvent velocity field do not play any role for a single particle in an unbounded fluid, they affect the motion of a particle near system boundaries [108–110] and at high concentrations. In particular hydrodynamic interactions between different swimmers will depend on these solvent flow fields [111–114]. Therefore the concept of effective forces and torques cannot straightforwardly be applied in these situations. How exactly the effective forces and torques would change near walls or other particles is a matter of future research.

6. Conclusions

We have shown that effective forces and torques can be used to model the self-propulsion of microswimmers and that this concept is an appropriate and consistent theoretical framework to describe the dynamics of anisotropic active particles and to understand related experimental results. We have provided general arguments as well as specific examples for the concept of effective forces and torques. In particular, we have presented a fine-grained derivation for an L-shaped mechanical microswimmer and an explicit hydrodynamic derivation for a biaxial diffusiophoretic microswimmer. Although the detailed processes responsible for the self-propulsion of some artificial microswimmers are not fully understood yet [115], our general theoretical approach constitutes a powerful tool to describe the dynamics of self-propelled particles of arbitrary shape and turns out to be in good agreement with experimental observations [12, 14].

Acknowledgments

This work was supported by the Deutsche Forschungsgemeinschaft (DFG) through the priority programme SPP 1726 on microswimmers under contracts BE 1788/13-1 and LO 418/17-1, by the Marie Curie-Initial Training Network Comploids funded by the European Union Seventh Framework Program (FP7), by the ERC Advanced Grant INTERCOCOS (Grant No. 267499), and by EPSRC (Grant No. EP/J007404). RW gratefully acknowledges financial support through a Postdoctoral Research Fellowship (WI 4170/1-2) from the DFG.

References

[1] Hansen J-P and McDonald I R 2006 *Theory of Simple Liquids* 3rd edn (London: Academic).

[2] Barrat J-L and Hansen J-P 2003 *Basic Concepts for Simple and Complex Liquids* (Cambridge: Cambridge University Press).

[3] Cates M E 2012 Diffusive transport without detailed balance in motile bacteria: does microbiology need statistical physics? *Rep. Prog. Phys.* **75** 042601.

[4] Marchetti M C, Joanny J F, Ramaswamy S, Liverpool T B, Prost J, Rao M, and Simha R A 2013 Hydrodynamics of soft active matter *Rev. Mod. Phys.* **85** 1143–89.

[5] Wensink H H, Dunkel J, Heidenreich S, Drescher K, Goldstein R E, Löwen H, and Yeomans J M 2012 Meso-scale turbulence in living fluids *Proc. Natl Acad. Sci. USA* **109** 14308–13.

[6] Schwarz-Linek J, Valeriani C, Cacciuto A, Cates M E, Marenduzzo D, Morozov A N, and Poon W C K 2012 Phase separation and rotor self-assembly in active particle suspensions *Proc. Natl Acad. Sci. USA* **109** 4052–7.

[7] Palacci J, Cottin-Bizonne C, Ybert C, and Bocquet L 2010 Sedimentation and effective temperature of active colloidal suspensions *Phys. Rev. Lett.* **105** 088304.

[8] Volpe G, Buttinoni I, Vogt D, Kümmerer H-J, and C Bechinger 2011 Microswimmers in patterned environments *Soft Matter* **7** 8810–5.

[9] Theurkauff I, Cottin-Bizonne C, Palacci J, Ybert C, and Bocquet L 2012 Dynamic clustering in active colloidal suspensions with chemical signaling *Phys. Rev. Lett.* **108** 268303.

[10] Buttinoni I, Volpe G, Kümmel F, Volpe G, and Bechinger C 2012 Active Brownian motion tunable by light *J. Phys.: Condens. Matter* **24** 284129.

[11] Buttinoni I, Bialké J, Kümmel F, Löwen H, Bechinger C, and Speck T 2013 Dynamical clustering and phase separation in suspensions of self-propelled colloidal particles *Phys. Rev. Lett.* **110** 238301.

[12] Kümmel F, ten Hagen B, Wittkowski R, Buttinoni I, Eichhorn R, Volpe G, Löwen H, and Bechinger C 2013 Circular motion of asymmetric self-propelling particles *Phys. Rev. Lett.* **110** 198302.

[13] Palacci J, Sacanna S, Preska Steinberg A, Pine D J, and Chaikin P M 2013 Living crystals of light-activated colloidal surfers *Science* **339** 936–40.

[14] ten Hagen B, Kümmel F, Wittkowski R, Takagi D, Löwen H, and Bechinger C 2014 Gravitaxis of asymmetric self-propelled colloidal particles *Nat. Commun.* **5** 4829.

[15] Takatori S C, Yan W, and Brady J F 2014 Swim pressure: stress generation in active matter *Phys. Rev. Lett.* **113** 028103.

[16] van Teeffelen S and Löwen H 2008 Dynamics of a Brownian circle swimmer *Phys. Rev. E* **78** 020101(R).

[17] ten Hagen B, Wittkowski R, and Löwen H 2011 Brownian dynamics of a self-propelled particle in shear flow *Phys. Rev. E* **84** 031105.

[18] Wittkowski R and Löwen H 2012 Self-propelled Brownian spinning top: dynamics of a biaxial swimmer at low Reynolds numbers *Phys. Rev. E* **85** 021406.

[19] Elgeti J and Gompper G 2013 Wall accumulation of self-propelled spheres *Europhys. Lett.* **101** 48003.

[20] Chen H-Y and Leung K-t 2006 Rotating states of self-propelling particles in two dimensions *Phys. Rev. E* **73** 056107.

[21] Li Y-X, Lukeman R, and Edelstein-Keshet L 2008 Minimal mechanisms for school formation in self-propelled particles *Physica D* **237** 699–720.

[22] Wensink H H and Löwen H 2012 Emergent states in dense systems of active rods: from swarming to turbulence *J. Phys.: Condens. Matter* **24** 464130.

[23] Peruani F, Deutsch A, and Bär M 2006 Nonequilibrium clustering of self-propelled rods *Phys. Rev. E* **74** 030904(R).

[24] Mehandia V and Nott P R 2008 The collective dynamics of self-propelled particles *J. Fluid Mech.* **595** 239–64.

[25] Wensink H H and Löwen H 2008 Aggregation of self-propelled colloidal rods near confining walls *Phys. Rev. E* **78** 031409.

[26] McCandlish S R, Baskaran A, and Hagan M F 2012 Spontaneous segregation of self-propelled particles with different motilities *Soft Matter* **8** 2527–34.

[27] Bialké J, Speck T, and Löwen H 2012 Crystallization in a dense suspension of self-propelled particles *Phys. Rev. Lett.* **108** 168301.

[28] Redner G S, Hagan M F, and Baskaran A 2013 Structure and dynamics of a phase-separating active colloidal fluid *Phys. Rev. Lett.* **110** 055701.

[29] Fily Y, Henkes S, and Marchetti M C 2014 Freezing and phase separation of self-propelled disks *Soft Matter* **10** 2132–40.

[30] Angelani L, Costanzo A, and Di Leonardo R 2011 Active ratchets *Europhys. Lett.* **96** 68002.

[31] Reichhardt C and Olson Reichhardt C J 2013 Active matter ratchets with an external drift *Phys. Rev. E* **88** 062310.

[32] Yang W, Misko V R, Nelissen K, Kong M, and Peeters F M 2012 Using self-driven microswimmers for particle separation *Soft Matter* **8** 5175–9.

[33] Costanzo A, Elgeti J, Auth T, Gompper G, and Ripoll M 2014 Motility-sorting of self-propelled particles in microchannels *Europhys. Lett.* **107** 36003.

[34] Kaiser A, Wensink H H, and Löwen H 2012 How to capture active particles *Phys. Rev. Lett.* **108** 268307.

[35] Kaiser A, Popowa K, Wensink H H, and Löwen H 2013 Capturing self-propelled particles in a moving microwedge *Phys. Rev. E* **88** 022311.

[36] Wang Z, Chen H-Y, Sheng Y-J, and Tsao H-K 2014 Diffusion, sedimentation equilibrium, and harmonic trapping of run-and-tumble nanoswimmers *Soft Matter* **10** 3209–17.

[37] Howse J R, Jones R A L, Ryan A J, Gough T, Vafabakhsh R, and Golestanian R 2007 Self-motile colloidal particles: from directed propulsion to random walk *Phys. Rev. Lett.* **99** 048102.

[38] ten Hagen B, van Teeffelen S, and Löwen H 2011 Brownian motion of a self-propelled particle *J. Phys.: Condens. Matter* **23** 194119.

[39] Happel J and Brenner H 1991 *Low Reynolds Number Hydrodynamics: With Special Applications to Particulate Media* (*Mechanics of Fluids and Transport Processes* vol 1) 2nd edn (Dordrecht: Kluwer).

[40] Kraft D J, Wittkowski R, ten Hagen B, Edmond K V, Pine D J, and Löwen H 2013 Brownian motion and the hydrodynamic friction tensor for colloidal particles of complex shape *Phys. Rev. E* **88** 050301(R).

[41] van Teeffelen S, Zimmermann U, and Löwen H 2009 Clockwise-directional circle swimmer moves counter-clockwise in Petri dish- and ring-like confinements *Soft Matter* **5** 4510–9.

[42] Cēbers A 2011 Diffusion of magnetotactic bacterium in rotating magnetic field *J. Magn. Magn. Mater.* **323** 279–82.

[43] Großmann R, Schimansky-Geier L, and Romanczuk P 2012 Active Brownian particles with velocity-alignment and active fluctuations *New J. Phys.* **14** 073033.

[44] Radtke P K and Schimansky-Geier L 2012 Directed transport of confined Brownian particles with torque *Phys. Rev. E* **85** 051110.

[45] Si T 2012 The equation of motion for the trajectory of a circling particle with an overdamped circle center *Physica A.* **391** 3054–60.

[46] Ganguly C and Chaudhuri D 2013 Stochastic thermodynamics of active Brownian particles *Phys. Rev. E* **88** 032102.

[47] Llopis I, Pagonabarraga I, Cosentino Lagomarsino M, and Lowe C P 2013 Cooperative motion of intrinsic and actuated semiflexible swimmers *Phys. Rev. E* **87** 032720.

[48] Marine N A, Wheat P M, Ault J, and Posner J D 2013 Diffusive behaviors of circle-swimming motors *Phys. Rev. E* **87** 052305.

[49] Takagi D, Braunschweig A B, Zhang J, and Shelley M J 2013 Dispersion of self-propelled rods undergoing fluctuation-driven flips *Phys. Rev. Lett.* **110** 038301.

[50] Nourhani A, Lammert P E, Borhan A, and Crespi V H 2013 Chiral diffusion of rotary nanomotors *Phys. Rev. E* **87** 050301(R).

[51] Kaiser A and Löwen H 2013 Vortex arrays as emergent collective phenomena for circle swimmers *Phys. Rev. E* **87** 032712.

[52] Mijalkov M and Volpe G 2013 Sorting of chiral microswimmers *Soft Matter* **9** 6376–81.

[53] Martinelli A 2014 Overdamped 2d Brownian motion for self-propelled and nonholonomic particles *J. Stat. Mech.* P03003.

[54] Mallory S A, Šarić A, Valeriani C, and Cacciuto A 2014 Anomalous thermomechanical properties of a self-propelled colloidal fluid *Phys. Rev. E* **89** 052303.

[55] Babel S, ten Hagen B, and Löwen H 2014 Swimming path statistics of an active Brownian particle with time-dependent self-propulsion *J. Stat. Mech.* P02011.

[56] Stenhammar J, Marenduzzo D, Allen R J, and Cates M E 2014 Phase behaviour of active Brownian particles: the role of dimensionality *Soft Matter* **10** 1489–99.

[57] Frymier P D, Ford R M, Berg H C, and Cummings P T 1995 Three-dimensional tracking of motile bacteria near a solid planar surface *Proc. Natl Acad. Sci. USA* **92** 6195–9.

[58] DiLuzio W R, Turner L, Mayer M, Garstecki P, Weibel D B, Berg H C, and Whitesides G M 2005 *Escherichia coli* swim on the right-hand side *Nature* **435** 1271–4.

[59] Sokolov A, Apodaca M M, Grzybowski B A, and Aranson I S 2010 Swimming bacteria power microscopic gears *Proc. Natl Acad. Sci. USA* **107** 969–74.

[60] Polin M, Tuval I, Drescher K, Gollub J P, and Goldstein R E 2009 *Chlamydomonas* swims with two 'gears' in a eukaryotic version of run-and-tumble locomotion *Science* **325** 487–90.

[61] Kantsler V, Dunkel J, Polin M, and Goldstein R E 2013 Ciliary contact interactions dominate surface scattering of swimming eukaryotes *Proc. Natl Acad. Sci. USA* **110** 1187–92.

[62] Riedel I H, Kruse K, and Howard J 2005 A self-organized vortex array of hydrodynamically entrained sperm cells *Science* **309** 300–3.

[63] Friedrich B M and Jülicher F 2008 The stochastic dance of circling sperm cells: sperm chemotaxis in the plane *New J. Phys.* **10** 123025.

[64] Elgeti J, Kaupp U B, and Gompper G 2010 Hydrodynamics of sperm cells near surfaces *Biophys. J.* **99** 1018–26.

[65] Friedrich B M, Riedel-Kruse I H, Howard J, and Jülicher F 2010 High-precision tracking of sperm swimming fine structure provides strong test of resistive force theory *J. Exp. Biol.* **213** 1226–34.

[66] Lighthill J 1976 Flagellar hydrodynamics *SIAM Rev.* **18** 161–230.

[67] Purcell E M 1997 The efficiency of propulsion by a rotating flagellum *Proc. Natl Acad. Sci. USA* **94** 11307–11.

[68] Lauga E and Powers T R 2009 The hydrodynamics of swimming microorganisms *Rep. Prog. Phys.* **72** 096601.

[69] Swan J W, Brady J F, Moore R S, and ChE 174 2011 Modeling hydrodynamic self-propulsion with Stokesian dynamics. Or teaching Stokesian dynamics to swim *Phys. Fluids* **23** 071901.

[70] Lobaskin V, Lobaskin D, and Kulić I M 2008 Brownian dynamics of a microswimmer *Eur. Phys. J. Spec. Top.* **157** 149–56.

[71] Wada H and Netz R R 2006 Non-equilibrium hydrodynamics of a rotating filament *Europhys. Lett.* **75** 645–51.

[72] Najafi A and Golestanian R 2004 Simple swimmer at low Reynolds number: three linked spheres *Phys. Rev. E* **69** 062901.

[73] Anderson J L 1989 Colloid transport by interfacial forces *Annu. Rev. Fluid Mech.* **21** 61–99.

[74] Ajdari A and Bocquet L 2006 Giant amplification of interfacially driven transport by hydrodynamic slip: diffusio-osmosis and beyond *Phys. Rev. Lett.* **96** 186102.

[75] Paxton W F, Sen A, and Mallouk T E 2005 Motility of catalytic nanoparticles through self-generated forces *Chem. Eur. J.* **11** 6462–70.

[76] Downton M T and Stark H 2009 Simulation of a model microswimmer *J. Phys.: Condens. Matter* **21** 204101.

[77] Götze I O and Gompper G 2010 Mesoscale simulations of hydrodynamic squirmer interactions *Phys. Rev. E* **82** 041921.

[78] Llopis I and Pagonabarraga I 2010 Hydrodynamic interactions in squirmer motion: swimming with a neighbour and close to a wall *J. Non-Newton. Fluid Mech.* **165** 946–52.

[79] Lighthill M J 1952 On the squirming motion of nearly spherical deformable bodies through liquids at very small Reynolds numbers *Commun. Pur. Appl. Math.* **5** 109–18.

[80] Ishikawa T, Simmonds M P, and Pedley T J 2006 Hydrodynamic interaction of two swimming model micro-organisms *J. Fluid Mech.* **568** 119.

[81] Ishikawa T and Pedley T J 2008 Coherent structures in monolayers of swimming particles *Phys. Rev. Lett.* **100** 088103.

[82] Alarcón F and Pagonabarraga I 2013 Spontaneous aggregation and global polar ordering in squirmer suspensions *J. Mol. Liq.* **185** 56–61.

[83] Matas-Navarro R, Golestanian R, Liverpool T B, and Fielding S M 2014 Hydrodynamic suppression of phase separation in active suspensions *Phys. Rev. E* **90** 032304.

[84] Purcell E M 1977 Life at low Reynolds number *Am. J. Phys.* **45** 3–11.

[85] Felderhof B U 2014 Comment on 'Circular motion of asymmetric self-propelling particles' *Phys. Rev. Lett.* **113** 029801.

[86] Kümmel F, ten Hagen B, Wittkowski R, Takagi D, Buttinoni I, Eichhorn R, Volpe G, Löwen H, and Bechinger C 2014 Reply to Comment on 'Circular motion of asymmetric self-propelling particles' *Phys. Rev. Lett.* **113** 029802.

[87] Córdova-Figueroa U M and Brady J F 2008 Osmotic propulsion: the osmotic motor *Phys. Rev. Lett.* **100** 158303.

[88] Jülicher F and Prost J 2009 Comment on 'Osmotic propulsion: the osmotic motor' *Phys. Rev. Lett.* **103** 079801.

[89] Golestanian R 2008 Three-sphere low-Reynolds-number swimmer with a cargo container *Eur. Phys. J. E* **25** 1–4.

[90] Wolff K, Hahn A M, and Stark H 2013 Sedimentation and polar order of active bottom-heavy particles *Eur. Phys. J. E* **36** 43.

[91] Campbell A I and Ebbens S J 2013 Gravitaxis in spherical Janus swimming devices *Langmuir* **29** 14066–73.

[92] Higdon J J L 1979 A hydrodynamic analysis of flagellar propulsion *J. Fluid Mech.* **90** 685–711.

[93] Kim S and Karrila S J 1991 *Microhydrodynamics: Principles and Selected Applications* (Boston: Butterworth-Heinemann).

[94] Batchelor G K 1970 Slender-body theory for particles of arbitrary cross-section in Stokes flow *J. Fluid Mech.* **44** 419–40.

[95] Cox R G 1970 The motion of long slender bodies in a viscous fluid part 1. General theory *J. Fluid Mech.* **44** 791–810.

[96] Becker L E, Koehler S A, and Stone H A 2003 On self-propulsion of micro-machines at low Reynolds number: Purcell's three-link swimmer *J. Fluid Mech.* **490** 15–35.

[97] Ramachandran S, Sunil Kumar P B, and Pagonabarraga I 2006 A Lattice-Boltzmann model for suspensions of self-propelling colloidal particles *Eur. Phys. J. E* **20** 151–8.

[98] Jülicher F and Prost J 2009 Generic theory of colloidal transport *Eur. Phys. J. E* **29** 27–36.

[99] Stone H A and Samuel A D T 1996 Propulsion of microorganisms by surface distortions *Phys. Rev. Lett.* **77** 4102–4.

[100] Gonzalez-Rodriguez D and Lauga E 2009 Reciprocal locomotion of dense swimmers in Stokes flow *J. Phys.: Condens. Matter* **21** 204103.

[101] Lauga E 2014 Locomotion in complex fluids: Integral theorems *Phys. Fluids* **26** 081902.

[102] Waterbury J B, Willey J M, Franks D G, Valois F W, and Watson S W 1985 A cyanobacterium capable of swimming motility *Science* **230** 74–6.

[103] Tailleur J and Cates M E 2008 Statistical mechanics of interacting run-and-tumble bacteria *Phys. Rev. Lett.* **100** 218103.

[104] Cates M E and Tailleur J 2013 When are active Brownian particles and run-and-tumble particles equivalent? Consequences for motility-induced phase separation *Europhys. Lett.* **101** 20010.

[105] Paoluzzi M, Di Leonardo R, and Angelani L 2013 Effective run-and-tumble dynamics of bacteria baths *J. Phys.: Condens. Matter* **25** 415102.

[106] Paxton W F, Kistler K C, Olmeda C C, Sen A, St. Angelo S K, Cao Y, Mallouk T E, Lammert P E, and Crespi V H 2004 Catalytic nanomotors: autonomous movement of striped nanorods *J. Am. Chem. Soc.* **126** 13424–31.

[107] Paxton W F, Baker P T, Kline T R, Wang Y, Mallouk T E, and Sen A 2006 Catalytically induced electrokinetics for motors and micropumps *J. Am. Chem. Soc.* **128** 14881–8.

[108] Kreuter C, Siems U, Nielaba P, Leiderer P, and Erbe A 2013 Transport phenomena and dynamics of externally and self-propelled colloids in confined geometry *Eur. Phys. J. Spec. Top.* **222** 2923–39.

[109] Takagi D, Palacci J, Braunschweig A B, Shelley M J, and Zhang J 2014 Hydrodynamic capture of microswimmers into sphere-bound orbits *Soft Matter* **10** 1784–9.

[110] Chilukuri S, Collins C H, and Underhill P T 2014 Impact of external flow on the dynamics of swimming microorganisms near surfaces *J. Phys.: Condens. Matter* **26** 115101.

[111] Kapral R 2008 Multiparticle collision dynamics: simulation of complex systems on mesoscales *Adv. Chem. Phys.* **140** 89–146.

[112] Gompper G, Ihle T, Kroll D M, and Winkler R G 2009 Multi-particle collision dynamics: a particle-based mesoscale simulation approach to the hydrodynamics of complex fluids *Adv. Polym. Sci.* **221** 1–87.

[113] Alexander G P, Pooley C M, and Yeomans J M 2009 Hydrodynamics of linked sphere model swimmers *J. Phys.: Condens. Matter* **21** 204108.

[114] Reigh S Y, Winkler R G, and Gompper G 2012 Synchronization and bundling of anchored bacterial flagella *Soft Matter* **8** 4363–72.

[115] Brown A and Poon W 2014 Ionic effects in self-propelled Pt-coated Janus swimmers *Soft Matter* **10** 4016–27.

Statement of the author: This article has been accepted for publication in *Journal of Physics: Condensed Matter*. It provides a detailed justification of the concept of effective forces and torques on a more general level than in the short Reply in reference [447].

I was mainly responsible for the paper and wrote substantial parts of it. The derivation in section 3 originates from Daisuke Takagi. The calculations were repeated by Raphael Wittkowski and me. I had the idea to use the example of the three-sphere-swimmer and performed the corresponding analysis. Hartmut Löwen established the contact with Daisuke Takagi and provided valuable input for both presented lines of reasoning. Felix Kümmel and Clemens Bechinger were involved in some general discussions about the concept of effective forces and torques.

CHAPTER 9

Gravitaxis of asymmetric self-propelled colloidal particles

The content of this chapter has been published in a similar form in *Nature Communications* **5**, 4829 (2014) by Borge ten Hagen, Felix Kümmel, Raphael Wittkowski, Daisuke Takagi, Hartmut Löwen, and Clemens Bechinger (see reference [220]).

Gravitaxis (known historically as geotaxis) describes the response of motile microorganisms to an external gravitational field and has been studied for several decades. In particular, negative gravitaxis, that is, a swimming motion opposed to a gravitational force $\mathbf{F}_{\mathrm{G}}$, is frequently observed for flagellates and ciliates such as *Chlamydomonas* [1] or *Paramecium* [2]. Since this ability enables microorganisms to counteract sedimentation, gravitaxis extends the range of their habitat and allows them to optimize their position in environments with spatial gradients [2]. In case of photosynthetic flagellates such as *Euglena gracilis*, gravitaxis (in addition to phototaxis) contributes to their vertical motion in water, which enables them to adjust the amount of exposure to solar radiation [3]. In order to achieve a gravitactic motion, in general, a stable orientation of the microorganism relative to the gravitational field is required. While in some organisms such an alignment is likely to be dominated by an active physiological mechanism [4, 5], its origin in other systems is still controversially discussed [1, 2, 6–13]. It has been suggested that gravitaxis can also result from purely passive effects like, for example, an inhomogeneous mass density within the organism (bottom-heaviness), which would lead to an alignment similar to that of a buoy in the ocean [14, 15]. (We use the term 'gravitaxis', which is not uniformly defined in the literature, also in the context of passive alignment effects.) Under such conditions, the organism's alignment should be the same regardless whether it sediments downward or upward in a hypo- or hyperdensity medium, respectively, as indeed observed for pluteus larvae [12]. However, corresponding experiments with immobilized *Paramecium* and gastrula larvae showed an alignment reversal for opposite sedimentation directions [12]. This suggests another mechanical alignment mechanism which is caused by the organism's fore-rear asymmetry [1, 2, 11, 12].

To explore the role of shape asymmetry on the gravitactic behaviour and to avoid the presence of additional physiological mechanisms, in our experiments, we study the motion of colloidal micron-sized swimmers with asymmetric shapes in a gravitational field. The self-propulsion mechanism is based on diffusiophoresis, where a local chemical gradient is induced in the solvent around the microswimmer [16]. It has been shown that the corresponding type of motion is similar to that of biological microorganisms [17–23]. On the basis of our experimental and theoretical results, we demonstrate that a shape anisotropy alone is sufficient to induce gravitactic motion with either upward or downward swimming. In addition to straight trajectories, also more complex swimming patterns are observed, where the swimmers perform a trochoid-like (that is, a generalized cycloid-like [24]) motion transversal to the direction of gravity.

Results

Experiments. For our experiments we use asymmetric L-shaped microswimmers with arm lengths of 9 and 6 μm, respectively, and 3 μm thickness that are obtained by soft lithography

[25, 26] (see Methods for details). To induce a self-diffusiophoretic motion, the particles are covered with a thin Au coating on the front side of the short arm, which leads to local heating upon laser illumination with intensity I (see Fig. 1d). When such particles are suspended in a binary mixture of water and 2,6-lutidine at critical composition, this heating causes a local demixing of the solvent that results in an intensity-dependent phoretic propulsion in the direction normal to the plane of the metal cap [26–28]. To restrict the particle's motion to two spatial dimensions, we use a sample cell with a height of 7 μm. Further details are provided in Methods. Variation of the gravitational force is achieved by mounting the sample cell on a microscope, which can be inclined by an angle α relative to the horizontal plane (see Fig. 1b).

Passive sedimentation. Figure 1a shows the measured orientational probability distribution $p(\phi)$ of a sedimenting passive L-shaped particle ($I = 0$) in a thin sample cell that was tilted by $\alpha = 10.67°$ relative to the horizontal plane. The data show a clear maximum at the orientation angle $\phi = -34°$, that is, the swimmer aligns slightly turned relative to the direction of gravity as schematically illustrated in Fig. 1b. A typical trajectory for such a sedimenting particle is plotted in Fig. 1c,1. To estimate the effect of the Au layer on the particle orientation, we also performed sedimentation experiments for non-coated particles and did not find measurable deviations. This suggests that the alignment cannot be attributed to an inhomogeneous mass distribution. The observed alignment with the *shorter* arm at the bottom (see Fig. 1b) is characteristic for sedimenting objects with homogeneous mass distribution and a fore-rear asymmetry [11]. This can easily be understood by considering an asymmetric dumbbell formed by two spheres with identical mass density but different radii R_1 and $R_2 > R_1$. The sedimentation speed of a single sphere with radius R due to gravitational and viscous forces is $v \propto R^2$. Therefore, if hydrodynamic interactions between the spheres are ignored, the dumbbell experiences a viscous torque resulting in an alignment where the bigger sphere is below the smaller one [11].

In Fig. 1c, we show the particle's centre-of-mass motion in the x-y plane as a consequence of the self-propulsion acting normally to the Au coating (see Fig. 1d). When the self-propulsion is sufficiently high to overcome sedimentation, the particle performs a rather rectilinear motion in upward direction, that is, against gravity (negative gravitaxis [11], see Fig. 1c,2-4). With increasing self-propulsion, that is, light intensity, the angle between the trajectory and the y axis increases until it exceeds 90°. For even higher intensities, interestingly, the particle performs an effective downward motion again (see Fig. 1c,5). It should be mentioned that in case of the above straight trajectories, the particle's orientation ϕ remains stable (apart from slight fluctuations) and shows a monotonic dependence on the illumination intensity (as will be discussed in more detail further below). Finally, for

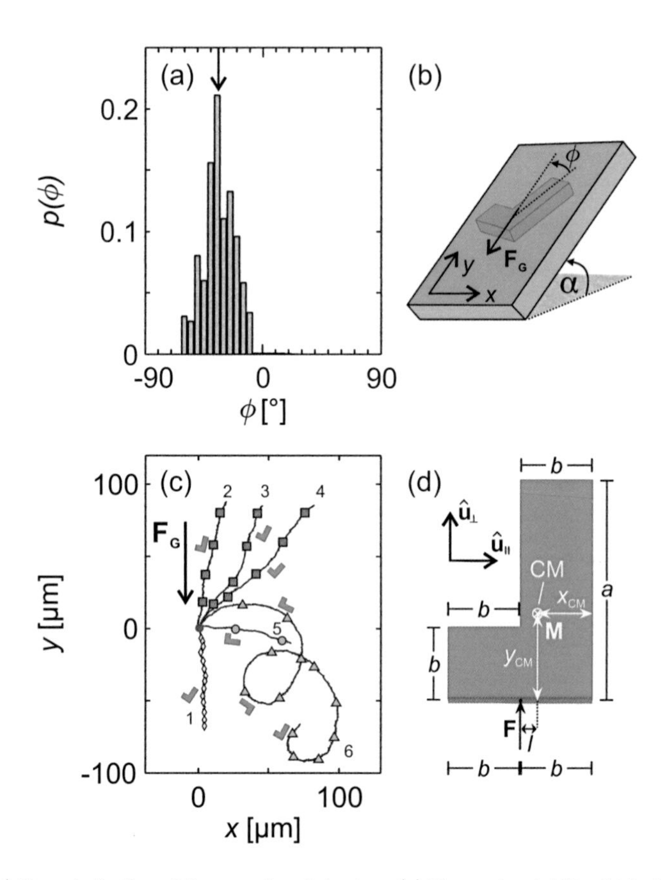

Figure 1 | Characterization of the experimental setup. (**a**) Measured probability distribution $p(\phi)$ of the orientation ϕ of a passive L-shaped particle during sedimentation for an inclination angle $\alpha = 10.67°$ (see sketch in **b**). (**c**) Experimental trajectories for the same inclination angle and increasing illumination intensity: (1) $I = 0$, (2-5) $0.6\ \mu\mathrm{W}\ \mu\mathrm{m}^{-2} < I < 4.8\ \mu\mathrm{W}\ \mu\mathrm{m}^{-2}$, (6) $I > 4.8\ \mu\mathrm{W}\ \mu\mathrm{m}^{-2}$. All trajectories start at the origin of the graph (red bullet). The particle positions after 1 min each are marked by yellow diamonds (passive straight downward trajectory), orange squares (straight upward trajectories), green bullets (active tilted straight downward trajectory), and blue triangles (trochoid-like trajectory). (**d**) Geometrical sketch of an ideal L-shaped particle (the Au coating is indicated by the yellow line) with dimensions $a = 9\ \mu\mathrm{m}$ and $b = 3\ \mu\mathrm{m}$ and coordinates $x_{\mathrm{CM}} = -2.25\ \mu\mathrm{m}$ and $y_{\mathrm{CM}} = 3.75\ \mu\mathrm{m}$ of the centre of mass (CM) (with the origin of coordinates in the bottom right corner of the particle and when the Au coating is neglected). The propulsion mechanism characterized by the effective force $\mathbf{F}$ and the effective torque $\mathbf{M}$ induces a rotation of the particle that depends on the length of the effective lever arm l. $\hat{\mathbf{u}}_{\parallel}(\phi)$ and $\hat{\mathbf{u}}_{\perp}(\phi)$ are particle-fixed unit vectors that denote the orientation of the particle (see Methods for details).

strong self-propulsion corresponding to high light intensity, the microswimmer performs a trochoid-like motion (see Fig. 1c,6).

To obtain a theoretical understanding of gravitaxis of asymmetric microswimmers, we first study the sedimentation of *passive* particles in a viscous solvent [29] based on the Langevin equations

$$\begin{aligned} \dot{\mathbf{r}} &= \beta \underline{\mathbf{D}}_{\mathrm{T}} \mathbf{F}_{\mathrm{G}} + \boldsymbol{\zeta}_{\mathbf{r}} \,, \\ \dot{\phi} &= \beta \mathbf{D}_{\mathrm{C}} \cdot \mathbf{F}_{\mathrm{G}} + \zeta_{\phi} \end{aligned} \tag{1}$$

for the time-dependent centre-of-mass position $\mathbf{r}(t) = (x(t), y(t))$ and orientation $\phi(t)$ of a particle. Here, the gravitational force $\mathbf{F}_{\mathrm{G}}$, the translational short-time diffusion tensor $\underline{\mathbf{D}}_{\mathrm{T}}$, the translational-rotational coupling vector $\mathbf{D}_{\mathrm{C}}$, the inverse effective thermal energy $\beta = 1/(k_{\mathrm{B}}T)$, and the Brownian noise terms $\boldsymbol{\zeta}_{\mathbf{r}}$ and ζ_{ϕ} are involved (see Methods for details).

Neglecting the stochastic contributions in equation (1), one obtains the stable long-time particle orientation angle

$$\phi_{\infty} \equiv \phi(t \to \infty) = -\arctan\left(\frac{D_{\mathrm{C}}^{\perp}}{D_{\mathrm{C}}^{\parallel}}\right), \tag{2}$$

which only depends on the two coupling coefficients $D_{\mathrm{C}}^{\parallel}$ and $D_{\mathrm{C}}^{\perp}$ determined by the particle's geometry.

Swimming patterns under gravity. Extending equation (1) to also account for the active motion of an asymmetric microswimmer yields (see Methods for a hydrodynamic derivation)

$$\dot{\mathbf{r}} = (P^{\star}/b)\big(\underline{\mathbf{D}}_{\mathrm{T}}\hat{\mathbf{u}}_{\perp} + l\mathbf{D}_{\mathrm{C}}\big) + \beta \underline{\mathbf{D}}_{\mathrm{T}} \mathbf{F}_{\mathrm{G}} + \boldsymbol{\zeta}_{\mathbf{r}} \,, \tag{3}$$

$$\dot{\phi} = (P^{\star}/b)\big(lD_{\mathrm{R}} + \mathbf{D}_{\mathrm{C}} \cdot \hat{\mathbf{u}}_{\perp}\big) + \beta \mathbf{D}_{\mathrm{C}} \cdot \mathbf{F}_{\mathrm{G}} + \zeta_{\phi} \,, \tag{4}$$

where the dimensionless number $P^{\star}$ is the strength of the self-propulsion, b is a characteristic length of the L-shaped particle (see Fig. 1d), l denotes an effective lever arm (see Fig. 1d), and D_{R} is the rotational diffusion coefficient of the particle. As shown in Methods, one can view $F = |\mathbf{F}| = P^{\star}/(b\beta)$ and $M = |\mathbf{M}| = Fl$ as an effective force (in $\hat{\mathbf{u}}_{\perp}$-direction) and an effective torque (perpendicular to the L-shaped particle), respectively, describing the self-propulsion of the particle (see Fig. 1d). This concept is in line with other theoretical work that has been presented recently [30–35]. The above equations of motion are fully compatible with the fact that, apart from gravity, a self-propelled swimmer is force-free and torque-free (see Methods).

The self-propulsion strength $P^{\star}$ is obtained from the experiments by measuring the particle velocity (see Methods for details). The rotational motion of the microswimmer depends on the detailed asymmetry of the particle shape and is characterized by the effective lever

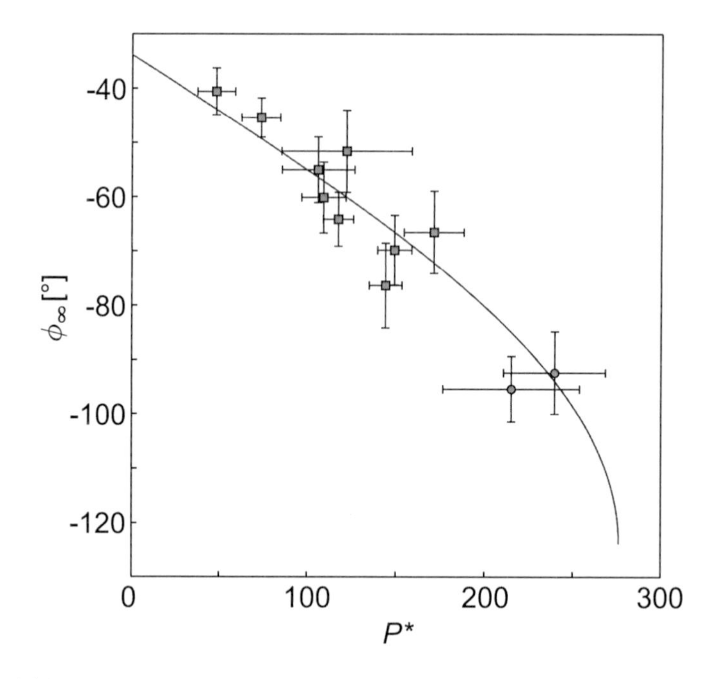

Figure 2 | Measured particle orientations. Long-time orientation ϕ_∞ of an L-shaped particle in the regime of straight motion as a function of the strength $P^\star$ of the self-propulsion for $\alpha = 10.67°$. The symbols with error bars representing the s.d. show the experimental data with the self-propulsion strength $P^\star$ determined using equation (29) and squares and bullets corresponding to upward and downward motion, respectively (see Fig. 1c). The solid curve is the theoretical prediction based on equation (5) with $D_{\mathrm{C}}^{\parallel}$ as fit parameter (see Methods for details).

arm l relative to the centre-of-mass position as the reference point (see Fig. 1d). Note that the choice of the reference point also changes the translational and coupling elements of the diffusion tensor. If the reference point does not coincide with the centre of mass of the particle, in the presence of a gravitational force an additional torque has to be considered in equation (4). In line with our experiments (see Fig. 1c), the noise-free asymptotic solutions of equations (3) and (4) are either straight (upward or downward) trajectories or periodic swimming paths. Up to a threshold value of $P^\star$, the effective torque originating from the self-propulsion (first term on the right-hand side of equation (4)) can be compensated by the gravitational torque (second term $\beta \mathbf{D}_{\mathrm{C}} \cdot \mathbf{F}_{\mathrm{G}}$ on the right-hand side of equation (4)) so that there is no net rotation and the trajectories are straight. In this case, after a transient

Table I | Experimentally determined values of the parameters used for the comparison with our theory.

translational diffusion coefficients	$D_{\parallel}$	$7.2 \times 10^{-3}\ \mu\mathrm{m}^2\,\mathrm{s}^{-1}$
	$D_{\perp}$	$8.1 \times 10^{-3}\ \mu\mathrm{m}^2\,\mathrm{s}^{-1}$
	$D_{\parallel}^{\perp}$	$0\ \mu\mathrm{m}^2\,\mathrm{s}^{-1}$
coupling coefficients	$D_{\mathrm{C}}^{\parallel}$	$5.7 \times 10^{-4}\ \mu\mathrm{m}\,\mathrm{s}^{-1}$
	$D_{\mathrm{C}}^{\perp}$	$3.8 \times 10^{-4}\ \mu\mathrm{m}\,\mathrm{s}^{-1}$
rotational diffusion coefficient	D_{R}	$6.2 \times 10^{-4}\ \mathrm{s}^{-1}$
effective lever arm	l	$-0.75\ \mu\mathrm{m}$
buoyant mass	m	$2.5 \times 10^{-14}\ \mathrm{kg}$

regime the particle orientation converges to

$$\phi_{\infty} = -\arctan\left(\frac{D_{\mathrm{C}}^{\perp}}{D_{\mathrm{C}}^{\parallel}}\right) + \arcsin\left(-\frac{P^{\star}}{\beta b F_{\mathrm{G}}}\frac{D_{\mathrm{C}}^{\perp} + l D_{\mathrm{R}}}{\sqrt{D_{\mathrm{C}}^{\parallel 2} + D_{\mathrm{C}}^{\perp 2}}}\right) \tag{5}$$

with the magnitude of the gravitational force $F_{\mathrm{G}} = |\mathbf{F}_{\mathrm{G}}|$. Obviously, ϕ_{∞} is a superposition of the passive case (first term on the right-hand side, cf. equation (2)) with a correction due to self-propulsion (second term on the right-hand side). The theoretical prediction given by equation (5) is visualized in Fig. 2 by the solid line, which is fitted to the experimental data (symbols) by using $D_{\mathrm{C}}^{\parallel}$ as fit parameter.

The restoring torque caused by gravity depends on the orientation of the particle and becomes maximal at the critical angle $\phi_{\mathrm{crit}} = \arctan(D_{\mathrm{C}}^{\parallel}/D_{\mathrm{C}}^{\perp}) - \pi$. According to equation (5), this corresponds to a critical self-propulsion strength

$$P_{\mathrm{crit}}^{\star} = \beta b m g \sin\alpha \frac{\sqrt{D_{\mathrm{C}}^{\parallel 2} + D_{\mathrm{C}}^{\perp 2}}}{D_{\mathrm{C}}^{\perp} + l D_{\mathrm{R}}} \tag{6}$$

with the particle's buoyant mass m and the gravity acceleration of earth $g = 9.81\ \mathrm{m\,s^{-2}}$. When $P^{\star}$ exceeds this critical value for a given inclination angle of the setup, the effective torque originating from the non-central drive can no longer be compensated by the restoring torque due to gravity (see equation (4)) so that $\dot{\phi} \neq 0$ and a periodic motion occurs.

To apply our theory to the experiments, we use the values of the various parameters as shown in Table I. All data are obtained from our measurements as described in Methods.

Dynamical state diagram. Depending on the self-propulsion strength $P^{\star}$ and the substrate inclination angle α, different types of motion occur (see Fig. 3a). (For clarity, we neglected the noise in Fig. 3, but we checked that noise changes the trajectories only marginally.) For very small values of $P^{\star}$, the particle performs straight downward swimming. The theoretical

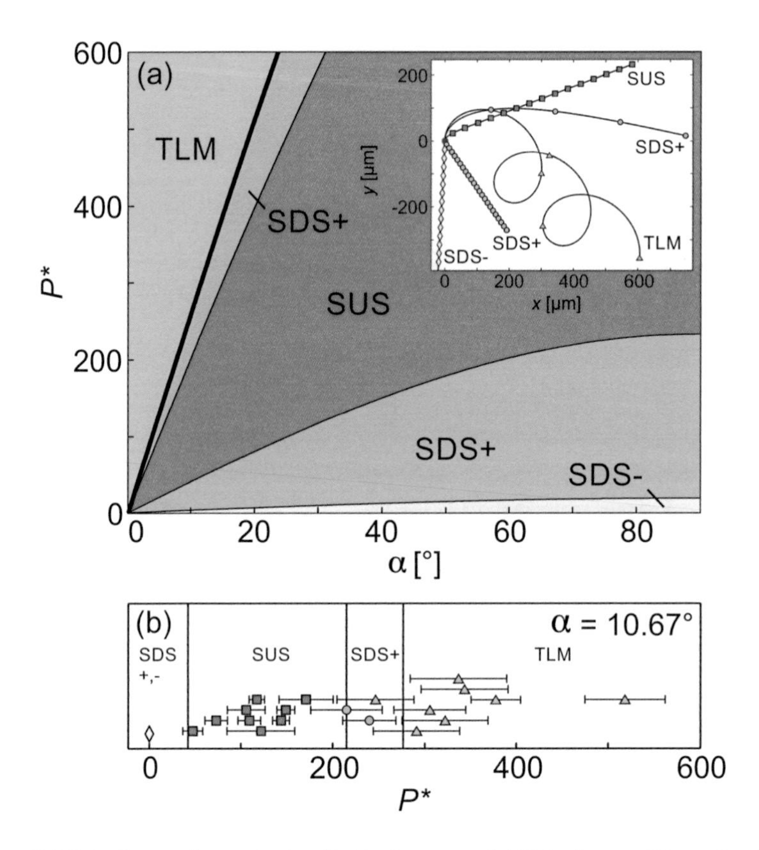

Figure 3 | State diagram for moderate effective lever arm. (**a**) State diagram of the motion of an active L-shaped particle with an effective lever arm $l = -0.75$ μm (see sketch in Fig. 1d) under gravity. The types of motion are straight downward swimming (SDS), straight upward swimming (SUS), and trochoid-like motion (TLM). Straight downward swimming is usually accompanied by a drift in negative (SDS-) or positive (SDS+) x direction. The transition from straight to circling motion is marked by a thick black line and determined analytically by equation (6). Theoretical noise-free example trajectories for the various states are shown in the inset. All trajectories start at the origin and the symbols (diamonds: SDS-, circles: SDS+, squares: SUS, triangles: TLM) indicate particle positions after 5 min each. (**b**) Experimentally observed types of motion for $\alpha = 10.67°$. The different symbols correspond to the various states as defined in **a** and are shifted in vertical direction for clarity. The error bars represent the s.d.

calculations reveal two regimes where on top of the downward motion either a small drift to the left or to the right is superimposed. For zero self-propulsion this additional lateral drift originates from the difference $D_{\parallel} - D_{\perp}$ between the translational diffusion coefficients, which is characteristic for non-spherical particles. Further increasing $P^{\star}$ results in negative gravitaxis, that is, straight upward swimming. Here, the vertical component of the self-propulsion counteracting gravity exceeds the strength of the gravitational force. For even higher $P^{\star}$, the velocity-dependent torque exerted on the particle further increases and leads to trajectories that are tilted more and more until a re-entrance to a straight downward motion with a drift to the right is observed. For the highest values of $P^{\star}$, the particle performs a trochoid-like motion. The critical self-propulsion $P^{\star}_{\mathrm{crit}}$ for the transition from straight to trochoid-like motion as obtained analytically from equation (6) is indicated by a thick black line in Fig. 3a.

Indeed, the experimentally observed types of motion taken from Fig. 1c correspond to those in the theoretical state diagram. This is shown in Fig. 3b, where we plotted the experimental data for $\alpha = 10.67°$ as a function of the self-propulsion strength $P^{\star}$. Apart from small deviations due to thermal fluctuations, quantitative agreement between experiment and theory is obtained.

According to equations (3) and (4), the state diagram should strongly depend on the effective lever arm l. To test this prediction experimentally, we tilted the silicon wafer with the L-particles about 25° relative to the Au source during the evaporation process. As a result, the Au coating slightly extends over the front face of the L-particles to one of the lateral planes which results in a shift of the effective propulsion force and a change in the lever arm. The value of l was experimentally determined to $l = -1.65\ \mu\mathrm{m}$ from the mean radius of the circular particle motion which is observed for $\alpha = 0°$ (see ref. [26]). The gravitactic behaviour of such particles is shown for different self-propulsion strengths in the inset of Fig. 4. Interestingly, under such conditions, we did not find evidence for negative gravitaxis. This is in good agreement with the corresponding theoretical state diagram shown in Fig. 4 and suggests that the occurrence of negative gravitaxis does not only depend on the strength of self-propulsion but also on the position where the effective force accounting for the self-propulsion mechanism acts on the body of the swimmer.

Discussion

Our model allows to derive general criteria for the existence of negative gravitaxis, which are applicable to arbitrary particle shapes. For a continuous upward motion, first, the self-propulsion $P^{\star}$ trivially has to be strong enough to overcome gravity. Secondly, the net rotation of the particle must vanish, which is the case for $P^{\star} \leqslant P^{\star}_{\mathrm{crit}}$ (see equation (6)). Otherwise, for example, trochoid-like trajectories are observed. As shown in the state diagrams in Figs. 3 and 4, the motional behaviour depends sensitively on several parameters

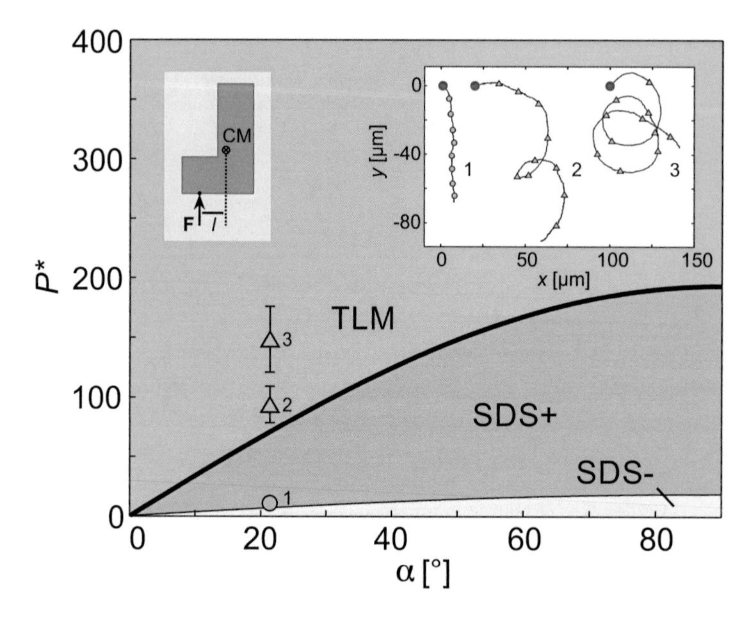

Figure 4 | State diagram for large effective lever arm. The same as in Fig. 3a but now for an active L-shaped particle with effective lever arm $l = -1.65$ µm (see left inset). Notice that the regime of upward swimming is missing here. Only straight downward swimming (with a drift in negative (SDS-) or positive (SDS+) x direction) and trochoid-like motion (TLM) are observed. The right inset shows three experimental trajectories whose corresponding positions in the state diagram are indicated together with error bars representing the s.d.

such as the strength of the self-propulsion and the length of the effective lever arm. This may also provide an explanation why gravitaxis is only observed for particular microorganisms as, for example, *Chlamydomonas reinhardtii* and why the gravitactic behaviour is subjected to large variations within a single population [7].

In our experiments, the particles had an almost homogeneous mass distribution in order to reveal pure shape-induced gravitaxis. For real biological systems, gravitaxis is a combination of the shape-induced mechanism studied here and bottom-heaviness resulting from an inhomogeneous mass distribution [36]. In particular for eukaryotic cells, small mass inhomogeneities due to the nucleus or organella are rather likely. Bottom-heaviness can straight-forwardly be included in our modelling by an additional torque contribution [37] and would support negative gravitaxis. Even for microorganisms with axial symmetry but fore-rear asymmetry [11] like *Paramecium* [2], both the associated shape-dependent hydrodynamic friction and bottom-heaviness contribute to gravitaxis so that the particle shape matters also in this special case.

In conclusion, our theory and experiments demonstrate that the presence of a gravitational field leads to straight downward and upward motion (gravitaxis) and also to trochoid-like trajectories of asymmetric self-propelled colloidal particles. On the basis of a set of Langevin equations, one can predict from the shape of a microswimmer, its mass distribution, and the geometry of the self-propulsion mechanism whether negative gravitaxis can occur. Although our study does not rule out additional physiological mechanisms for gravitaxis [4, 5], our results suggest that a swimming motion of biological microswimmers opposed to gravity can be entirely caused by a fore-rear asymmetry, that is, passive effects [11]. Such passive alignment mechanisms may also be useful in situations where directed motion of autonomous self-propelled objects is required as, for example, in applications where they serve as microshuttles for directed cargo delivery. Furthermore, their different response to a gravitational field could be utilized to sort microswimmers with respect to their shape and activity [33].

Methods

Experimental details. Asymmetric L-shaped microswimmers were fabricated from photoresist SU-8 by photolithography. A 3-µm-thick layer of SU-8 is spin coated onto a silicon wafer, soft baked for 80 s at 95 °C, and then exposed to ultraviolet light through a photomask. After a postexposure bake at 95 °C for 140 s, the entire wafer with the attached particles is covered by a several-nm-thick Au coating. During this process, the substrate is aligned by a specific angle relative to the evaporation source. Depending on the chosen angle, the Au coating can selectively be applied to specific regions of the front sides of the short arms of the particles. Afterwards, the particles are released from the substrate by an ultrasonic treatment and suspended in a binary mixture of water and 2,6-lutidine at critical composition (28.6 mass percent of lutidine) that is kept several degrees below its lower critical point $T_{\mathrm{C}} = 34.1$ °C. Time-dependent particle positions and orientations were acquired by video microscopy at a frame rate of 7.5 f.p.s. and stored for further analysis.

Langevin equations for a sedimenting passive particle. The Langevin equations for the centre-of-mass position $\mathbf{r}(t)$ and orientation $\phi(t)$ of an arbitrarily shaped passive particle that sediments under the gravitational force $\mathbf{F}_{\mathrm{G}} = (0, -mg\sin\alpha)$ are given by equation (1). The key quantities are the translational short-time diffusion tensor $\underline{\mathbf{D}}_{\mathrm{T}} = D_{\|}\hat{\mathbf{u}}_{\|} \otimes \hat{\mathbf{u}}_{\|} + D_{\|}^{\perp}(\hat{\mathbf{u}}_{\|} \otimes \hat{\mathbf{u}}_{\perp} + \hat{\mathbf{u}}_{\perp} \otimes \hat{\mathbf{u}}_{\|}) + D_{\perp}\hat{\mathbf{u}}_{\perp} \otimes \hat{\mathbf{u}}_{\perp}$ with $\otimes$ denoting the dyadic product and the translational-rotational coupling vector $\mathbf{D}_{\mathrm{C}} = D_{\mathrm{C}}^{\|}\hat{\mathbf{u}}_{\|} + D_{\mathrm{C}}^{\perp}\hat{\mathbf{u}}_{\perp}$ with the centre-of-mass position as reference point. The orientation vectors $\hat{\mathbf{u}}_{\|} = (\cos\phi, \sin\phi)$ and $\hat{\mathbf{u}}_{\perp} = (-\sin\phi, \cos\phi)$

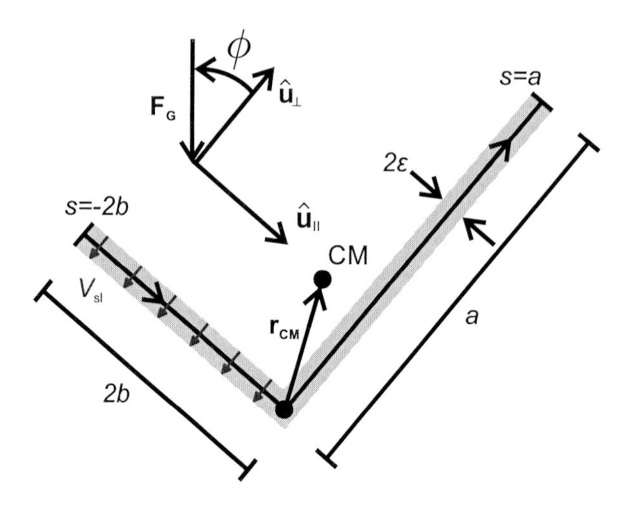

Figure 5 | Sketch for the slender-body model. Illustration of a rigid L-shaped particle with arm lengths a and $2b$. The position $\mathbf{r}(t)$ of the centre of mass (CM) and the orientation angle $\phi(t)$ evolve over time due to the gravitational force $\mathbf{F}_\mathrm{G}$ and fluid slip velocity V_sl on the particle surface. The unit vectors $\hat{\mathbf{u}}_\parallel$ and $\hat{\mathbf{u}}_\perp$ indicate the particle-fixed coordinate system.

are fixed in the body frame of the particle. (Notice that equation (1) could equivalently be written with friction coefficients instead of diffusion coefficients [38]. In this article, we use the given representation since the diffusion coefficients can directly be obtained from our experiments.) Finally, $\boldsymbol{\zeta}_\mathbf{r}(t)$ and $\zeta_\phi(t)$ are Gaussian noise terms of zero mean and variances $\langle \boldsymbol{\zeta}_\mathbf{r}(t_1) \otimes \boldsymbol{\zeta}_\mathbf{r}(t_2) \rangle = 2\,\underline{\mathbf{D}}_\mathrm{T}\,\delta(t_1 - t_2)$, $\langle \boldsymbol{\zeta}_\mathbf{r}(t_1)\,\zeta_\phi(t_2) \rangle = 2\,\mathbf{D}_\mathrm{C}\,\delta(t_1 - t_2)$, and $\langle \zeta_\phi(t_1)\zeta_\phi(t_2) \rangle = 2D_\mathrm{R}\delta(t_1 - t_2)$. Within this formalism, the specific particle shape enters via the translational diffusion coefficients $D_\parallel$, $D_\parallel^\perp$, and $D_\perp$, the coupling coefficients $D_\mathrm{C}^\parallel$ and $D_\mathrm{C}^\perp$, and the rotational diffusion coefficient D_R.

Langevin equations for an active particle under gravity. In the following, we provide a hydrodynamic derivation of our equations of motion (3) and (4) to justify their validity. Our derivation is based on slender-body theory for Stokes flow [39]. This approach has been applied successfully to model, for example, flagellar locomotion [40, 41]. The theory is adapted to describe the movement of a rigid L-shaped particle in three-dimensional Stokes flow as sketched in Fig. 5. A key assumption is that the width 2ϵ of the arms of the particle is much smaller than the total arc length $L = a + 2b$ of the particle, where a and $2b$ are its arm lengths. Though $\epsilon = 1.5$ µm is only one order of magnitude smaller than $L = 15$ µm in the experiments, the slenderness approximation $\epsilon \ll L$ offers valuable fundamental insight into the effects of self-propulsion and gravity on the particle's motion, as examined below.

The centreline position $\mathbf{x}(s)$ of the slender L-shaped particle is described by a parameter s with $-2b \leq s \leq a$:

$$\mathbf{x}(s) = \mathbf{r} - \mathbf{r}_{\mathrm{CM}} + \begin{cases} s\hat{\mathbf{u}}_{\parallel}\,, & \text{if } -2b \leq s \leq 0\,, \\ s\hat{\mathbf{u}}_{\perp}\,, & \text{if } 0 < s \leq a\,. \end{cases} \tag{7}$$

Here, $\mathbf{r}$ is the centre-of-mass position of the particle, $\hat{\mathbf{u}}_{\parallel}$ and $\hat{\mathbf{u}}_{\perp}$ are unit vectors defining the particle's frame of reference, and $\mathbf{r}_{\mathrm{CM}} = (a^2\hat{\mathbf{u}}_{\perp} - 4b^2\hat{\mathbf{u}}_{\parallel})/(2L)$ is a vector from the point where the two arms meet at right angles to the centre of mass (CM) (see Fig. 5). The fluid velocity on the particle surface is approximated by $\dot{\mathbf{x}} + \mathbf{v}_{\mathrm{sl}}(s)$, where $\mathbf{v}_{\mathrm{sl}}(s)$ is a prescribed slip velocity averaged over the intersection line of the particle surface and the plane through the point $\mathbf{x}(s)$ adjacent to the centreline. Motivated by the slip flow generated near the Au coating in the experiments, we set $\mathbf{v}_{\mathrm{sl}} = -V_{\mathrm{sl}}\hat{\mathbf{u}}_{\perp}$ along the shorter arm $(-2b \leq s \leq 0)$ and no slip (that is, $\mathbf{v}_{\mathrm{sl}} = \mathbf{0}$) along the other arm. According to the leading-order terms in slender-body theory [39], the fluid velocity is related to the local force per unit length $\mathbf{f}(s)$ on the particle surface and given by

$$\dot{\mathbf{x}} + \mathbf{v}_{\mathrm{sl}} = c(\underline{\mathbf{I}} + \mathbf{x}' \otimes \mathbf{x}')\mathbf{f} \tag{8}$$

with the constant $c = \log(L/\epsilon)/(4\pi\eta)$, the viscosity of the surrounding fluid η, the identity matrix $\underline{\mathbf{I}}$, the vector $\mathbf{x}' = \partial\mathbf{x}/\partial s$ that is locally tangent to the centreline, and the dyadic product $\otimes$. The force density $\mathbf{f}$ satisfies the integral constraints that the net force on the particle is the gravitational force,

$$\mathbf{F}_{\mathrm{G}} = \int_{-2b}^{a} \mathbf{f}\,\mathrm{d}s\,, \tag{9}$$

and that the net torque relative to the centre of mass of the particle vanishes,

$$\begin{aligned} &\hat{\mathbf{e}}_z \cdot \int_{-2b}^{a} (-\mathbf{r}_{\mathrm{CM}} + s\mathbf{x}') \times \mathbf{f}\,\mathrm{d}s \\ &= \int_{-2b}^{0} s\hat{\mathbf{u}}_{\perp} \cdot \mathbf{f}\,\mathrm{d}s - \int_{0}^{a} s\hat{\mathbf{u}}_{\parallel} \cdot \mathbf{f}\,\mathrm{d}s + \frac{1}{2L}(a^2\hat{\mathbf{u}}_{\parallel} \cdot \mathbf{F}_{\mathrm{G}} + 4b^2\hat{\mathbf{u}}_{\perp} \cdot \mathbf{F}_{\mathrm{G}}) = 0\,, \end{aligned} \tag{10}$$

where $\hat{\mathbf{e}}_z$ is the unit vector parallel to the z axis. Thus, in the absence of gravity, the self-propelled particle is force-free and torque-free as required.

We differentiate equation (7) with respect to time and insert it into equation (8). This leads to

$$\dot{\mathbf{r}} + \mathbf{v}_{\mathrm{sl}} + \frac{\dot{\phi}}{2L}(a^2\hat{\mathbf{u}}_{\parallel} + 4b^2\hat{\mathbf{u}}_{\perp}) = \begin{cases} c(\underline{\mathbf{I}} + \hat{\mathbf{u}}_{\parallel} \otimes \hat{\mathbf{u}}_{\parallel})\mathbf{f} - s\dot{\phi}\hat{\mathbf{u}}_{\perp}\,, & \text{if } -2b \leq s \leq 0\,, \\ c(\underline{\mathbf{I}} + \hat{\mathbf{u}}_{\perp} \otimes \hat{\mathbf{u}}_{\perp})\mathbf{f} + s\dot{\phi}\hat{\mathbf{u}}_{\parallel}\,, & \text{if } 0 < s \leq a\,. \end{cases} \tag{11}$$

The $\hat{\mathbf{u}}_{\parallel}$ and $\hat{\mathbf{u}}_{\perp}$ components of equation (11) are

$$\hat{\mathbf{u}}_{\parallel} \cdot (\dot{\mathbf{r}} + \mathbf{v}_{\mathrm{sl}}) + \frac{a^2\dot{\phi}}{2L} = \begin{cases} 2c\hat{\mathbf{u}}_{\parallel} \cdot \mathbf{f}\,, & \text{if } -2b \leq s \leq 0\,, \\ c\hat{\mathbf{u}}_{\parallel} \cdot \mathbf{f} + s\dot{\phi}\,, & \text{if } 0 < s \leq a\,, \end{cases} \tag{12}$$

and

$$\hat{\mathbf{u}}_{\perp} \cdot (\dot{\mathbf{r}} + \mathbf{v}_{\mathrm{sl}}) + 2\frac{b^2\dot{\phi}}{L} = \begin{cases} c\hat{\mathbf{u}}_{\perp} \cdot \mathbf{f} - s\dot{\phi}\,, & \text{if } -2b \leq s \leq 0\,, \\ 2c\hat{\mathbf{u}}_{\perp} \cdot \mathbf{f}\,, & \text{if } 0 < s \leq a\,. \end{cases} \tag{13}$$

After integrating equations (12) and (13) over s from $-2b$ to 0 and separately from 0 to a, the unknown $\mathbf{f}$ can be eliminated using equations (9) and (10). This results in the system of ordinary differential equations

$$\begin{aligned} \eta\underline{\mathbf{H}} \begin{pmatrix} \hat{\mathbf{u}}_{\parallel} \cdot \dot{\mathbf{r}} \\ \hat{\mathbf{u}}_{\perp} \cdot \dot{\mathbf{r}} \\ \dot{\phi} \end{pmatrix} &= \frac{1}{2c} \begin{pmatrix} 2(a+b) & 0 & -a^2b/L \\ 0 & a+4b & -2ab^2/L \\ -a^2b/L & -2ab^2/L & A \end{pmatrix} \begin{pmatrix} \hat{\mathbf{u}}_{\parallel} \cdot \dot{\mathbf{r}} \\ \hat{\mathbf{u}}_{\perp} \cdot \dot{\mathbf{r}} \\ \dot{\phi} \end{pmatrix} \\ &= \begin{pmatrix} \hat{\mathbf{u}}_{\parallel} \cdot \mathbf{F}_{\mathrm{G}} \\ \hat{\mathbf{u}}_{\perp} \cdot \mathbf{F}_{\mathrm{G}} \\ 0 \end{pmatrix} + \begin{pmatrix} 0 \\ 2bV_{\mathrm{sl}}/c \\ -2ab^2V_{\mathrm{sl}}/(cL) \end{pmatrix} \end{aligned} \tag{14}$$

with the constant $A = ((8L^2 - 6ab)(a^3 + 8b^3) - 6L(a^4 + 16b^4))/(12L^2)$. It is important to note that the matrix $\underline{\mathbf{H}}$ is identical to the grand resistance matrix (also called 'hydrodynamic friction tensor' [42]) for a passive particle (see further below). Since equation (14) is given in the particle's frame of reference, we next transform it to the laboratory frame by means of the rotation matrix

$$\underline{\mathbf{M}}(\phi) = \begin{pmatrix} \cos\phi & -\sin\phi & 0 \\ \sin\phi & \cos\phi & 0 \\ 0 & 0 & 1 \end{pmatrix}. \tag{15}$$

As obtained from the hydrodynamic derivation, the generalized diffusion tensor

$$\underline{\mathbf{D}} = \frac{1}{\beta\eta}\underline{\mathbf{H}}^{-1} = \begin{pmatrix} D_{\parallel} & D_{\parallel}^{\perp} & D_{\mathrm{C}}^{\parallel} \\ D_{\parallel}^{\perp} & D_{\perp} & D_{\mathrm{C}}^{\perp} \\ D_{\mathrm{C}}^{\parallel} & D_{\mathrm{C}}^{\perp} & D_{\mathrm{R}} \end{pmatrix}, \tag{16}$$

where $\beta = 1/(k_B T)$ is the inverse effective thermal energy, is determined by the parameters

$$D_{\parallel} = \left((a+4b)A - \frac{4a^2b^4}{L^2}\right)K\,, \tag{17}$$

$$D_{\parallel}^{\perp} = \frac{2a^3b^3K}{L^2}\,, \tag{18}$$

$$D_{\perp} = \left(2(a+b)A - \frac{a^4b^2}{L^2}\right)K\,, \tag{19}$$

$$D_{\mathrm{C}}^{\parallel} = \frac{a^2b(a+4b)K}{L}\,, \tag{20}$$

$$D_{\mathrm{C}}^{\perp} = \frac{4ab^2(a+b)K}{L}\,, \tag{21}$$

$$D_{\mathrm{R}} = 2(a+b)(a+4b)K \tag{22}$$

with the constant

$$K = \frac{2c}{\beta}\left(2(a+b)(a+4b)A - \frac{a^2b^2}{L^2}\left(4abL + a^3 + 8b^3\right)\right)^{-1}. \tag{23}$$

Defining

$$P^{\star} = \frac{2\beta b^2 V_{\mathrm{sl}}}{c}\,, \tag{24}$$

$$l = -\frac{ab}{L} \tag{25}$$

one finally obtains

$$\dot{\mathbf{r}} = (P^{\star}/b)\big(\underline{\mathbf{D}}_{\mathrm{T}}\hat{\mathbf{u}}_{\perp} + l\mathbf{D}_{\mathrm{C}}\big) + \beta\underline{\mathbf{D}}_{\mathrm{T}}\mathbf{F}_{\mathrm{G}}\,, \tag{26}$$

$$\dot{\phi} = (P^{\star}/b)\big(lD_{\mathrm{R}} + \mathbf{D}_{\mathrm{C}}\cdot\hat{\mathbf{u}}_{\perp}\big) + \beta\mathbf{D}_{\mathrm{C}}\cdot\mathbf{F}_{\mathrm{G}}\,, \tag{27}$$

where $\underline{\mathbf{D}}_{\mathrm{T}}$ is the translational short-time diffusion tensor, $\mathbf{D}_{\mathrm{C}}$ the translational-rotational coupling vector, and D_{R} the rotational diffusion coefficient of the particle. Up to the noise terms, which follow directly from the fluctuation-dissipation theorem and can easily be added, equations (26) and (27) are the same as our equations of motion (3) and (4). This proves that our equations of motion are physically justified and that they provide an appropriate theoretical framework to understand our experimental results.

Finally, we justify the concept of effective forces and torques by comparing the equations of motion of a self-propelled particle with the equations of motion of a passive particle that is driven by an external force $\mathbf{F}_{\mathrm{ext}}$, which is constant in the particle's frame of reference, and an external torque M_{ext}. To derive the equations of motion of an externally driven passive particle, we assume no-slip conditions for the fluid on the entire particle surface. In analogy to the derivation for a self-propelled particle (see further above) one obtains

$$\begin{aligned}\eta\underline{\mathbf{H}}\begin{pmatrix}\hat{\mathbf{u}}_{\parallel}\cdot\dot{\mathbf{r}}\\ \hat{\mathbf{u}}_{\perp}\cdot\dot{\mathbf{r}}\\ \dot{\phi}\end{pmatrix} &= \frac{1}{2c}\begin{pmatrix}2(a+b) & 0 & -a^2b/L\\ 0 & a+4b & -2ab^2/L\\ -a^2b/L & -2ab^2/L & A\end{pmatrix}\begin{pmatrix}\hat{\mathbf{u}}_{\parallel}\cdot\dot{\mathbf{r}}\\ \hat{\mathbf{u}}_{\perp}\cdot\dot{\mathbf{r}}\\ \dot{\phi}\end{pmatrix}\\ &= \begin{pmatrix}\hat{\mathbf{u}}_{\parallel}\cdot\mathbf{F}_{\mathrm{G}}\\ \hat{\mathbf{u}}_{\perp}\cdot\mathbf{F}_{\mathrm{G}}\\ 0\end{pmatrix}+\begin{pmatrix}\hat{\mathbf{u}}_{\parallel}\cdot\mathbf{F}_{\mathrm{ext}}\\ \hat{\mathbf{u}}_{\perp}\cdot\mathbf{F}_{\mathrm{ext}}\\ M_{\mathrm{ext}}\end{pmatrix}.\end{aligned} \tag{28}$$

We emphasize that the grand resistance matrix $\underline{\mathbf{H}}$ in equation (28) for an externally driven passive particle is identical to that in equation (14) for a self-propelled particle. Formally, the two equations are exactly the same if $\hat{\mathbf{u}}_{\parallel}\cdot\mathbf{F}_{\mathrm{ext}} = 0$, $\hat{\mathbf{u}}_{\perp}\cdot\mathbf{F}_{\mathrm{ext}} = 2bV_{\mathrm{sl}}/c$, and $M_{\mathrm{ext}} = -2ab^2V_{\mathrm{sl}}/(cL)$. This shows that the motion of a self-propelled particle with fluid slip $\mathbf{v}_{\mathrm{sl}} = -V_{\mathrm{sl}}\hat{\mathbf{u}}_{\perp}$ along the shorter arm is identical to the motion of a passive particle driven by a net external force $\mathbf{F}_{\mathrm{ext}} = F\hat{\mathbf{u}}_{\perp} = (2bV_{\mathrm{sl}}/c)\hat{\mathbf{u}}_{\perp}$ and torque $M_{\mathrm{ext}} = -(ab/L)F = lF$ with the effective self-propulsion force $F = 2bV_{\mathrm{sl}}/c$ and the effective lever arm $l = -ab/L$. (Note that while the equations of motion of a self-propelled particle and an externally driven passive particle are the same in this case, the flow and pressure fields generated by the particles are different.)

Consequently, the concept of effective forces and torques, where formally external ('effective') forces and torques that move with the self-propelled particle are used to model its self-propulsion, is justified and can be applied to facilitate the understanding of the dynamics of self-propelled particles.

Analysis of the experimental trajectories. For the interpretation of the experimental results in the context of the theoretical model, it is necessary to determine the self-propulsion strength $P^\star$ for the various observed trajectories. This is achieved by means of the relation

$$P^\star = b\frac{v_x - \beta F_{\mathrm{G}}(D_{\parallel} - D_{\perp})\sin\phi\cos\phi}{lD_{\mathrm{C}}^{\parallel}\cos\phi - (D_{\perp} + lD_{\mathrm{C}}^{\perp})\sin\phi}, \tag{29}$$

which is obtained from equation (3). It provides $P^\star$ as a function of the measured centre-of-mass velocity v_x in x direction. On the other hand, the gravitational force $F_{\mathrm{G}} = mg\sin\alpha$ is directly obtained from the easily adjustable inclination angle α of the experimental setup.

Determination of the experimental parameters. For the comparison of the theory with our measurements, specific values of a number of parameters have to be determined. These are the various diffusion and coupling coefficients, the effective lever arm, and the buoyant mass of the particles. The translational and rotational diffusion coefficients

$D_{\parallel} = 7.2 \times 10^{-3}\ \mu\mathrm{m}^2\,\mathrm{s}^{-1}$, $D_{\perp} = 8.1 \times 10^{-3}\ \mu\mathrm{m}^2\,\mathrm{s}^{-1}$, and $D_{\mathrm{R}} = 6.2 \times 10^{-4}\ \mathrm{s}^{-1}$ are obtained experimentally from short-time correlations of the particle trajectories for zero gravity [26]. Since $D_{\parallel}^{\perp}$ is negligible compared to $D_{\parallel}$ and $D_{\perp}$ here, we set $D_{\parallel}^{\perp} = 0$. The relation between the two coupling coefficients $D_{\mathrm{C}}^{\perp}$ and $D_{\mathrm{C}}^{\parallel}$ is determined from sedimentation experiments with passive L-particles. The peak position $\phi = -34°$ of the measured probability distribution $p(\phi)$ (see arrow in Fig. 1a) implies $D_{\mathrm{C}}^{\perp} = 0.67 D_{\mathrm{C}}^{\parallel}$ via equation (2). The only remaining open parameter is the absolute value of the coupling coefficient $D_{\mathrm{C}}^{\parallel}$, which is used as fit parameter in Fig. 2. Best agreement of the experimental data with the theoretical prediction (see equation (5)) is achieved for $D_{\mathrm{C}}^{\parallel} = 5.7 \times 10^{-4}\ \mu\mathrm{m}\,\mathrm{s}^{-1}$. The diffusion coefficients have also been calculated numerically by solving the Stokes equation [43] and good agreement with the experimental values has been found [26]. The influence of the substrate has been taken into account by applying the Stokeslet close to a no-slip boundary [44]. The effective lever arm of the L-shaped particles is determined to $l = -0.75\ \mu\mathrm{m}$ by assuming an ideally shaped swimmer with homogeneous mass distribution and the effective self-propulsion force acting vertically on the centre of the front side of the short arm (see Fig. 1d). Finally, the buoyant mass $m = 2.5 \times 10^{-14}$ kg is obtained by measuring the sedimentation velocity of passive particles.

References

[1] Roberts, A. M. Mechanisms of gravitaxis in *Chlamydomonas*. *Biol. Bull.* **210**, 78–80 (2006).

[2] Roberts, A. M. The mechanics of gravitaxis in *Paramecium*. *J. Exp. Biol.* **213**, 4158–4162 (2010).

[3] Ntefidou, M., Iseki, M., Watanabe, M., Lebert, M. & Häder, D.-P. Photoactivated adenylyl cyclase controls phototaxis in the flagellate *Euglena gracilis*. *Plant Physiol.* **133**, 1517–1521 (2003).

[4] Häder, D.-P. & Hemmersbach, R. Graviperception and graviorientation in flagellates. *Planta* **203**, S7–S10 (1997).

[5] Häder, D.-P., Richter, P. R., Schuster, M., Daiker, V. & Lebert, M. Molecular analysis of the graviperception signal transduction in the flagellate *Euglena gracilis*: involvement of a transient receptor potential-like channel and a calmodulin. *Adv. Space Res.* **43**, 1179–1184 (2009).

[6] Richter, P. R., Schuster, M., Lebert, M., Streb, C. & Häder, D.-P. Gravitaxis of *Euglena gracilis* depends only partially on passive buoyancy. *Adv. Space Res.* **39**, 1218–1224 (2007).

[7] Hemmersbach, R., Volkmann, D. & Häder, D.-P. Graviorientation in protists and plants. *J. Plant Physiol.* **154**, 1–15 (1999).

[8] Häder, D.-P. Gravitaxis in unicellular microorganisms. *Adv. Space Res.* **24**, 843–850 (1999).

[9] Nagel, U. & Machemer, H. Physical and physiological components of the graviresponses of wild-type and mutant *Paramecium tetraurelia*. *J. Exp. Biol.* **203**, 1059–1070 (2000).

[10] Machemer, H. & Braucker, R. Gravitaxis screened for physical mechanism using g-modulated cellular orientational behaviour. *Microgravity Sci. Technol.* **9**, 2–9 (1996).

[11] Roberts, A. M. & Deacon, F. M. Gravitaxis in motile micro-organisms: the role of fore-aft body asymmetry. *J. Fluid Mech.* **452**, 405–423 (2002).

[12] Mogami, Y., Ishii, J. & Baba, S. A. Theoretical and experimental dissection of gravity-dependent mechanical orientation in gravitactic microorganisms. *Biol. Bull.* **201**, 26–33 (2001).

[13] Streb, C., Richter, P., Ntefidou, M., Lebert, M. & Häder, D.-P. Sensory transduction of gravitaxis in *Euglena gracilis*. *J. Plant Physiol.* **159**, 855–862 (2002).

[14] Durham, W. M., Kessler, J. O. & Stocker, R. Disruption of vertical motility by shear triggers formation of thin phytoplankton layers. *Science* **323**, 1067–1070 (2009).

[15] Kessler, J. O. Hydrodynamic focusing of motile algal cells. *Nature* **313**, 218–220 (1985).

[16] Anderson, J. L. Colloid transport by interfacial forces. *Annu. Rev. Fluid Mech.* **21**, 61–99 (1989).

[17] Howse, J. R. *et al.* Self-motile colloidal particles: from directed propulsion to random walk. *Phys. Rev. Lett.* **99**, 048102 (2007).

[18] Hong, Y., Blackman, N. M., Kopp, N. D., Sen, A. & Velegol, D. Chemotaxis of nonbiological colloidal rods. *Phys. Rev. Lett.* **99**, 178103 (2007).

[19] Gibbs, J. G. & Zhao, Y.-P. Autonomously motile catalytic nanomotors by bubble propulsion. *Appl. Phys. Lett.* **94**, 163104 (2009).

[20] Jiang, H.-R., Yoshinaga, N. & Sano, M. Active motion of a Janus particle by self-thermophoresis in a defocused laser beam. *Phys. Rev. Lett.* **105**, 268302 (2010).

[21] Golestanian, R., Liverpool, T. B. & Ajdari, A. Propulsion of a molecular machine by asymmetric distribution of reaction products. *Phys. Rev. Lett.* **94**, 220801 (2005).

[22] Theurkauff, I., Cottin-Bizonne, C., Palacci, J., Ybert, C. & Bocquet, L. Dynamic clustering in active colloidal suspensions with chemical signaling. *Phys. Rev. Lett.* **108**, 268303 (2012).

[23] Palacci, J., Cottin-Bizonne, C., Ybert, C. & Bocquet, L. Sedimentation and effective temperature of active colloidal suspensions. *Phys. Rev. Lett.* **105**, 088304 (2010).

[24] Yates, R. C. *A Handbook on Curves and Their Properties* (J. W. Edwards, Ann Arbor, 1947).

[25] Badaire, S., Cottin-Bizonne, C., Woody, J. W., Yang, A. & Stroock, A. D. Shape selectivity in the assembly of lithographically designed colloidal particles. *J. Am. Chem. Soc.* **129**, 40–41 (2007).

[26] Kümmel, F. *et al.* Circular motion of asymmetric self-propelling particles. *Phys. Rev. Lett.* **110**, 198302 (2013).

[27] Volpe, G., Buttinoni, I., Vogt, D., Kümmerer, H.-J. & Bechinger, C. Microswimmers in patterned environments. *Soft Matter* **7**, 8810–8815 (2011).

[28] Buttinoni, I., Volpe, G., Kümmel, F., Volpe, G. & Bechinger, C. Active Brownian motion tunable by light. *J. Phys. Condens. Matter* **24**, 284129 (2012).

[29] Brenner, H. & Condiff, D. W. Transport mechanics in systems of orientable particles. III. Arbitary particles. *J. Colloid Interf. Sci.* **41**, 228–274 (1972).

[30] Friedrich, B. M. & Jülicher, F. The stochastic dance of circling sperm cells: sperm chemotaxis in the plane. *New J. Phys.* **10**, 123025 (2008).

[31] Jékely, G. *et al.* Mechanism of phototaxis in marine zooplankton. *Nature* **456**, 395–399 (2008).

[32] Radtke, P. K. & Schimansky-Geier, L. Directed transport of confined Brownian particles with torque. *Phys. Rev. E* **85**, 051110 (2012).

[33] Mijalkov, M. & Volpe, G. Sorting of chiral microswimmers. *Soft Matter* **9**, 6376–6381 (2013).

[34] Nourhani, A., Lammert, P. E., Borhan, A. & Crespi, V. H. Chiral diffusion of rotary nanomotors. *Phys. Rev. E* **87**, 050301(R) (2013).

[35] Marine, N. A., Wheat, P. M., Ault, J. & Posner, J. D. Diffusive behaviors of circle-swimming motors. *Phys. Rev. E* **87**, 052305 (2013).

[36] Campbell, A. I. & Ebbens, S. J. Gravitaxis in spherical Janus swimming devices. *Langmuir* **29**, 14066–14073 (2013).

[37] Wolff, K., Hahn, A. M. & Stark, H. Sedimentation and polar order of active bottom-heavy particles. *Eur. Phys. J. E* **36**, 43 (2013).

[38] Makino, M. & Doi, M. Brownian motion of a particle of general shape in Newtonian fluid. *J. Phys. Soc. Jpn.* **73**, 2739–2745 (2004).

[39] Batchelor, G. K. Slender-body theory for particles of arbitrary cross-section in Stokes flow. *J. Fluid Mech.* **44**, 419–440 (1970).

[40] Lighthill, J. Flagellar hydrodynamics. *SIAM Rev.* **18**, 161–230 (1976).

[41] Lauga, E. & Powers, T. R. The hydrodynamics of swimming microorganisms. *Rep. Prog. Phys.* **72**, 096601 (2009).

[42] Kraft, D. J. *et al.* Brownian motion and the hydrodynamic friction tensor for colloidal particles of complex shape. *Phys. Rev. E* **88**, 050301(R) (2013).

[43] Carrasco, B. & Garcia de la Torre, J. Improved hydrodynamic interaction in macromolecular bead models. *J. Chem. Phys.* **111**, 4817–4826 (1999).

[44] Blake, J. R. A note on the image system for a stokeslet in a no-slip boundary. *Math. Proc. Cambridge* **70**, 303–310 (1971).

Acknowledgements

This work was supported by the Marie Curie-Initial Training Network Comploids funded by the European Union Seventh Framework Program (FP7), by the ERC Advanced Grant INTERCOCOS (Grant No. 267499), and by EPSRC (Grant No. EP/J007404). R.W. gratefully acknowledges financial support through a Postdoctoral Research Fellowship (WI 4170/1-1) from the Deutsche Forschungsgemeinschaft (DFG).

Author contributions

B.t.H., F.K., R.W., H.L., and C.B. designed the research, analysed the data, and wrote the paper; B.t.H. and R.W. carried out the analytical calculations and the simulations; F.K. performed the experiments; D.T. provided the hydrodynamic derivation in the Methods section.

Additional information

The authors declare no competing financial interests.

Statement of the author: This article, published in *Nature Communications*, is the second centerpiece of my doctorate. While Felix Kümmel performed the experiments under the supervision of Clemens Bechinger, I analyzed most of the experimental data, derived the theoretical expressions in the main part of the paper, and calculated the state diagrams. Raphael Wittkowski and Hartmut Löwen provided valuable input in many fruitful discussions. I wrote significant parts of the paper, though all authors helped to improve the manuscript in several revisions. Hartmut Löwen established the contact with Daisuke Takagi, who provided the hydrodynamic derivation in the Methods section.

Conclusions

Active colloidal particles have been widely studied during the last decade. Biological microswimmers as well as artificial self-propelled particles have been topic of intense scientific research. Such systems have been addressed both theoretically and experimentally. In most cases, theoretical models and corresponding computer simulations have been developed for spherical or other highly symmetric particle shapes. However, on a more fine-grained level of description, shape anisotropy must not be neglected as it is the generic case for nearly all natural microswimmers and at least to some extent also in most artificial self-propelled particles. Therefore, the main goal of this thesis was to provide a theoretical framework for the description of asymmetric microswimmers. As demonstrated by the presented results, their motional behavior indeed largely deviates from the dynamics of idealized isotropic self-propelled particles.

Starting with the most simple model situation, first spherical self-propelled Janus particles were addressed. A theory based on translational and rotational Langevin equations has been applied to a specific experimental setup consisting of Pt-silica Janus spheres moving in hydrogen peroxide solutions. In addition to the mean square displacement, which is a standard quantity for characterizing different dynamical regimes, higher moments of the displacement probability distribution function have been considered. In particular, the skewness and the excess kurtosis have been calculated analytically and compared to experimental data. Thus, a characteristic non-Gaussian behavior could be identified and quantitatively analyzed. This complements the classification of different diffusive and superdiffusive regimes of motion. Additional insights about the dynamics of self-propelled Janus particles have been obtained by analyzing the full probability distributions for the magnitude and the direction of particle displacements. By comparing simulation results with experimental data, it could be shown that the rotational motion of the Janus spheres under study is restricted by a higher hydrogen peroxide concentration of the solvent.

The first step towards a more realistic description of biological microswimmers was achieved by explicitly including a time-dependent self-propulsion in the equations of motion. This is particularly relevant with regard to microorganisms which perform a characteristic run-and-tumble motion or exhibit other variations in their swimming velocity. But also in the context

of artificial self-propelled particles it is sometimes desired to prescribe certain propulsion protocols and thus to induce an explicit time dependence in the system under study. In this thesis, analytical solutions have been provided both for the mean position and the mean square displacement of microswimmers with various types of time-dependent self-propulsion. These results can serve as a reference for future experimental studies. In particular, it has been shown that the mean trajectories of microswimmers with time-periodic self-propulsion are self-similar curves if their motion is affected by an additional torque.
Based on a different model, some results have also been presented for deformable active particles. Surface deformations are especially important in the context of active droplets but also for certain biological microorganisms such as amoebae and cyanobacteria. In this thesis, the motion of such deformable particles exposed to a linear shear flow has been studied. By numerical calculations, a great variety of different types of motion has been observed and systematically classified by means of comprehensive dynamical phase diagrams. In the two special cases of either vanishing shear flow or vanishing particle deformability, previously reported theoretical models could be recovered.

When dealing with rigid colloidal particles of complex shape, important quantities for characterizing the dynamics are the grand resistance matrix or, equivalently, the diffusion tensor. For a reliable theoretical description of the occurring motile behavior, the various friction or diffusion coefficients have to be determined accurately. Here, it has been discussed in detail how this can be achieved either based on experimental tracking data of the Brownian motion of such particles or by means of numerical calculations for the specific particle shape. According to the experimental route, appropriate short-time correlation functions obtained from the recorded particle trajectories are analyzed. The theoretical method relies on a hydrodynamic calculation for the Stokes flow around a particle.
In a first application, this theoretical procedure of predicting the diffusion coefficients was used to set up a computer simulation for a mesoscopic wedgelike carrier embedded in a suspension of rodlike bacteria. Based on this system, it could be shown that it is possible to extract mechanical energy from a bath of bacteria moving in a turbulent state. Thus, a passive carrier can be transported through the suspension by exploiting the active motion of the swimming microorganisms. The two effects mainly responsible for this phenomenon are swirl shielding and polar ordering of the bacteria inside the cusp region of the carrier.

Regarding the theoretical description of rigid asymmetric microswimmers, which is the core issue of this thesis, the diffusion and coupling coefficients do not only determine the Brownian motion but are also responsible for the degree of coupling between translational and rotational active motion. Instead of one single translational and one rotational diffusion coefficient as for spherical particles, usually the full generalized diffusion tensor enters the overdamped equations of motion when asymmetric particles are considered. These Langevin equations have been derived explicitly for L-shaped microswimmers, for which the theoretical predictions could directly be compared to experimental observations. It has been demonstrated that such particles move in general on circular trajectories with a constant radius, which does not depend on the swimming velocity but only on the shape of a particle. This is due to a velocity-dependent torque, which is coupled to the strength of self-propulsion via an effective lever arm. Moreover, the noise-averaged trajectory has been analyzed and shown to be a logarithmic spiral. In order to quantitatively compare the theoretical prediction to experimental data, the measured trajectories were divided into segments of equal length and subsequently averaged. As an additional result, for the special case of an L-shaped self-propelled particle interacting with a solid wall, two qualitatively different types of motion have been observed. Depending on the approaching angle, a particle either enters a stable sliding regime or it is reflected by the wall.

The general applicability of the theoretical model provided in this thesis is demonstrated by the fact that different experimental setups have been used to test the predictions. While the experiments discussed in chapters 7 and 9 are based on self-diffusiophoretic microswimmers in a critical water-lutidine mixture [115, 116], the measurements analyzed in chapter 2 have been performed with Janus particles moving by a catalytic propulsion mechanism. The universality of the model using effective forces and torques has also been demonstrated from a theoretical point of view. Detailed derivations have been provided both for mechanical swimming and for self-propulsion induced by a slip velocity on the particle surface. Thus, the theory provides a consistent and powerful framework for the description of self-propelled particles of complex shape although it does not explicitly include the flow and pressure fields in the vicinity of the particle. In many respects, this is even an important advantage of the model because the detailed self-propulsion mechanisms of artificial microswimmers are often not completely understood yet [111, 112].

In the final chapter of the main part of this thesis, the gravitactic behavior of asymmetric self-propelled particles has been investigated. Using the theory based on a set of Langevin equations in combination with experimental results for L-shaped microswimmers, various types of motion under gravity have been identified. Depending on the ratio of the self-propulsion strength and the gravitational force, the particles move either on straight downward or straight upward trajectories, or they perform a trochoidlike motion. When moving on a straight swimming path, the particle assumes a constant orientation after an initial transient regime. This long-time orientation has been predicted theoretically and compared with experimental observations. In addition to a detailed interpretation of the different trajectories, the full state diagram has been calculated. It sensitively depends on the detailed shape of the microswimmers.
The results are also important in the context of gravitaxis of biological microorganisms. It has been demonstrated that a swimming motion opposed to the direction of gravity can entirely result from an asymmetric particle shape. Thus, a physiological gravity sensor is not necessary. Most of the existing studies on gravitaxis of biological microswimmers already date back more than a decade and are controversially discussed (see references [189, 190, 200, 202, 207]

and references therein). Probably, more accurate measurements can be performed with new technological methods that have become available during the last few years. With regard to the discussed findings for the gravitactic behavior of artificial asymmetric self-propelled particles, it would be very interesting to see whether the origin of gravitaxis can be definitely clarified for certain microorganisms in the near future.

For many possible applications of self-propelled particles one has to be able to control their motional behavior. This can relatively easily be achieved if particles whose self-propulsion is triggered by illumination with light are used. A theory for time-dependent self-propulsion protocols has already been presented in this thesis. On the other hand, applications may also require a position-dependent swimming velocity. Light-activated microswimmers can be tuned in such a way by using appropriate spatially varying light fields [448, 449]. For the setup described in references [115, 116], the propulsion velocity is typically proportional to the local light intensity [218]. Thus, the particles move fast in brightly illuminated areas and become passive in the limit of completely dark regions. This can also be exploited for the trapping of self-propelled particles [450] and hence represents an alternative to trapping by means of external potentials [256, 451, 452] or in geometrical devices [379, 380]. From a theoretical point of view, it is an interesting question how exactly the behavior of light-activated self-propelled particles under illumination with spatially varying intensity deviates from setups with symmetric or asymmetric energy barriers. For the latter case, a concept of effective temperatures has been introduced recently [453].

This thesis clearly illustrates the strong influence of the particle shape on the motion of microswimmers. For the future, a next logical step would be to explicitly consider asymmetric shapes not only in the theoretical description of the single-particle dynamics but also for the investigation of interacting multi-particle systems. Even in the simple case of isotropic particle shapes many interesting collective phenomena have been observed. Examples that have been studied in detail include clustering, swarming, and lane formation. Such phenomena are expected to be clearly influenced by shape asymmetry. The behavior will probably be significantly more sophisticated than for spherical particles. The theory developed in the previous chapters of this thesis can be used as a starting point for appropriate modeling and computer simulations. Recent calculations for self-propelled particle systems have shown that even a small fore-rear asymmetry may lead to substantially different collective behavior [454]. When L-shaped or C-shaped swimmers are considered, for example, the individual particles can get caught into each other so that the formation of long chains and clusters is boosted. Thus, even completely new collective effects may arise. Similar emergent phenomena have already been studied in simulations for rotationally driven particles consisting of several disks [455]. Here, the specific shape leads to an effective attraction between neighboring particles rotating in the same direction [455]. Another situation which has only rarely been addressed up to now are mixtures of self-propelled particles [456]. A rich phase behavior is expected when the individual particle species have different shapes.

New types of interaction are also likely to occur when asymmetric microswimmers get in contact with obstacles. While the motion of isotropic particles in confined geometries such as channels or near symmetric or asymmetric barriers has often been considered in computer simulations [381, 457, 458], much less is known about the dynamics of asymmetric self-propelled particles under such conditions. As opposed to spherical particles, microswimmers with more complicated shapes could also get stuck at corners of solid obstacles and are thus expected to show different behavior in transport processes [459]. While a constant torque has already been included in theoretical modeling [269, 458], this has not been attributed to an asymmetric particle shape so far. A detailed study of the interaction of microswimmers with surfaces is also interesting in the context of biological microorganisms, where often additional direct ciliary or flagellar contact interactions play an important role as has recently been shown both for sperm cells and green algae [57].

One reason for the interest in the dynamics of active particles moving through channels or near obstacles is the fact that such geometries can be used for various sorting [382, 460, 461], separation [462, 463], or pumping applications [464]. Different techniques have been proposed for circle swimmers and particles moving on rather straight lines [382, 383, 465]. Particle sorting can be performed with respect to shape or velocity, for example [465]. For particles moving on circular trajectories due to the presence of an internal torque, another relevant sorting criterion is their chirality [460]. The results for asymmetric self-propelled particles presented in this thesis might contribute to the improvement of such applications. While up to now the majority of related studies is purely theoretical, corresponding systems are likely to be realized in experimental setups in the near future. For such purposes, it is necessary to describe the real conditions as precisely as possible. Therefore, the assumption of idealized spherical particle shapes is often not appropriate. A more accurate theory can probably give valuable input how certain features should be designed in order to obtain efficient setups. It would also be desirable to include hydrodynamic interactions [407, 466] in the theoretical modeling of multi-particle systems, although this is clearly a nontrivial topic when asymmetric shapes are considered.

While the control of systems with large numbers of self-propelled particles is one area of possible applications, an even more promising field is the construction of specific micromachines which are able to perform certain highly sophisticated tasks, such as cargo transport on a microscale [72, 73, 467, 468], targeted drug delivery [74, 75, 469], or more specialized functions like cleaning polluted water [470, 471]. Apart from applications based on artificial micro- and nanodevices [472], ongoing progress is also expected in the field of biological micromachines [473–475]. To use the green alga *Chlamydomonas reinhardtii* as a microtransporter, for example, various mechanisms based on surface chemistry, phototaxis, and photochemistry can be utilized [476]. Moreover, it has recently been illustrated how swimming bacteria can fulfill

the task of targeted delivery of colloids without any external control fields [477]. New ideas for the design of artificial microdevices may also evolve from the investigation of molecular motors [478–480], which enable intracellular cargo transport [481,482].
Up to now, cargo transportation is the mostly studied application of micromachines. However, by making use of specific asymmetric shapes, such devices can also be engineered for drilling themselves into biological tissue, for example [76,483]. This is clearly favored by screw-like trajectories as typically observed in three dimensions for microswimmers with an asymmetry around their propulsion axis.

Undoubtedly, the investigation of self-propelled particles remains a fascinating topic, both from a fundamental point of view and with regard to new technological applications. It will be exciting to see how theory and experiment will continue to mutually complement each other in putting forward new ideas and making substantial progress in research in the coming years.

Appendix: Numerical calculation of diffusion coefficients

In this appendix, it is described in detail how the diffusion coefficients for a specific particle shape can be calculated numerically. While in previous studies on self-propelled particles usually highly symmetric shapes have been assumed, one of the main questions addressed in this thesis is how the theoretical modeling can be transferred to asymmetric microswimmers. In this context, the grand resistance matrix and the diffusion tensor, respectively, provide essential information. The most important advantage of the presented calculation method is that it can also be applied to particles with arbitrarily complicated shape.

We calculate the translational and rotational diffusion coefficients based on a bead model using the software HYDRO++ [349, 377, 484]. This software has originally been designed for the prediction of hydrodynamic properties of macromolecules [349], but it can also be used for colloidal particles modeled by a large number of beads. To calculate the diffusion coefficients of such particles, first the bead positions have to be determined. In order to achieve a high accuracy, it is important that the bead model closely represents the specific particle shape. As the number of beads that can be used for the calculations with the software HYDRO++ is limited, it is most efficient to place the beads only on the surface of the particle to be modeled. This method is referred to as *shell model* and has originally been introduced by Bloomfield [485, 486]. It enables a much better resolution of the particle shapes than filling models with a full-volume bead representation, as beads with a significantly smaller radius can be used. Moreover, in many situations shell models provide even more accurate results than filling models [487]. Best results are obtained if the beads are placed on the surface without larger voids so that they ideally touch each other. However, according to the HYDRO++ userguide it is important that the beads do not overlap because the software may provide wrong results in that case. If bead representations with overlapping spheres are required, the program HYDROSUB can be used to perform the hydrodynamic calculations [216, 348]. This software also allows for the implementation of ellipsoidal and cylindrical subunits [348]. In order to include the effect of a wall or a substrate, it is in principle possible to modify the calculations by using the Stokeslet close to a no-slip boundary [408]. This is in particular relevant to experimental situations with a quasi-two-dimensional setup [218, 220].

In the following, we illustrate how the method can easily be applied to rotationally symmetric particles. Explicitly, cylinders, spherocylinders, ellipsoids, spindles, double cones, and lenses are considered. Both for prolate and oblate solids of revolution an efficient implementation of the calculation of the bead positions that are required as an input for HYDRO++ is possible. The beads are placed on several rings centered on the z axis, which coincides with the symmetry axis of the particles. These rings have a distance of $2r$ each, where r is the radius of a single bead. We always use the same value of r for all beads representing a particle. The radius $R(z)$ of the rings is given for cylinders by

$$R_{\mathrm{cylinder}}(z) = \frac{L}{2\nu}\,, \tag{A.1}$$

for spherocylinders by

$$R_{\mathrm{spherocylinder}}(z) = \begin{cases} \sqrt{\frac{L(L/2+z)}{\nu} - (\frac{L}{2}+z)^2} & \text{for } -\frac{L}{2} \leq z < \frac{L}{2\nu} - \frac{L}{2}\,, \\ \frac{L}{2\nu} & \text{for } -\frac{L}{2} + \frac{L}{2\nu} \leq z \leq \frac{L}{2} - \frac{L}{2\nu}\,, \\ \sqrt{\frac{L(L/2-z)}{\nu} - (\frac{L}{2}-z)^2} & \text{for } \frac{L}{2} - \frac{L}{2\nu} < z \leq \frac{L}{2}\,, \end{cases} \tag{A.2}$$

for ellipsoids by

$$R_{\mathrm{ellipsoid}}(z) = \frac{1}{2\nu}\sqrt{L^2 - 4z^2}\,, \tag{A.3}$$

for spindles by

$$R_{\mathrm{spindle}}(z) = \frac{L}{4\nu}(1-\nu^2) + \sqrt{\frac{L^2}{16\nu^2}(1+\nu^2)^2 - z^2}\,, \tag{A.4}$$

and for double cones by

$$R_{\mathrm{doublecone}}(z) = \frac{L}{2\nu} - \frac{1}{\nu}|z|\,. \tag{A.5}$$

In these formulas, L is the length of the particle and ν is the aspect ratio. For cylinders, it is necessary to add beads on the top and bottom areas, as equation (A.1) places beads only on the lateral surface area of the cylinder. Therefore, additional rings with radii equidistantly distributed between zero and $R_{\mathrm{max}} = \sigma/2$, where $\sigma = L/\nu$ is the equatorial diameter of the particle, are considered in this case.

Once the radii of the rings are known, the explicit coordinates of the beads are calculated. The maximum number of beads that can be placed on a ring with radius R is determined by means of the formula $s = 2R\sin(\alpha/2)$ for the chord length s. Here, α is the corresponding central angle. Using the condition $s = 2r$, one obtains that the minimal ring radius

$$R_{\mathrm{min}}(N) = \frac{r}{\sin(\pi/N)} \tag{A.6}$$

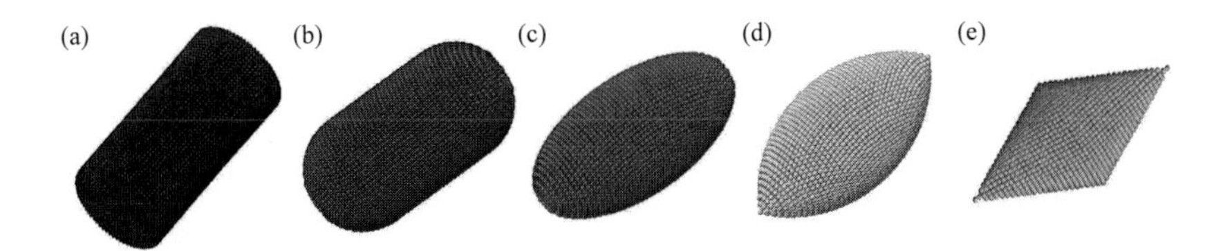

Figure A.1: *Bead model representation of the considered prolate particle shapes with aspect ratio $\nu = 2$: (a) cylinder, (b) spherocylinder, (c) ellipsoid, (d) spindle, and (e) double cone.*

is required in order to place N beads of radius r on the ring. Based on equation (A.6), for each value of $R(z)$ the maximal number of beads can easily be determined. Using polar coordinates, the beads are positioned equidistantly on the ring. The resulting bead model representations for the various considered particle shapes are visualized in figure A.1.

The method described so far is efficient for prolate particle shapes. However, for oblate shapes the method has to be adapted in order to achieve a high accuracy. In that case, it is more reasonable to use the radius R of the rings for the non-overlap condition instead of the z value. Accordingly, the radii are defined in equidistant steps between zero and R_{max}, similarly to the occupation of the top and bottom areas of the cylinder. Based on these radii, the corresponding z values are calculated in order to determine the positions of the beads. While cylinders, spherocylinders, ellipsoids, spindles, and double cones were considered as prolate particle shapes, spherocylinders and spindles are not well defined as oblate objects. Therefore, only cylinders, ellipsoids, double cones, and additionally lenses are considered in the following (see figure A.2 for the respective bead model representations). A lens can also be interpreted as an abbreviated spherocylinder with missing intermediate part. For ellipsoids, lenses, and double cones the equations for the z positions of the rings with beads are

$$|z_{\text{ellipsoid}}(R)| = \nu\sqrt{\frac{\sigma^2}{4} - R^2}\,, \tag{A.7}$$

$$|z_{\text{lens}}(R)| = \frac{\sigma}{4\nu}(\nu^2 - 1) + \sqrt{\frac{\sigma^2}{16\nu^2}(\nu^2+1)^2 - R^2}\,, \tag{A.8}$$

and

$$|z_{\text{doublecone}}(R)| = \frac{\sigma\nu}{2} - \nu R\,, \tag{A.9}$$

respectively. The z values can be positive or negative with $-L/2 \le z \le L/2$. For an oblate cylinder, the same procedure as for a prolate cylinder (see equation (A.1)) can be used.

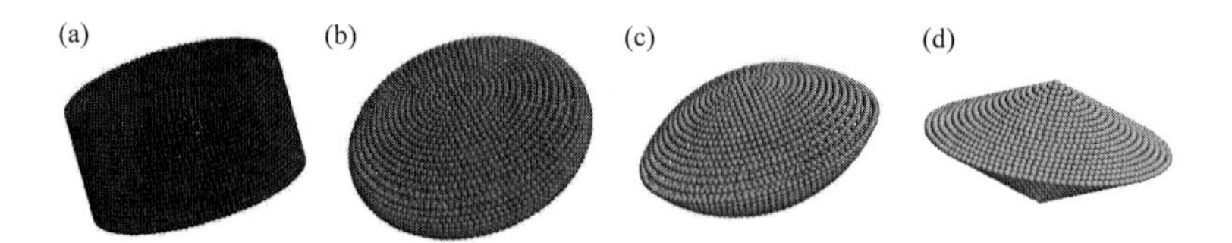

Figure A.2: *Bead model representation of the considered oblate particle shapes with aspect ratio $\nu = 1/2$: (a) cylinder, (b) ellipsoid, (c) lens, and (d) double cone.*

The method for oblate particle shapes can further be improved by adding rings with beads if the distance Δz between the centers of two adjacent rings is larger than $4r$. For ellipsoids and double cones, the radii of these additional rings are given by equations (A.3) and (A.5) presented above in the context of prolate particle shapes. For lenses the corresponding expression is

$$R_{\text{lens}}(z) = \sqrt{\frac{\sigma^2}{4} + \frac{z\sigma}{2\nu}(\nu^2 - 1) - z^2}\,. \tag{A.10}$$

While the additional rings may lead to a remarkable improvement of the numerical results for the diffusion coefficients if spheres or nearly spherical particle shapes with aspect ratio $\nu \approx 1$ are considered, they are not necessary for strongly oblate particles.

The previously described method is specifically designed for solids of revolution. When more complicated particle shapes are considered, the positioning of the beads has to be adapted accordingly. Another situation where a relatively simple implementation is possible are shapes consisting of several cuboids. An example are the L-shaped microswimmers studied in detail in chapters 7–9 of this thesis. In that case, the spheres are distributed homogeneously on the rectangular surfaces.

Bibliography

[1] T. Vicsek and A. Zafeiris. Collective motion. *Phys. Rep.* **517**, 71–140 (2012).

[2] I. D. Couzin and J. Krause. Self-organization and collective behavior in vertebrates. *Adv. Stud. Behav.* **32**, 1–75 (2003).

[3] N. Abaid and M. Porfiri. Fish in a ring: spatio-temporal pattern formation in one-dimensional animal groups. *J. R. Soc. Interface* **7**, 1441–1453 (2010).

[4] Y. Katz, K. Tunstrøm, C. C. Ioannou, C. Huepe, and I. D. Couzin. Inferring the structure and dynamics of interactions in schooling fish. *Proc. Natl. Acad. Sci. USA* **108**, 18720–18725 (2011).

[5] M. Ballerini, N. Cabibbo, R. Candelier, A. Cavagna, E. Cisbani, I. Giardina, V. Lecomte, A. Orlandi, G. Parisi, A. Procaccini, M. Viale, and V. Zdravkovic. Interaction ruling animal collective behavior depends on topological rather than metric distance: evidence from a field study. *Proc. Natl. Acad. Sci. USA* **105**, 1232–1237 (2008).

[6] R. Lukeman, Y.-X. Li, and L. Edelstein-Keshet. Inferring individual rules from collective behavior. *Proc. Natl. Acad. Sci. USA* **107**, 12576–12580 (2010).

[7] M. Nagy, Z. Ákos, D. Biro, and T. Vicsek. Hierarchical group dynamics in pigeon flocks. *Nature* **464**, 890–893 (2010).

[8] G. Theraulaz, E. Bonabeau, S. C. Nicolis, R. V. Solé, V. Fourcassié, S. Blanco, R. Fournier, J.-L. Joly, P. Fernández, A. Grimal, P. Dalle, and J.-L. Deneubourg. Spatial patterns in ant colonies. *Proc. Natl. Acad. Sci. USA* **99**, 9645–9649 (2002).

[9] J. Buhl, D. J. T. Sumpter, I. D. Couzin, J. J. Hale, E. Despland, E. R. Miller, and S. J. Simpson. From disorder to order in marching locusts. *Science* **312**, 1402–1406 (2006).

[10] S. Bazazi, J. Buhl, J. J. Hale, M. L. Anstey, G. A. Sword, S. J. Simpson, and I. D. Couzin. Collective motion and cannibalism in locust migratory bands. *Curr. Biol.* **18**, 735–739 (2008).

[11] J. G. Puckett and N. T. Ouellette. Determining asymptotically large population sizes in insect swarms. *J. R. Soc. Interface* **11**, 20140710 (2014).

[12] D. Helbing, I. Farkas, and T. Vicsek. Simulating dynamical features of escape panic. *Nature* **407**, 487–490 (2000).

[13] J. Zhang, W. Klingsch, A. Schadschneider, and A. Seyfried. Ordering in bidirectional pedestrian flows and its influence on the fundamental diagram. *J. Stat. Mech.* (2012) P02002.

[14] J. L. Silverberg, M. Bierbaum, J. P. Sethna, and I. Cohen. Collective motion of humans in mosh and circle pits at heavy metal concerts. *Phys. Rev. Lett.* **110**, 228701 (2013).

[15] D. Helbing. Traffic and related self-driven many-particle systems. *Rev. Mod. Phys.* **73**, 1067–1141 (2001).

[16] T. Obata, T. Shimizu, H. Osaki, H. Oshima, and H. Hara. Fluctuations in human's walking (II). *J. Korean Phys. Soc.* **46**, 713–718 (2005).

[17] A. M. Menzel. Tuned, driven, and active soft matter. *Phys. Rep.* **554**, 1–45 (2015).

[18] Y. Sumino, K. H. Nagai, Y. Shitaka, D. Tanaka, K. Yoshikawa, H. Chaté, and K. Oiwa. Large-scale vortex lattice emerging from collectively moving microtubules. *Nature* **483**, 448–452 (2012).

[19] E. M. Purcell. Life at low Reynolds number. *Am. J. Phys.* **45**, 3–11 (1977).

[20] G. Fuchs and H. G. Schlegel. *Allgemeine Mikrobiologie*, 9th ed. (Thieme, Stuttgart, 2014).

[21] H. C. Berg and R. A. Anderson. Bacteria swim by rotating their flagellar filaments. *Nature* **245**, 380–382 (1973).

[22] M. Silverman and M. Simon. Flagellar rotation and the mechanism of bacterial motility. *Nature* **249**, 73–74 (1974).

[23] J. B. Waterbury, J. M. Willey, D. G. Franks, F. W. Valois, and S. W. Watson. A cyanobacterium capable of swimming motility. *Science* **230**, 74–76 (1985).

[24] K. M. Ehlers, A. D. T. Samuel, H. C. Berg, and R. Montgomery. Do cyanobacteria swim using traveling surface waves? *Proc. Natl. Acad. Sci. USA* **93**, 8340–8343 (1996).

[25] K. M. Ehlers and J. Koiller. Micro-swimming without flagella: propulsion by internal structures. *Regul. Chaotic Dyn.* **16**, 623–652 (2011).

[26] K. G. Grell. *Protozoologie*, 2nd ed. (Springer, Berlin, 1968).

[27] J. J. Molina, Y. Nakayama, and R. Yamamoto. Hydrodynamic interactions of self-propelled swimmers. *Soft Matter* **9**, 4923–4936 (2013).

[28] A. Zöttl and H. Stark. Hydrodynamics determines collective motion and phase behavior of active colloids in quasi-two-dimensional confinement. *Phys. Rev. Lett.* **112**, 118101 (2014).

[29] G.-J. Li and A. M. Ardekani. Hydrodynamic interaction of microswimmers near a wall. *Phys. Rev. E* **90**, 013010 (2014).

[30] G. Gaines and F. J. R. Taylor. Form and function of the dinoflagellate transverse flagellum. *J. Protozool.* **32**, 290–296 (1985).

[31] J. R. Blake and M. A. Sleigh. Mechanics of ciliary locomotion. *Biol. Rev.* **49**, 85–125 (1974).

[32] H. Streble and D. Krauter. *Das Leben im Wassertropfen: Mikroflora und Mikrofauna des Süßwassers. Ein Bestimmungsbuch*, 11th ed. (Kosmos, Stuttgart, 2009).

[33] A. N. Bragg. Clockwise rotation in *Paramecium trichium*. *Science* **79**, 524 (1934).

[34] H. C. Berg. *E. coli in Motion* (Springer, New York, 2004).

[35] H. C. Berg. Motile behavior of bacteria. *Phys. Today* **53**, 24–29 (2000).

[36] P. D. Frymier, R. M. Ford, H. C. Berg, and P. T. Cummings. Three-dimensional tracking of motile bacteria near a solid planar surface. *Proc. Natl. Acad. Sci. USA* **92**, 6195–6199 (1995).

[37] J. Saragosti, P. Silberzan, and A. Buguin. Modeling *E. coli* tumbles by rotational diffusion. Implications for chemotaxis. *PLoS ONE* **7**, e35412 (2012).

[38] H. C. Berg and D. A. Brown. Chemotaxis in *Escherichia coli* analysed by three-dimensional tracking. *Nature* **239**, 500–504 (1972).

[39] J. Saragosti, V. Calvez, N. Bournaveas, B. Perthame, A. Buguin, and P. Silberzan. Directional persistence of chemotactic bacteria in a traveling concentration wave. *Proc. Natl. Acad. Sci. USA* **108**, 16235–16240 (2011).

[40] T. Ishikawa, N. Yoshida, H. Ueno, M. Wiedeman, Y. Imai, and T. Yamaguchi. Energy transport in a concentrated suspension of bacteria. *Phys. Rev. Lett.* **107**, 028102 (2011).

[41] J. Schwarz-Linek, C. Valeriani, A. Cacciuto, M. E. Cates, D. Marenduzzo, A. N. Morozov, and W. C. K. Poon. Phase separation and rotor self-assembly in active particle suspensions. *Proc. Natl. Acad. Sci. USA* **109**, 4052–4057 (2012).

[42] J. Gachelin, A. Rousselet, A. Lindner, and E. Clement. Collective motion in an active suspension of *Escherichia coli* bacteria. *New J. Phys.* **16**, 025003 (2014).

[43] P. Galajda, J. Keymer, P. Chaikin, and R. Austin. A wall of funnels concentrates swimming bacteria. *J. Bacteriol.* **189**, 8704–8707 (2007).

[44] W. W. Navarre and O. Schneewind. Surface proteins of Gram-positive bacteria and mechanisms of their targeting to the cell wall envelope. *Microbiol. Mol. Biol. Rev.* **63**, 174–229 (1999).

[45] T. J. Beveridge. Structures of Gram-negative cell walls and their derived membrane vesicles. *J. Bacteriol.* **181**, 4725–4733 (1999).

[46] M. T. Cabeen and C. Jacobs-Wagner. Bacterial cell shape. *Nat. Rev. Microbiol.* **3**, 601–610 (2005).

[47] C. Dombrowski, L. Cisneros, S. Chatkaew, R. E. Goldstein, and J. O. Kessler. Self-concentration and large-scale coherence in bacterial dynamics. *Phys. Rev. Lett.* **93**, 098103 (2004).

[48] A. Sokolov, I. S. Aranson, J. O. Kessler, and R. E. Goldstein. Concentration dependence of the collective dynamics of swimming bacteria. *Phys. Rev. Lett.* **98**, 158102 (2007).

[49] A. Sokolov, M. M. Apodaca, B. A. Grzybowski, and I. S. Aranson. Swimming bacteria power microscopic gears. *Proc. Natl. Acad. Sci. USA* **107**, 969–974 (2010).

[50] X. Chen, X. Dong, A. Be'er, H. L. Swinney, and H. P. Zhang. Scale-invariant correlations in dynamic bacterial clusters. *Phys. Rev. Lett.* **108**, 148101 (2012).

[51] H. H. Wensink, J. Dunkel, S. Heidenreich, K. Drescher, R. E. Goldstein, H. Löwen, and J. M. Yeomans. Meso-scale turbulence in living fluids. *Proc. Natl. Acad. Sci. USA* **109**, 14308–14313 (2012).

[52] G. Li, L.-K. Tam, and J. X. Tang. Amplified effect of Brownian motion in bacterial near-surface swimming. *Proc. Natl. Acad. Sci. USA* **105**, 18355–18359 (2008).

[53] R. E. Goldstein. Green algae as model organisms for biological fluid dynamics. *Annu. Rev. Fluid Mech.* **47**, 343–375 (2015).

[54] S. Rafaï, L. Jibuti, and P. Peyla. Effective viscosity of microswimmer suspensions. *Phys. Rev. Lett.* **104**, 098102 (2010).

[55] M. Polin, I. Tuval, K. Drescher, J. P. Gollub, and R. E. Goldstein. *Chlamydomonas* swims with two "gears" in a eukaryotic version of run-and-tumble locomotion. *Science* **325**, 487–490 (2009).

[56] K. Drescher, R. E. Goldstein, N. Michel, M. Polin, and I. Tuval. Direct measurement of the flow field around swimming microorganisms. *Phys. Rev. Lett.* **105**, 168101 (2010).

[57] V. Kantsler, J. Dunkel, M. Polin, and R. E. Goldstein. Ciliary contact interactions dominate surface scattering of swimming eukaryotes. *Proc. Natl. Acad. Sci. USA* **110**, 1187–1192 (2013).

[58] M. Eisenbach. Sperm chemotaxis. *Rev. Reprod.* **4**, 56–66 (1999).

[59] U. B. Kaupp, J. Solzin, E. Hildebrand, J. E. Brown, A. Helbig, V. Hagen, M. Beyermann, F. Pampaloni, and I. Weyand. The signal flow and motor response controling chemotaxis of sea urchin sperm. *Nat. Cell Biol.* **5**, 109–117 (2003).

[60] U. B. Kaupp, N. D. Kashikar, and I. Weyand. Mechanisms of sperm chemotaxis. *Annu. Rev. Physiol.* **70**, 93–117 (2008).

[61] A. Guidobaldi, Y. Jeyaram, I. Berdakin, V. V. Moshchalkov, C. A. Condat, V. I. Marconi, L. Giojalas, and A. V. Silhanek. Geometrical guidance and trapping transition of human sperm cells. *Phys. Rev. E* **89**, 032720 (2014).

[62] L. Rothschild. Non-random distribution of bull spermatozoa in a drop of sperm suspension. *Nature* **198**, 1221–1222 (1963).

[63] B. M. Friedrich, I. H. Riedel-Kruse, J. Howard, and F. Jülicher. High-precision tracking of sperm swimming fine structure provides strong test of resistive force theory. *J. Exp. Biol.* **213**, 1226–1234 (2010).

[64] D. M. Woolley. Motility of spermatozoa at surfaces. *Reproduction* **216**, 259–270 (2003).

[65] I. H. Riedel, K. Kruse, and J. Howard. A self-organized vortex array of hydrodynamically entrained sperm cells. *Science* **309**, 300–303 (2005).

[66] J. K. G. Dhont. *An Introduction to Dynamics of Colloids* (Elsevier, Amsterdam, 1996).

[67] S. Ramaswamy. The mechanics and statistics of active matter. *Annu. Rev. Condens. Matter Phys.* **1**, 323–345 (2010).

[68] M. C. Marchetti, J. F. Joanny, S. Ramaswamy, T. B. Liverpool, J. Prost, M. Rao, and R. A. Simha. Hydrodynamics of soft active matter. *Rev. Mod. Phys.* **85**, 1143–1189 (2013).

[69] H. Löwen. Colloidal soft matter under external control. *J. Phys.: Condens. Matter* **13**, R415–R432 (2001).

[70] H. Löwen. Introduction to colloidal dispersions in external fields. *Eur. Phys. J. Spec. Top.* **222**, 2727–2737 (2013).

[71] P. Romanczuk, M. Bär, W. Ebeling, B. Lindner, and L. Schimansky-Geier. Active Brownian particles. From individual to collective stochastic dynamics. *Eur. Phys. J. Spec. Top.* **202**, 1–162 (2012).

[72] L. Baraban, M. Tasinkevych, M. N. Popescu, S. Sanchez, S. Dietrich, and O. G. Schmidt. Transport of cargo by catalytic Janus micro-motors. *Soft Matter* **8**, 48–52 (2012).

[73] J. Palacci, S. Sacanna, A. Vatchinsky, P. M. Chaikin, and D. J. Pine. Photoactivated colloidal dockers for cargo transportation. *J. Am. Chem. Soc.* **135**, 15978–15981 (2013).

[74] D. Kagan, R. Laocharoensuk, M. Zimmerman, C. Clawson, S. Balasubramanian, D. Kang, D. Bishop, S. Sattayasamitsathit, L. Zhang, and J. Wang. Rapid delivery of drug carriers propelled and navigated by catalytic nanoshuttles. *Small* **6**, 2741–2747 (2010).

[75] D. J. Irvine. Drug delivery: one nanoparticle, one kill. *Nat. Mater.* **10**, 342–343 (2011).

[76] A. A. Solovev, W. Xi, D. H. Gracias, S. M. Harazim, C. Deneke, S. Sanchez, and O. G. Schmidt. Self-propelled nanotools. *ACS Nano* **6**, 1751–1756 (2012).

[77] S. J. Ebbens and J. R. Howse. In pursuit of propulsion at the nanoscale. *Soft Matter* **6**, 726–738 (2010).

[78] J. L. Anderson. Colloid transport by interfacial forces. *Annu. Rev. Fluid Mech.* **21**, 61–99 (1989).

[79] A. Ajdari and L. Bocquet. Giant amplification of interfacially driven transport by hydrodynamic slip: diffusio-osmosis and beyond. *Phys. Rev. Lett.* **96**, 186102 (2006).

[80] R. A. Rica and M. Z. Bazant. Electrodiffusiophoresis: particle motion in electrolytes under direct current. *Phys. Fluids* **22**, 112109 (2010).

[81] R. Golestanian, T. B. Liverpool, and A. Ajdari. Propulsion of a molecular machine by asymmetric distribution of reaction products. *Phys. Rev. Lett.* **94**, 220801 (2005).

[82] R. Golestanian, T. B. Liverpool, and A. Ajdari. Designing phoretic micro- and nano-swimmers. *New J. Phys.* **9**, 126 (2007).

[83] W. F. Paxton, K. C. Kistler, C. C. Olmeda, A. Sen, S. K. St. Angelo, Y. Cao, T. E. Mallouk, P. E. Lammert, and V. H. Crespi. Catalytic nanomotors: autonomous movement of striped nanorods. *J. Am. Chem. Soc.* **126**, 13424–13431 (2004).

[84] W. F. Paxton, P. T. Baker, T. R. Kline, Y. Wang, T. E. Mallouk, and A. Sen. Catalytically induced electrokinetics for motors and micropumps. *J. Am. Chem. Soc.* **128**, 14881–14888 (2006).

[85] Y. Wang, R. M. Hernandez, D. J. Bartlett, J. M. Bingham, T. R. Kline, A. Sen, and T. E. Mallouk. Bipolar electrochemical mechanism for the propulsion of catalytic nanomotors in hydrogen peroxide solutions. *Langmuir* **22**, 10451–10456 (2006).

[86] J. L. Moran, P. M. Wheat, and J. D. Posner. Locomotion of electrocatalytic nanomotors due to reaction induced charge autoelectrophoresis. *Phys. Rev. E* **81**, 065302(R) (2010).

[87] J. L. Moran and J. D. Posner. Electrokinetic locomotion due to reaction-induced charge auto-electrophoresis. *J. Fluid Mech.* **680**, 31–66 (2011).

[88] B. Sabass and U. Seifert. Nonlinear, electrocatalytic swimming in the presence of salt. *J. Chem. Phys.* **136**, 214507 (2012).

[89] P. Mitchell. Hypothetical thermokinetic and electrokinetic mechanisms of locomotion in micro-organisms. *Proc. R. Phys. Soc. (Edinburgh)* **25**, 32–34 (1956).

[90] P. Mitchell. Self-electrophoretic locomotion in microorganisms: bacterial flagella as giant ionophores. *FEBS Lett.* **28**, 1–4 (1972).

[91] T. P. Pitta and H. C. Berg. Self-electrophoresis is not the mechanism for motility in swimming cyanobacteria. *J. Bacteriol.* **177**, 5701–5703 (1995).

[92] M. N. Popescu, S. Dietrich, and G. Oshanin. Confinement effects on diffusiophoretic self-propellers. *J. Chem. Phys.* **130**, 194702 (2009).

[93] J. R. Howse, R. A. L. Jones, A. J. Ryan, T. Gough, R. Vafabakhsh, and R. Golestanian. Self-motile colloidal particles: from directed propulsion to random walk. *Phys. Rev. Lett.* **99**, 048102 (2007).

[94] M. Braibanti, D. Vigolo, and R. Piazza. Does thermophoretic mobility depend on particle size? *Phys. Rev. Lett.* **100**, 108303 (2008).

[95] H. R. Jiang, N. Yoshinaga, and M. Sano. Active motion of a Janus particle by self-thermophoresis in a defocused laser beam. *Phys. Rev. Lett.* **105**, 268302 (2010).

[96] L. Baraban, R. Streubel, D. Makarov, L. Han, D. Karnaushenko, O. G. Schmidt, and G. Cuniberti. Fuel-free locomotion of Janus motors: magnetically induced thermophoresis. *ACS Nano* **7**, 1360–1367 (2013).

[97] R. Golestanian. Collective behavior of thermally active colloids. *Phys. Rev. Lett.* **108**, 038303 (2012).

[98] R. F. Ismagilov, A. Schwartz, N. Bowden, and G. M. Whitesides. Autonomous movement and self-assembly. *Angew. Chem. Int. Ed.* **41**, 652–654 (2002).

[99] W. F. Paxton, A. Sen, and T. E. Mallouk. Motility of catalytic nanoparticles through self-generated forces. *Chem. Eur. J.* **11**, 6462–6470 (2005).

[100] S. Fournier-Bidoz, A. C. Arsenault, I. Manners, and G. A. Ozin. Synthetic self-propelled nanorotors. *Chem. Commun.* **2005**, 441–443 (2005).

[101] W. F. Paxton, S. Sundararajan, T. E. Mallouk, and A. Sen. Chemical locomotion. *Angew. Chem. Int. Ed.* **45**, 5420–5429 (2006).

[102] P. Dhar, T. M. Fischer, Y. Wang, T. E. Mallouk, W. F. Paxton, and A. Sen. Autonomously moving nanorods at a viscous interface. *Nano Lett.* **6**, 66–72 (2006).

[103] R. Laocharoensuk, J. Burdick, and J. Wang. Carbon-nanotube-induced acceleration of catalytic nanomotors. *ACS Nano* **2**, 1069–1075 (2008).

[104] U. K. Demirok, R. Laocharoensuk, K. M. Manesh, and J. Wang. Ultrafast catalytic alloy nanomotors. *Angew. Chem. Int. Ed.* **47**, 9349–9351 (2008).

[105] G. Rückner and R. Kapral. Chemically powered nanodimers. *Phys. Rev. Lett.* **98**, 150603 (2007).

[106] L. F. Valadares, Y.-G. Tao, N. S. Zacharia, V. Kitaev, F. Galembeck, R. Kapral, and G. A. Ozin. Catalytic nanomotors: self-propelled sphere dimers. *Small* **6**, 565–572 (2010).

[107] H. Ke, S. Ye, R. L. Carroll, and K. Showalter. Motion analysis of self-propelled Pt-silica particles in hydrogen peroxide solutions. *J. Phys. Chem. A.* **114**, 5462–5467 (2010).

[108] S. J. Ebbens and J. R. Howse. Direct observation of the direction of motion for spherical catalytic swimmers. *Langmuir* **27**, 12293–12296 (2011).

[109] S. Ebbens, M.-H. Tu, J. R. Howse, and R. Golestanian. Size dependence of the propulsion velocity for catalytic Janus-sphere swimmers. *Phys. Rev. E* **85**, 020401(R) (2012).

[110] S. Ebbens, R. A. L. Jones, A. J. Ryan, R. Golestanian, and J. R. Howse. Self-assembled autonomous runners and tumblers. *Phys. Rev. E* **82**, 015304(R) (2010).

[111] A. Brown and W. Poon. Ionic effects in self-propelled Pt-coated Janus swimmers. *Soft Matter* **10**, 4016–4027 (2014).

[112] S. Ebbens, D. A. Gregory, G. Dunderdale, J. R. Howse, Y. Ibrahim, T. B. Liverpool, and R. Golestanian. Electrokinetic effects in catalytic platinum-insulator Janus swimmers. *Europhys. Lett.* **106**, 58003 (2014).

[113] J. Palacci, S. Sacanna, A. Preska Steinberg, D. J. Pine, and P. M. Chaikin. Living crystals of light-activated colloidal surfers. *Science* **339**, 936–940 (2013).

[114] J. Palacci, S. Sacanna, S.-H. Kim, G.-R. Yi, D. J. Pine, and P. M. Chaikin. Light-activated self-propelled colloids. *Phil. Trans. R. Soc. A* **372**, 20130372 (2014).

[115] G. Volpe, I. Buttinoni, D. Vogt, H.-J. Kümmerer, and C. Bechinger. Microswimmers in patterned environments. *Soft Matter* **7**, 8810–8815 (2011).

[116] I. Buttinoni, G. Volpe, F. Kümmel, G. Volpe, and C. Bechinger. Active Brownian motion tunable by light. *J. Phys.: Condens. Matter* **24**, 284129 (2012).

[117] L. Baraban, D. Makarov, R. Streubel, I. Mönch, D. Grimm, S. Sanchez, and O. G. Schmidt. Catalytic Janus motors on microfluidic chip: deterministic motion for targeted cargo delivery. *ACS Nano* **6**, 3383–3389 (2012).

[118] L. Baraban, D. Makarov, O. G. Schmidt, G. Cuniberti, P. Leiderer, and A. Erbe. Control over Janus micromotors by the strength of a magnetic field. *Nanoscale* **5**, 1332–1336 (2013).

[119] P. Tierno, R. Golestanian, I. Pagonabarraga, and F. Sagués. Controlled swimming in confined fluids of magnetically actuated colloidal rotors. *Phys. Rev. Lett.* **101**, 218304 (2008).

[120] P. Tierno, R. Golestanian, I. Pagonabarraga, and F. Sagués. Magnetically actuated colloidal microswimmers. *J. Phys. Chem. B* **112**, 16525–16528 (2008).

[121] A. Snezhko, M. Belkin, I. S. Aranson, and W.-K. Kwok. Self-assembled magnetic surface swimmers. *Phys. Rev. Lett.* **102**, 118103 (2009).

[122] A. Snezhko and I. S. Aranson. Magnetic manipulation of self-assembled colloidal asters. *Nat. Mater.* **10**, 698–703 (2011).

[123] P. Tierno, O. Güell, F. Sagués, R. Golestanian, and I. Pagonabarraga. Controlled propulsion in viscous fluids of magnetically actuated colloidal doublets. *Phys. Rev. E* **81**, 011402 (2010).

[124] R. Dreyfus, J. Baudry, M. L. Roper, M. Fermigier, H. A. Stone, and J. Bibette. Microscopic artificial swimmers. *Nature* **437**, 862–865 (2005).

[125] L. Zhang, J. J. Abbott, L. Dong, K. E. Peyer, B. E. Kratochvil, H. Zhang, C. Bergeles, and B. J. Nelson. Characterizing the swimming properties of artificial bacterial flagella. *Nano Lett.* **9**, 3663–3667 (2009).

[126] B. J. Williams, S. V. Anand, J. Rajagopalan, and M. T. A. Saif. A self-propelled biohybrid swimmer at low Reynolds number. *Nat. Commun.* **5**, 3081 (2014).

[127] T. Qiu, T.-C. Lee, A. G. Mark, K. I. Morozov, R. Münster, O. Mierka, S. Turek, A. M. Leshansky, and P. Fischer. Swimming by reciprocal motion at low Reynolds number. *Nat. Commun.* **5**, 5119 (2014).

[128] N. Mano and A. Heller. Bioelectrochemical propulsion. *J. Am. Chem. Soc.* **127**, 11574–11575 (2005).

[129] A. Reinmüller, H. J. Schöpe, and T. Palberg. Self-organized cooperative swimming at low Reynolds numbers. *Langmuir* **29**, 1738–1742 (2013).

[130] W. Wang, L. A. Castro, M. Hoyos, and T. E. Mallouk. Autonomous motion of metallic microrods propelled by ultrasound. *ACS Nano* **6**, 6122–6132 (2012).

[131] S. Ahmed, W. Wang, L. O. Mair, R. D. Fraleigh, S. Li, L. A. Castro, M. Hoyos, T. J. Huang, and T. E. Mallouk. Steering acoustically propelled nanowire motors toward cells in a biologically compatible environment using magnetic fields. *Langmuir* **29**, 16113–16118 (2013).

[132] J. Adler. Chemotaxis in bacteria. *Science* **153**, 708–716 (1966).

[133] N. Mittal, E. O. Budrene, M. P. Brenner, and A. van Oudenaarden. Motility of *Escherichia coli* cells in clusters formed by chemotactic aggregation. *Proc. Natl. Acad. Sci. USA* **100**, 13259–13263 (2003).

[134] B. M. Friedrich and F. Jülicher. Chemotaxis of sperm cells. *Proc. Natl. Acad. Sci. USA* **104**, 13256–13261 (2007).

[135] B. M. Friedrich and F. Jülicher. The stochastic dance of circling sperm cells: sperm chemotaxis in the plane. *New J. Phys.* **10**, 123025 (2008).

[136] R. D. Sjoblad and P. H. Frederikse. Chemotactic responses of *Chlamydomonas reinhardtii*. *Mol. Cell Biol.* **1**, 1057–1060 (1981).

[137] E. G. Govorunova and O. A. Sineshchekov. Chemotaxis in the green flagellate alga *Chlamydomonas*. *Biochemistry (Moscow)* **70**, 717–725 (2005).

[138] E. F. Keller and L. A. Segel. Initiation of slime mold aggregation viewed as an instability. *J. Theor. Biol.* **26**, 399–415 (1970).

[139] A. Levchenko and P. A. Iglesias. Models of eukaryotic gradient sensing: application to chemotaxis of amoebae and neutrophils. *Biophys. J.* **82**, 50–63 (2002).

[140] P. J. M. van Haastert and P. N. Devreotes. Chemotaxis: signalling the way forward. *Nat. Rev. Mol. Cell Biol.* **5**, 626–634 (2004).

[141] M. Kollmann, L. Løvdok, K. Bartholomé, J. Timmer, and V. Sourjik. Design principles of a bacterial signalling network. *Nature* **438**, 504–507 (2005).

[142] A. Celani and M. Vergassola. Bacterial strategies for chemotaxis response. *Proc. Natl. Acad. Sci. USA* **107**, 1391–1396 (2010).

[143] R. Stocker, J. R. Seymour, A. Samadani, D. E. Hunt, and M. F. Polz. Rapid chemotactic response enables marine bacteria to exploit ephemeral microscale nutrient patches. *Proc. Natl. Acad. Sci. USA* **105**, 4209–4214 (2008).

[144] Y. Hong, N. M. K. Blackman, N. D. Kopp, A. Sen, and D. Velegol. Chemotaxis of nonbiological colloidal rods. *Phys. Rev. Lett.* **99**, 178103 (2007).

[145] J. Taktikos, V. Zaburdaev, and H. Stark. Collective dynamics of model microorganisms with chemotactic signaling. *Phys. Rev. E* **85**, 051901 (2012).

[146] S. Saha, R. Golestanian, and S. Ramaswamy. Clusters, asters, and collective oscillations in chemotactic colloids. *Phys. Rev. E* **89**, 062316 (2014).

[147] A. Sengupta, T. Kruppa, and H. Löwen. Chemotactic predator-prey dynamics. *Phys. Rev. E* **83**, 031914 (2011).

[148] R. Grima. Strong-coupling dynamics of a multicellular chemotactic system. *Phys. Rev. Lett.* **95**, 128103 (2005).

[149] A. Sengupta, S. van Teeffelen, and H. Löwen. Dynamics of a microorganism moving by chemotaxis in its own secretion. *Phys. Rev. E* **80**, 031122 (2009).

[150] J. Taktikos, V. Zaburdaev, and H. Stark. Modeling a self-propelled autochemotactic walker. *Phys. Rev. E* **84**, 041924 (2011).

[151] B. L. Taylor, I. B. Zhulin, and M. S. Johnson. Aerotaxis and other energy-sensing behavior in bacteria. *Annu. Rev. Microbiol.* **53**, 103–128 (1999).

[152] A. J. Hillesdon and T. J. Pedley. Bioconvection in suspensions of oxytactic bacteria: linear theory. *J. Fluid Mech.* **324**, 223–259 (1996).

[153] D. M. Porterfield. Orientation of motile unicellular algae to oxygen: oxytaxis in *Euglena*. *Biol. Bull.* **193**, 229–230 (1997).

[154] A. V. Kuznetsov. Thermo-bioconvection in a suspension of oxytactic bacteria. *Int. Commun. Heat Mass* **32**, 991–999 (2005).

[155] E. Paster and W. S. Ryu. The thermal impulse response of *Escherichia coli*. *Proc. Natl. Acad. Sci. USA* **105**, 5373–5377 (2008).

[156] A. Bahat, I. Tur-Kaspa, A. Gakamsky, L. C. Giojalas, H. Breitbart, and M. Eisenbach. Thermotaxis of mammalian sperm cells: a potential navigation mechanism in the female genital tract. *Nat. Med.* **9**, 149–150 (2003).

[157] A. Bahat and M. Eisenbach. Sperm thermotaxis. *Mol. Cell. Endocrinol.* **252**, 115–119 (2006).

[158] M. Ntefidou, M. Iseki, M. Watanabe, M. Lebert, and D.-P. Häder. Photoactivated adenylyl cyclase controls phototaxis in the flagellate *Euglena gracilis*. *Plant Physiol.* **133**, 1517–1521 (2003).

[159] V. Daiker, D.-P. Häder, P. R. Richter, and M. Lebert. The involvement of a protein kinase in phototaxis and gravitaxis of *Euglena gracilis*. *Planta* **233**, 1055–1062 (2011).

[160] K.-i. Wakabayashi, Y. Misawa, S. Mochiji, and R. Kamiya. Reduction-oxidation poise regulates the sign of phototaxis in *Chlamydomonas reinhardtii*. *Proc. Natl. Acad. Sci. USA* **108**, 11280–11284 (2011).

[161] K. Josef, J. Saranak, and K. W. Foster. Ciliary behavior of a negatively phototactic *Chlamydomonas reinhardtii*. *Cell Motil. Cytoskel.* **61**, 97–111 (2005).

[162] K. Josef, J. Saranak, and K. W. Foster. Linear systems analysis of the ciliary steering behavior associated with negative-phototaxis in *Chlamydomonas reinhardtii*. *Cell Motil. Cytoskel.* **63**, 758–777 (2006).

[163] G. Jékely, J. Colombelli, H. Hausen, K. Guy, E. Stelzer, F. Nédélec, and D. Arendt. Mechanism of phototaxis in marine zooplankton. *Nature* **456**, 395–399 (2008).

[164] W. D. Hoff, M. A. van der Horst, C. B. Nudel, and K. J. Hellingwerf. Prokaryotic phototaxis. *Methods Mol. Biol.* **571**, 25–49 (2009).

[165] R. Blakemore. Magnetotactic bacteria. *Science* **190**, 377–379 (1975).

[166] R. P. Blakemore, R. B. Frankel, and A. J. Kalmijn. South-seeking magnetotactic bacteria in the southern hemisphere. *Nature* **286**, 384–385 (1980).

[167] S. L. Simmons, D. A. Bazylinski, and K. J. Edwards. South-seeking magnetotactic bacteria in the northern hemisphere. *Science* **311**, 371–374 (2006).

[168] F. F. Torres de Araujo, M. A. Pires, R. B. Frankel, and C. E. M. Bicudo. Magnetite and magnetotaxis in algae. *Biophys. J.* **50**, 375–378 (1986).

[169] E. F. DeLong, R. B. Frankel, and D. A. Bazylinski. Multiple evolutionary origins of magnetotaxis in bacteria. *Science* **259**, 803–806 (1993).

[170] R. B. Frankel, D. A. Bazylinski, M. S. Johnson, and B. L. Taylor. Magneto-aerotaxis in marine coccoid bacteria. *Biophys. J.* **73**, 994–1000 (1997).

[171] D. H. Kim, U. K. Cheang, L. Kőhidai, D. Byun, and M. J. Kim. Artificial magnetotactic motion control of *Tetrahymena pyriformis* using ferromagnetic nanoparticles: a tool for fabrication of microbiorobots. *Appl. Phys. Lett.* **97**, 173702 (2010).

[172] J. C. Montgomery, C. F. Baker, and A. G. Carton. The lateral line can mediate rheotaxis in fish. *Nature* **389**, 960–963 (1997).

[173] F. P. Bretherton and L. Rothschild. Rheotaxis of spermatozoa. *Proc. R. Soc. London B. Biol. Sci.* **153**, 490–502 (1961).

[174] A. M. Roberts. Motion of spermatozoa in fluid streams. *Nature* **228**, 375–376 (1970).

[175] V. Kantsler, J. Dunkel, M. Blayney, and R. E. Goldstein. Rheotaxis facilitates upstream navigation of mammalian sperm cells. *eLife* **3**, e02403 (2014).

[176] J. Hill, O. Kalkanci, J. L. McMurry, and H. Koser. Hydrodynamic surface interactions enable *Escherichia coli* to seek efficient routes to swim upstream. *Phys. Rev. Lett.* **98**, 068101 (2007).

[177] Marcos, H. C. Fu, T. R. Powers, and R. Stocker. Bacterial rheotaxis. *Proc. Natl. Acad. Sci. USA* **109**, 4780–4785 (2012).

[178] T. M. El-Sherry, M. Elsayed, H. K. Abdelhafez, and M. Abdelgawad. Characterization of rheotaxis of bull sperm using microfluidics. *Integr. Biol.* **6**, 1111–1121 (2014).

[179] K. Miki and D. E. Clapham. Rheotaxis guides mammalian sperm. *Curr. Biol.* **23**, 443–452 (2013).

[180] J. O. Kessler. Hydrodynamic focusing of motile algal cells. *Nature* **313**, 218–220 (1985).

[181] T. J. Pedley, N. A. Hill, and J. O. Kessler. The growth of bioconvection patterns in a uniform suspension of gyrotactic micro-organisms. *J. Fluid Mech.* **195**, 223–237 (1988).

[182] T. J. Pedley and J. O. Kessler. Hydrodynamic phenomena in suspensions of swimming microorganisms. *Annu. Rev. Fluid Mech.* **24**, 313–358 (1992).

[183] W. M. Durham, J. O. Kessler, and R. Stocker. Disruption of vertical motility by shear triggers formation of thin phytoplankton layers. *Science* **323**, 1067–1070 (2009).

[184] U. Timm and A. Okubo. Gyrotaxis: a plume model for self-focusing micro-organisms. *Bull. Math. Biol.* **56**, 187–206 (1994).

[185] W. M. Durham, E. Climent, and R. Stocker. Gyrotaxis in a steady vortical flow. *Phys. Rev. Lett.* **106**, 238102 (2011).

[186] F. De Lillo, M. Cencini, W. M. Durham, M. Barry, R. Stocker, E. Climent, and G. Boffetta. Turbulent fluid acceleration generates clusters of gyrotactic microorganisms. *Phys. Rev. Lett.* **112**, 044502 (2014).

[187] C. R. Williams and M. A. Bees. A tale of three taxes: photo-gyro-gravitactic bioconvection. *J. Exp. Biol.* **214**, 2398–2408 (2011).

[188] C. R. Williams and M. A. Bees. Photo-gyrotactic bioconvection. *J. Fluid Mech.* **678**, 41–86 (2011).

[189] D.-P. Häder, R. Hemmersbach, and M. Lebert. *Gravity and the Behavior of Unicellular Organisms* (Cambridge University Press, Cambridge, 2005).

[190] A. M. Roberts. The mechanics of gravitaxis in *Paramecium*. *J. Exp. Biol.* **213**, 4158–4162 (2010).

[191] A. I. Campbell and S. J. Ebbens. Gravitaxis in spherical Janus swimming devices. *Langmuir* **29**, 14066–14073 (2013).

[192] R. N. Bearon. Helical swimming can provide robust upwards transport for gravitactic single-cell algae; a mechanistic model. *J. Math. Biol.* **66**, 1341–1359 (2013).

[193] K. Wolff, A. M. Hahn, and H. Stark. Sedimentation and polar order of active bottom-heavy particles. *Eur. Phys. J. E* **36**, 43 (2013).

[194] R. Wittkowski and H. Löwen. Self-propelled Brownian spinning top: dynamics of a biaxial swimmer at low Reynolds numbers. *Phys. Rev. E* **85**, 021406 (2012).

[195] H. C. Crenshaw. A new look at locomotion in microorganisms: rotating and translating. *Am. Zool.* **36**, 608–618 (1996).

[196] E. Gurarie, D. Grünbaum, and M. T. Nishizaki. Estimating 3D movements from 2D observations using a continuous model of helical swimming. *Bull. Math. Biol.* **73**, 1358–1377 (2011).

[197] D. E. Boakes, E. A. Codling, G. J. Thorn, and M. Steinke. Analysis and modelling of swimming behaviour in *Oxyrrhis marina*. *J. Plankton Res.* **33**, 641–649 (2011).

[198] D.-P. Häder and R. Hemmersbach. Graviperception and graviorientation in flagellates. *Planta* **203**, S7–S10 (1997).

[199] D.-P. Häder. Gravitaxis in unicellular microorganisms. *Adv. Space Res.* **24**, 843–850 (1999).

[200] R. Hemmersbach, D. Volkmann, and D.-P. Häder. Graviorientation in protists and plants. *J. Plant Physiol.* **154**, 1–15 (1999).

[201] H. Machemer and R. Braucker. Gravitaxis screened for physical mechanism using g-modulated cellular orientational behaviour. *Microgravity Sci. Technol.* **9**, 2–9 (1996).

[202] U. Nagel and H. Machemer. Physical and physiological components of the graviresponses of wild-type and mutant *Paramecium tetraurelia*. *J. Exp. Biol.* **203**, 1059–1070 (2000).

[203] C. Streb, P. Richter, M. Ntefidou, M. Lebert, and D.-P. Häder. Sensory transduction of gravitaxis in *Euglena gracilis*. *J. Plant Physiol.* **159**, 855–862 (2002).

[204] D.-P. Häder, P. R. Richter, M. Schuster, V. Daiker, and M. Lebert. Molecular analysis of the graviperception signal transduction in the flagellate *Euglena gracilis*: involvement of a transient receptor potential-like channel and a calmodulin. *Adv. Space Res.* **43**, 1179–1184 (2009).

[205] P. R. Richter, M. Schuster, M. Lebert, C. Streb, and D.-P. Häder. Gravitaxis of *Euglena gracilis* depends only partially on passive buoyancy. *Adv. Space Res.* **39**, 1218–1224 (2007).

[206] A. M. Roberts. Geotaxis in motile micro-organisms. *J. Exp. Biol.* **53**, 687–699 (1970).

[207] A. M. Roberts and F. M. Deacon. Gravitaxis in motile micro-organisms: the role of fore-aft body asymmetry. *J. Fluid Mech.* **452**, 405–423 (2002).

[208] K. Fukui and H. Asai. The most probable mechanism of the negative geotaxis of *Paramecium caudatum*. *Proc. Jpn. Acad. B* **56**, 172–177 (1980).

[209] K. Fukui and H. Asai. Negative geotactic behavior of *Paramecium caudatum* is completely described by the mechanism of buoyancy-oriented upward swimming. *Biophys. J.* **47**, 479–482 (1985).

[210] Y. Mogami, J. Ishii, and S. A. Baba. Theoretical and experimental dissection of gravity-dependent mechanical orientation in gravitactic microorganisms. *Biol. Bull.* **201**, 26–33 (2001).

[211] V. Kam, N. Moseyko, J. Nemson, and L. J. Feldman. Gravitaxis in *Chlamydomonas reinhardtii*: characterization using video microscopy and computer analysis. *Int. J. Plant Sci.* **160**, 1093–1098 (1999).

[212] A. M. Roberts. Mechanisms of gravitaxis in *Chlamydomonas*. *Biol. Bull.* **210**, 78–80 (2006).

[213] X. Zheng, B. ten Hagen, A. Kaiser, M. Wu, H. Cui, Z. Silber-Li, and H. Löwen. Non-Gaussian statistics for the motion of self-propelled Janus particles: experiment versus theory. *Phys. Rev. E* **88**, 032304 (2013).

[214] S. Babel, B. ten Hagen, and H. Löwen. Swimming path statistics of an active Brownian particle with time-dependent self-propulsion. *J. Stat. Mech.* (2014) P02011.

[215] M. Tarama, A. M. Menzel, B. ten Hagen, R. Wittkowski, T. Ohta, and H. Löwen. Dynamics of a deformable active particle under shear flow. *J. Chem. Phys.* **139**, 104906 (2013).

[216] D. J. Kraft, R. Wittkowski, B. ten Hagen, K. V. Edmond, D. J. Pine, and H. Löwen. Brownian motion and the hydrodynamic friction tensor for colloidal particles of complex shape. *Phys. Rev. E* **88**, 050301(R) (2013).

[217] A. Kaiser, A. Peshkov, A. Sokolov, B. ten Hagen, H. Löwen, and I. S. Aranson. Transport powered by bacterial turbulence. *Phys. Rev. Lett.* **112**, 158101 (2014).

[218] F. Kümmel, B. ten Hagen, R. Wittkowski, I. Buttinoni, R. Eichhorn, G. Volpe, H. Löwen, and C. Bechinger. Circular motion of asymmetric self-propelling particles. *Phys. Rev. Lett.* **110**, 198302 (2013).

[219] B. ten Hagen, R. Wittkowski, D. Takagi, F. Kümmel, C. Bechinger, and H. Löwen. Can the self-propulsion of anisotropic microswimmers be described by using forces and torques? arXiv:1410.6707 (2014).

[220] B. ten Hagen, F. Kümmel, R. Wittkowski, D. Takagi, H. Löwen, and C. Bechinger. Gravitaxis of asymmetric self-propelled colloidal particles. *Nat. Commun.* **5**, 4829 (2014).

[221] A. Walther and A. H. E. Müller. Janus particles: synthesis, self-assembly, physical properties, and applications. *Chem. Rev.* **113**, 5194–5261 (2013).

[222] M. Doi and S. F. Edwards. *The Theory of Polymer Dynamics* (Oxford Science Publications, Oxford, 1986).

[223] B. ten Hagen, S. van Teeffelen, and H. Löwen. Brownian motion of a self-propelled particle. *J. Phys.: Condens. Matter* **23**, 194119 (2011).

[224] M. B. Wan, C. J. Olson Reichhardt, Z. Nussinov, and C. Reichhardt. Rectification of swimming bacteria and self-driven particle systems by arrays of asymmetric barriers. *Phys. Rev. Lett.* **101**, 018102 (2008).

[225] O. Chepizhko and F. Peruani. Diffusion, subdiffusion, and trapping of active particles in heterogeneous media. *Phys. Rev. Lett.* **111**, 160604 (2013).

[226] N. Bartolo, E. Komatsu, S. Matarrese, and A. Riotto. Non-Gaussianity from inflation: theory and observations. *Phys. Rep.* **402**, 103–266 (2004).

[227] M. Alishahiha, E. Silverstein, and D. Tong. DBI in the sky: non-Gaussianity from inflation with a speed limit. *Phys. Rev. D* **70**, 123505 (2004).

[228] D. H. Lyth and Y. Rodríguez. Inflationary prediction for primordial non-Gaussianity. *Phys. Rev. Lett.* **95**, 121302 (2005).

[229] D. Koga, A. C.-L. Chian, R. A. Miranda, and E. L. Rempel. Intermittent nature of solar wind turbulence near the Earth's bow shock: phase coherence and non-Gaussianity. *Phys. Rev. E* **75**, 046401 (2007).

[230] W. Kob, C. Donati, S. J. Plimpton, P. H. Poole, and S. C. Glotzer. Dynamical heterogeneities in a supercooled Lennard-Jones liquid. *Phys. Rev. Lett.* **79**, 2827–2830 (1997).

[231] A. Arbe, J. Colmenero, F. Alvarez, M. Monkenbusch, D. Richter, B. Farago, and B. Frick. Non-Gaussian nature of the α relaxation of glass-forming polyisoprene. *Phys. Rev. Lett.* **89**, 245701 (2002).

[232] K. Vollmayr-Lee, W. Kob, K. Binder, and A. Zippelius. Dynamical heterogeneities below the glass transition. *J. Chem. Phys.* **116**, 5158–5166 (2002).

[233] A. M. Puertas, M. Fuchs, and M. E. Cates. Simulation study of nonergodicity transitions: gelation in colloidal systems with short-range attractions. *Phys. Rev. E* **67**, 031406 (2003).

[234] D. Lavallée and R. J. Archuleta. Stochastic modeling of slip spatial complexities for the 1979 Imperial Valley, California, earthquake. *Geophys. Res. Lett.* **30**, 1245 (2003).

[235] J.-P. Bouchaud and M. Potters. *Theory of Financial Risk and Derivative Pricing: From Statistical Physics to Risk Management*, 2nd ed. (Cambridge University Press, Cambridge, 2009).

[236] K. Kiyono, Z. R. Struzik, and Y. Yamamoto. Criticality and phase transition in stock-price fluctuations. *Phys. Rev. Lett.* **96**, 068701 (2006).

[237] B. ten Hagen, S. van Teeffelen, and H. Löwen. Non-Gaussian behaviour of a self-propelled particle on a substrate. *Condens. Matter Phys.* **12**, 725–738 (2009).

[238] F. J. Sevilla and L. A. Gómez Nava. Theory of diffusion of active particles that move at constant speed in two dimensions. *Phys. Rev. E* **90**, 022130 (2014).

[239] A. Martinelli. Overdamped 2D Brownian motion for self-propelled and nonholonomic particles. *J. Stat. Mech.* (2014) P03003.

[240] D. M. Phillips. Comparative analysis of mammalian sperm motility. *J. Cell Biol.* **53**, 561–573 (1972).

[241] D. L. Ringo. Flagellar motion and fine structure of the flagellar apparatus in *Chlamydomonas*. *J. Cell Biol.* **33**, 543–571 (1967).

[242] C. J. Brokaw, D. J. L. Luck, and B. Huang. Analysis of the movement of *Chlamydomonas* flagella: the function of the radial-spoke system is revealed by comparison of wild-type and mutant flagella. *J. Cell Biol.* **92**, 722–732 (1982).

[243] B. M. Friedrich and F. Jülicher. Flagellar synchronization independent of hydrodynamic interactions. *Phys. Rev. Lett.* **109**, 138102 (2012).

[244] R. R. Bennett and R. Golestanian. Emergent run-and-tumble behavior in a simple model of *Chlamydomonas* with intrinsic noise. *Phys. Rev. Lett.* **110**, 148102 (2013).

[245] S. Dodson and C. Ramcharan. Size-specific swimming behavior of *Daphnia pulex*. *J. Plankton Res.* **13**, 1367–1379 (1991).

[246] A. Ordemann, G. Balazsi, and F. Moss. Pattern formation and stochastic motion of the zooplankton *Daphnia* in a light field. *Physica A* **325**, 260–266 (2003).

[247] N. Komin, U. Erdmann, and L. Schimansky-Geier. Random walk theory applied to *Daphnia* motion. *Fluct. Noise Lett.* **4**, L151–L159 (2004).

[248] L. Schimansky-Geier, U. Erdmann, and N. Komin. Advantages of hopping on a zig-zag course. *Physica A* **351**, 51–59 (2005).

[249] L. Haeggqwist, L. Schimansky-Geier, I. M. Sokolov, and F. Moss. Hopping on a zig-zag course. *Eur. Phys. J. Spec. Top.* **157**, 33–42 (2008).

[250] A. C. H. Tsang and E. Kanso. Flagella-induced transitions in the collective behavior of confined microswimmers. *Phys. Rev. E* **90**, 021001(R) (2014).

[251] J. Strefler, W. Ebeling, E. Gudowska-Nowak, and L. Schimansky-Geier. Dynamics of individuals and swarms with shot noise induced by stochastic food supply. *Eur. Phys. J. B* **72**, 597–606 (2009).

[252] C. Hoell and H. Löwen. Theory of microbe motion in a poisoned environment. *Phys. Rev. E* **84**, 042903 (2011).

[253] G. Rosser, R. E. Baker, J. P. Armitage, and A. G. Fletcher. Modelling and analysis of bacterial tracks suggest an active reorientation mechanism in *Rhodobacter sphaeroides*. *J. R. Soc. Interface* **11**, 20140320 (2014).

[254] J. Tailleur and M. E. Cates. Statistical mechanics of interacting run-and-tumble bacteria. *Phys. Rev. Lett.* **100**, 218103 (2008).

[255] M. E. Cates. Diffusive transport without detailed balance in motile bacteria: does microbiology need statistical physics? *Rep. Prog. Phys.* **75**, 042601 (2012).

[256] R. W. Nash, R. Adhikari, J. Tailleur, and M. E. Cates. Run-and-tumble particles with hydrodynamics: sedimentation, trapping, and upstream swimming. *Phys. Rev. Lett.* **104**, 258101 (2010).

[257] N. Uchida and R. Golestanian. Generic conditions for hydrodynamic synchronization. *Phys. Rev. Lett.* **106**, 058104 (2011).

[258] P. Cremer and H. Löwen. Scaling of cluster growth for coagulating active particles. *Phys. Rev. E* **89**, 022307 (2014).

[259] R. Arditi, Y. Tyutyunov, A. Morgulis, V. Govorukhin, and I. Senina. Directed movement of predators and the emergence of density-dependence in predator-prey models. *Theor. Popul. Biol.* **59**, 207–221 (2001).

[260] D. Takagi, A. B. Braunschweig, J. Zhang, and M. J. Shelley. Dispersion of self-propelled rods undergoing fluctuation-driven flips. *Phys. Rev. Lett.* **110**, 038301 (2013).

[261] H.-O. Peitgen, H. Jürgens, and D. Saupe. *Chaos and Fractals: New Frontiers of Science* (Springer, New York, 1992).

[262] J. Hutchinson. Fractals and self-similarity. *Indiana Univ. Math. J.* **30**, 713–747 (1981).

[263] W. Leland, M. Taqqu, W. Willinger, and D. Wilson. On the self-similar nature of ethernet traffic (extended version). *IEEE/ACM Trans. Network.* **2**, 1–15 (1994).

[264] C. Song, S. Havlin, and H. A. Makse. Self-similarity of complex networks. *Nature* **433**, 392–395 (2005).

[265] P. Meakin. *Fractals, Scaling and Growth Far from Equilibrium* (Cambridge University Press, Cambridge, 1998).

[266] M. Prähofer and H. Spohn. Statistical self-similarity of one-dimensional growth processes. *Physica A* **279**, 342–352 (2000).

[267] R. Benzi, S. Ciliberto, R. Tripiccione, C. Baudet, F. Massaioli, and S. Succi. Extended self-similarity in turbulent flows. *Phys. Rev. E* **48**, R29–R32 (1993).

[268] C. R. Evans and J. S. Coleman. Critical phenomena and self-similarity in the gravitational collapse of radiation fluid. *Phys. Rev. Lett.* **72**, 1782–1785 (1994).

[269] S. van Teeffelen and H. Löwen. Dynamics of a Brownian circle swimmer. *Phys. Rev. E* **78**, 020101(R) (2008).

[270] L. Li, S. F. Nørrelykke, and E. C. Cox. Persistent cell motion in the absence of external signals: a search strategy for eukaryotic cells. *PLoS ONE* **3**, e2093 (2008).

[271] Y. T. Maeda, J. Inose, M. Y. Matsuo, S. Iwaya, and M. Sano. Ordered patterns of cell shape and orientational correlation during spontaneous cell migration. *PLoS ONE* **3**, e3734 (2008).

[272] N. P. Barry and M. S. Bretscher. *Dictyostelium* amoebae and neutrophils can swim. *Proc. Natl. Acad. Sci. USA* **107**, 11376–11380 (2010).

[273] A. J. Bae and E. Bodenschatz. On the swimming of *Dictyostelium* amoebae. *Proc. Natl. Acad. Sci. USA* **107**, E165–E166 (2010).

[274] T. Lämmermann and M. Sixt. Mechanical modes of 'amoeboid' cell migration. *Curr. Opin. Cell Biol.* **21**, 636–644 (2009).

[275] A. Farutin, S. Rafaï, D. K. Dysthe, A. Duperray, P. Peyla, and C. Misbah. Amoeboid swimming: a generic self-propulsion of cells in fluids by means of membrane deformations. *Phys. Rev. Lett.* **111**, 228102 (2013).

[276] M. Arroyo, L. Heltai, D. Millán, and A. DeSimone. Reverse engineering the euglenoid movement. *Proc. Natl. Acad. Sci. USA* **109**, 17874–17879 (2012).

[277] H. A. Stone and A. D. T. Samuel. Propulsion of microorganisms by surface distortions. *Phys. Rev. Lett.* **77**, 4102–4104 (1996).

[278] L. E. Scriven and C. V. Sternling. The Marangoni effects. *Nature* **187**, 186–188 (1960).

[279] Y. Sumino, N. Magome, T. Hamada, and K. Yoshikawa. Self-running droplet: emergence of regular motion from nonequilibrium noise. *Phys. Rev. Lett.* **94**, 068301 (2005).

[280] F. Takabatake, N. Magome, M. Ichikawa, and K. Yoshikawa. Spontaneous mode-selection in the self-propelled motion of a solid/liquid composite driven by interfacial instability. *J. Chem. Phys.* **134**, 114704 (2011).

[281] K. Nagai, Y. Sumino, H. Kitahata, and K. Yoshikawa. Mode selection in the spontaneous motion of an alcohol droplet. *Phys. Rev. E* **71**, 065301(R) (2005).

[282] H. Boukellal, O. Campás, J.-F. Joanny, J. Prost, and C. Sykes. Soft *Listeria*: actin-based propulsion of liquid drops. *Phys. Rev. E* **69**, 061906 (2004).

[283] T. Toyota, N. Maru, M. M. Hanczyc, T. Ikegami, and T. Sugawara. Self-propelled oil droplets consuming "fuel" surfactant. *J. Am. Chem. Soc.* **131**, 5012–5013 (2009).

[284] N. Yoshinaga. Spontaneous motion and deformation of a self-propelled droplet. *Phys. Rev. E* **89**, 012913 (2014).

[285] M. Schmitt and H. Stark. Swimming active droplet: a theoretical analysis. *Europhys. Lett.* **101**, 44008 (2013).

[286] S. Thutupalli, R. Seemann, and S. Herminghaus. Swarming behavior of simple model squirmers. *New J. Phys.* **13**, 073021 (2011).

[287] S. Herminghaus, C. C. Maass, C. Krüger, S. Thutupalli, L. Goehring, and C. Bahr. Interfacial mechanisms in active emulsions. *Soft Matter* **10**, 7008–7022 (2014).

[288] T. Ohta and T. Ohkuma. Deformable self-propelled particles. *Phys. Rev. Lett.* **102**, 154101 (2009).

[289] T. Ohkuma and T. Ohta. Deformable self-propelled particles with a global coupling. *Chaos* **20**, 023101 (2010).

[290] T. Hiraiwa, M. Y. Matsuo, T. Ohkuma, T. Ohta, and M. Sano. Dynamics of a deformable self-propelled domain. *Europhys. Lett.* **91**, 20001 (2010).

[291] Y. Itino, T. Ohkuma, and T. Ohta. Collective dynamics of deformable self-propelled particles with repulsive interaction. *J. Phys. Soc. Jpn.* **80**, 033001 (2011).

[292] T. Hiraiwa, K. Shitara, and T. Ohta. Dynamics of a deformable self-propelled particle in three dimensions. *Soft Matter* **7**, 3083–3086 (2011).

[293] M. Tarama and T. Ohta. Dynamics of a deformable self-propelled particle under external forcing. *Eur. Phys. J. B* **83**, 391–400 (2011).

[294] M. Tarama and T. Ohta. Spinning motion of a deformable self-propelled particle in two dimensions. *J. Phys.: Condens. Matter* **24**, 464129 (2012).

[295] A. M. Menzel and T. Ohta. Soft deformable self-propelled particles. *Europhys. Lett.* **99**, 58001 (2012).

[296] M. Tarama and T. Ohta. Dynamics of a deformable self-propelled particle with internal rotational force. *Prog. Theor. Exp. Phys.* (2013) 013A01.

[297] M. Tarama and T. Ohta. Oscillatory motions of an active deformable particle. *Phys. Rev. E* **87**, 062912 (2013).

[298] M. Tarama, Y. Itino, A. Menzel, and T. Ohta. Individual and collective dynamics of self-propelled soft particles. *Eur. Phys. J. Spec. Top.* **223**, 121–139 (2014).

[299] A. A. Golovin, Y. P. Gupalo, and Y. S. Ryazantsev. Change in shape of drop moving due to the chemithermocapillary effect. *J. Appl. Mech. Tech. Phys.* **30**, 602–609 (1989).

[300] W. Alt and R. T. Tranquillo. Basic morphogenetic system modeling shape changes of migrating cells: how to explain fluctuating lamellipodial dynamics. *J. Biol. Syst.* **3**, 905–916 (1995).

[301] A. Stéphanou, E. Mylona, M. Chaplain, and P. Tracqui. A computational model of cell migration coupling the growth of focal adhesions with oscillatory cell protrusions. *J. Theor. Biol.* **253**, 701–716 (2008).

[302] H. Wada and R. R. Netz. Hydrodynamics of helical-shaped bacterial motility. *Phys. Rev. E* **80**, 021921 (2009).

[303] D. Shao, W.-J. Rappel, and H. Levine. Computational model for cell morphodynamics. *Phys. Rev. Lett.* **105**, 108104 (2010).

[304] K. Doubrovinski and K. Kruse. Cell motility resulting from spontaneous polymerization waves. *Phys. Rev. Lett.* **107**, 258103 (2011).

[305] F. Ziebert, S. Swaminathan, and I. S. Aranson. Model for self-polarization and motility of keratocyte fragments. *J. R. Soc. Interface* **9**, 1084–1092 (2011).

[306] B. ten Hagen, R. Wittkowski, and H. Löwen. Brownian dynamics of a self-propelled particle in shear flow. *Phys. Rev. E* **84**, 031105 (2011).

[307] M. Sandoval, N. K. Marath, G. Subramanian, and E. Lauga. Stochastic dynamics of active swimmers in linear flows. *J. Fluid Mech.* **742**, 50–70 (2014).

[308] Y.-G. Tao and R. Kapral. Swimming upstream: self-propelled nanodimer motors in a flow. *Soft Matter* **6**, 756–761 (2010).

[309] A. Zöttl and H. Stark. Nonlinear dynamics of a microswimmer in Poiseuille flow. *Phys. Rev. Lett.* **108**, 218104 (2012).

[310] A. Zöttl and H. Stark. Periodic and quasiperiodic motion of an elongated microswimmer in Poiseuille flow. *Eur. Phys. J. E* **36**, 4 (2013).

[311] R. Chacón. Chaotic dynamics of a microswimmer in Poiseuille flow. *Phys. Rev. E* **88**, 052905 (2013).

[312] A. E. Frankel and A. S. Khair. Dynamics of a self-diffusiophoretic particle in shear flow. *Phys. Rev. E* **90**, 013030 (2014).

[313] P. Talkner, G.-L. Ingold, and P. Hänggi. Transport of flexible chiral objects in a uniform shear flow. *New J. Phys.* **14**, 073006 (2012).

[314] A. Chengala, M. Hondzo, and J. Sheng. Microalga propels along vorticity direction in a shear flow. *Phys. Rev. E* **87**, 052704 (2013).

[315] M. Tarama, A. M. Menzel, and H. Löwen. Deformable microswimmer in a swirl: capturing and scattering dynamics. *Phys. Rev. E* **90**, 032907 (2014).

[316] A. Einstein. Über die von der molekularkinetischen Theorie der Wärme geforderte Bewegung von in ruhenden Flüssigkeiten suspendierten Teilchen. *Ann. Phys. (Berlin)* **322**, 549–560 (1905).

[317] M. J. Perrin. Mouvement brownien et réalité moléculaire. *Ann. Chim. Phys.* **18**, 5–114 (1909).

[318] Y. Sun and Y. Xia. Shape-controlled synthesis of gold and silver nanoparticles. *Science* **298**, 2176–2179 (2002).

[319] S. C. Glotzer and M. J. Solomon. Anisotropy of building blocks and their assembly into complex structures. *Nat. Mater.* **6**, 557–562 (2007).

[320] D. J. Kraft, W. S. Vlug, C. M. van Kats, A. van Blaaderen, A. Imhof, and W. K. Kegel. Self-assembly of colloids with liquid protrusions. *J. Am. Chem. Soc.* **131**, 1182–1186 (2009).

[321] D. J. Kraft, J. Groenewold, and W. K. Kegel. Colloidal molecules with well-controlled bond angles. *Soft Matter* **5**, 3823–3826 (2009).

[322] H. Brenner. Coupling between the translational and rotational Brownian motions of rigid particles of arbitrary shape: I. Helicoidally isotropic particles. *J. Colloid Sci.* **20**, 104–122 (1965).

[323] H. Brenner. Coupling between the translational and rotational Brownian motions of rigid particles of arbitrary shape: II. General theory. *J. Colloid Interface Sci.* **23**, 407–436 (1967).

[324] H. Z. Cummins, F. D. Carlson, T. J. Herbert, and G. Woods. Translational and rotational diffusion constants of tobacco mosaic virus from Rayleigh linewidths. *Biophys. J.* **9**, 518–546 (1969).

[325] J. García de la Torre and V. A. Bloomfield. Hydrodynamic properties of complex, rigid, biological macromolecules: theory and applications. *Q. Rev. Biophys.* **14**, 81–139 (1981).

[326] Y. Han, A. M. Alsayed, M. Nobili, J. Zhang, T. C. Lubensky, and A. G. Yodh. Brownian motion of an ellipsoid. *Science* **314**, 626–630 (2006).

[327] S. Harvey and J. García de la Torre. Coordinate systems for modeling the hydrodynamic resistance and diffusion coefficients of irregularly shaped rigid macromolecules. *Macromolecules* **13**, 960–964 (1980).

[328] W. A. Wegener. Diffusion coefficients for rigid macromolecules with irregular shapes that allow rotational-translational coupling. *Biopolymers* **20**, 303–326 (1981).

[329] M. X. Fernandes and J. García de la Torre. Brownian dynamics simulation of rigid particles of arbitrary shape in external fields. *Biophys. J.* **83**, 3039–3048 (2002).

[330] M. Makino and M. Doi. Brownian motion of a particle of general shape in Newtonian fluid. *J. Phys. Soc. Jpn.* **73**, 2739–2745 (2004).

[331] F. Perrin. Mouvement brownien d'un ellipsoide (I). Dispersion diélectrique pour des molécules ellipsoidales. *J. Phys. Radium* **5**, 497–511 (1934).

[332] F. Perrin. Mouvement brownien d'un ellipsoide (II). Rotation libre et dépolarisation des fluorescence. Translation et diffusion de molécules ellipsoidales. *J. Phys. Radium* **7**, 1–11 (1936).

[333] A. Nir and A. Acrivos. On the creeping motion of two arbitrary-sized touching spheres in a linear shear field. *J. Fluid Mech.* **59**, 209–223 (1973).

[334] A. Chakrabarty, A. Konya, F. Wang, J. V. Selinger, K. Sun, and Q.-H. Wei. Brownian motion of boomerang colloidal particles. *Phys. Rev. Lett.* **111**, 160603 (2013).

[335] A. Chakrabarty, F. Wang, C.-Z. Fan, K. Sun, and Q.-H. Wei. High-precision tracking of Brownian boomerang colloidal particles confined in quasi two dimensions. *Langmuir* **29**, 14396–14402 (2013).

[336] A. Chakrabarty, A. Konya, F. Wang, J. V. Selinger, K. Sun, and Q.-H. Wei. Brownian motion of arbitrarily shaped particles in two dimensions. *Langmuir* **30**, 13844–13853 (2014).

[337] J. Happel and H. Brenner. *Low Reynolds Number Hydrodynamics: With Special Applications to Particulate Media*, 2nd ed. (Kluwer Academic Publishers, Dordrecht, 1991).

[338] J. C. Crocker and D. G. Grier. Methods of digital video microscopy for colloidal studies. *J. Colloid Interface Sci.* **179**, 298–310 (1996).

[339] R. H. Webb. Confocal optical microscopy. *Rep. Prog. Phys.* **59**, 427–471 (1996).

[340] M. C. Jenkins and S. U. Egelhaaf. Confocal microscopy of colloidal particles: towards reliable, optimum coordinates. *Adv. Colloid Interface Sci.* **136**, 65–92 (2008).

[341] G. L. Hunter, K. V. Edmond, M. T. Elsesser, and E. R. Weeks. Tracking rotational diffusion of colloidal clusters. *Opt. Express* **19**, 17189–17202 (2011).

[342] P. Marquet, B. Rappaz, P. J. Magistretti, E. Cuche, Y. Emery, T. Colomb, and C. Depeursinge. Digital holographic microscopy: a noninvasive contrast imaging tech-

nique allowing quantitative visualization of living cells with subwavelength axial accuracy. *Opt. Lett.* **30**, 468–470 (2005).

[343] S.-H. Lee, Y. Roichman, G.-R. Yi, S.-H. Kim, S.-M. Yang, A. van Blaaderen, P. van Oostrum, and D. G. Grier. Characterizing and tracking single colloidal particles with video holographic microscopy. *Opt. Express* **15**, 18275–18282 (2007).

[344] J. Fung, K. E. Martin, R. W. Perry, D. M. Kaz, R. McGorty, and V. N. Manoharan. Measuring translational, rotational, and vibrational dynamics in colloids with digital holographic microscopy. *Opt. Express* **19**, 8051–8065 (2011).

[345] J. Fung and V. N. Manoharan. Holographic measurements of anisotropic three-dimensional diffusion of colloidal clusters. *Phys. Rev. E* **88**, 020302(R) (2013).

[346] A. Wang, T. G. Dimiduk, J. Fung, S. Razavi, I. Kretzschmar, K. Chaudhary, and V. N. Manoharan. Using the discrete dipole approximation and holographic microscopy to measure rotational dynamics of non-spherical colloidal particles. *J. Quant. Spectrosc. Ra.* **146**, 499–509 (2014).

[347] V. N. Manoharan, M. T. Elsesser, and D. J. Pine. Dense packing and symmetry in small clusters of microspheres. *Science* **301**, 483–487 (2003).

[348] J. García de la Torre and B. Carrasco. Hydrodynamic properties of rigid macromolecules composed of ellipsoidal and cylindrical subunits. *Biopolymers* **63**, 163–167 (2002).

[349] J. García de la Torre, S. Navarro, M. C. López Martínez, F. G. Díaz, and J. J. López Cascales. HYDRO: a computer program for the prediction of hydrodynamic properties of macromolecules. *Biophys. J.* **67**, 530–531 (1994).

[350] R. A. Simha and S. Ramaswamy. Hydrodynamic fluctuations and instabilities in ordered suspensions of self-propelled particles. *Phys. Rev. Lett.* **89**, 058101 (2002).

[351] I. Llopis and I. Pagonabarraga. Dynamic regimes of hydrodynamically coupled self-propelling particles. *Europhys. Lett.* **75**, 999–1005 (2006).

[352] H. H. Wensink and H. Löwen. Aggregation of self-propelled colloidal rods near confining walls. *Phys. Rev. E* **78**, 031409 (2008).

[353] V. Mehandia and P. R. Nott. The collective dynamics of self-propelled particles. *J. Fluid Mech.* **595**, 239–264 (2008).

[354] F. Ginelli, F. Peruani, M. Bär, and H. Chaté. Large-scale collective properties of self-propelled rods. *Phys. Rev. Lett.* **104**, 184502 (2010).

[355] H. P. Zhang, A. Be'er, E.-L. Florin, and H. L. Swinney. Collective motion and density fluctuations in bacterial colonies. *Proc. Natl. Acad. Sci. USA* **107**, 13626–13630 (2010).

[356] F. Peruani, T. Klauss, A. Deutsch, and A. Voss-Boehme. Traffic jams, gliders, and bands in the quest for collective motion of self-propelled particles. *Phys. Rev. Lett.* **106**, 128101 (2011).

[357] J. Bialké, T. Speck, and H. Löwen. Crystallization in a dense suspension of self-propelled particles. *Phys. Rev. Lett.* **108**, 168301 (2012).

[358] L. Berthier. Nonequilibrium glassy dynamics of self-propelled hard disks. *Phys. Rev. Lett.* **112**, 220602 (2014).

[359] A. Wysocki, R. G. Winkler, and G. Gompper. Cooperative motion of active Brownian spheres in three-dimensional dense suspensions. *Europhys. Lett.* **105**, 48004 (2014).

[360] F. Peruani, A. Deutsch, and M. Bär. Nonequilibrium clustering of self-propelled rods. *Phys. Rev. E* **74**, 030904(R) (2006).

[361] S. R. McCandlish, A. Baskaran, and M. F. Hagan. Spontaneous segregation of self-propelled particles with different motilities. *Soft Matter* **8**, 2527–2534 (2012).

[362] I. Buttinoni, J. Bialké, F. Kümmel, H. Löwen, C. Bechinger, and T. Speck. Dynamical clustering and phase separation in suspensions of self-propelled colloidal particles. *Phys. Rev. Lett.* **110**, 238301 (2013).

[363] G. S. Redner, M. F. Hagan, and A. Baskaran. Structure and dynamics of a phase-separating active colloidal fluid. *Phys. Rev. Lett.* **110**, 055701 (2013).

[364] J. Bialké, H. Löwen, and T. Speck. Microscopic theory for the phase separation of self-propelled repulsive disks. *Europhys. Lett.* **103**, 30008 (2013).

[365] Y. Fily, S. Henkes, and M. C. Marchetti. Freezing and phase separation of self-propelled disks. *Soft Matter* **10**, 2132–2140 (2014).

[366] D. Levis and L. Berthier. Clustering and heterogeneous dynamics in a kinetic Monte Carlo model of self-propelled hard disks. *Phys. Rev. E* **89**, 062301 (2014).

[367] T. Speck, J. Bialké, A. M. Menzel, and H. Löwen. Effective Cahn-Hilliard equation for the phase separation of active Brownian particles. *Phys. Rev. Lett.* **112**, 218304 (2014).

[368] H.-Y. Chen and K.-t. Leung. Rotating states of self-propelling particles in two dimensions. *Phys. Rev. E* **73**, 056107 (2006).

[369] Y.-X. Li, R. Lukeman, and L. Edelstein-Keshet. Minimal mechanisms for school formation in self-propelled particles. *Physica D* **237**, 699–720 (2008).

[370] Y. Yang, V. Marceau, and G. Gompper. Swarm behavior of self-propelled rods and swimming flagella. *Phys. Rev. E* **82**, 031904 (2010).

[371] H. H. Wensink and H. Löwen. Emergent states in dense systems of active rods: from swarming to turbulence. *J. Phys.: Condens. Matter* **24**, 464130 (2012).

[372] S. Henkes, Y. Fily, and M. C. Marchetti. Active jamming: self-propelled soft particles at high density. *Phys. Rev. E* **84**, 040301(R) (2011).

[373] X. Yang, M. L. Manning, and M. C. Marchetti. Aggregation and segregation of confined active particles. *Soft Matter* **10**, 6477–6484 (2014).

[374] F. D. C. Farrell, M. C. Marchetti, D. Marenduzzo, and J. Tailleur. Pattern formation in self-propelled particles with density-dependent motility. *Phys. Rev. Lett.* **108**, 248101 (2012).

[375] A. M. Menzel. Unidirectional laning and migrating cluster crystals in confined self-propelled particle systems. *J. Phys.: Condens. Matter* **25**, 505103 (2013).

[376] M. M. Tirado, C. López Martínez, and J. García de la Torre. Comparison of theories for the translational and rotational diffusion coefficients of rod-like macromolecules. Application to short DNA fragments. *J. Chem. Phys.* **81**, 2047–2052 (1984).

[377] J. García de la Torre, G. del Rio Echenique, and A. Ortega. Improved calculation of rotational diffusion and intrinsic viscosity of bead models for macromolecules and nanoparticles. *J. Phys. Chem. B* **111**, 955–961 (2007).

[378] R. Di Leonardo, L. Angelani, D. Dell'Arciprete, G. Ruocco, V. Iebba, S. Schippa, M. P. Conte, F. Mecarini, F. De Angelis, and E. Di Fabrizio. Bacterial ratchet motors. *Proc. Natl. Acad. Sci. USA* **107**, 9541–9545 (2010).

[379] A. Kaiser, H. H. Wensink, and H. Löwen. How to capture active particles. *Phys. Rev. Lett.* **108**, 268307 (2012).

[380] A. Kaiser, K. Popowa, H. H. Wensink, and H. Löwen. Capturing self-propelled particles in a moving microwedge. *Phys. Rev. E* **88**, 022311 (2013).

[381] C. Reichhardt and C. J. Olson Reichhardt. Active matter ratchets with an external drift. *Phys. Rev. E* **88**, 062310 (2013).

[382] J. A. Drocco, C. J. Olson Reichhardt, and C. Reichhardt. Bidirectional sorting of flocking particles in the presence of asymmetric barriers. *Phys. Rev. E* **85**, 056102 (2012).

[383] C. Reichhardt and C. J. Olson Reichhardt. Dynamics and separation of circularly moving particles in asymmetrically patterned arrays. *Phys. Rev. E* **88**, 042306 (2013).

[384] A. Kaiser and H. Löwen. Unusual swelling of a polymer in a bacterial bath. *J. Chem. Phys.* **141**, 044903 (2014).

[385] A. Kaiser, A. Sokolov, I. S. Aranson, and H. Löwen. Mechanisms of carrier transport induced by a microswimmer bath. arXiv:1408.1883 (2014).

[386] C. R. Calladine. Construction of bacterial flagella. *Nature* **255**, 121–124 (1975).

[387] L. Turner, W. S. Ryu, and H. C. Berg. Real-time imaging of fluorescent flagellar filaments. *J. Bacteriol.* **182**, 2793–2801 (2000).

[388] R. Vogel and H. Stark. Force-extension curves of bacterial flagella. *Eur. Phys. J. E* **33**, 259–271 (2010).

[389] S. Y. Reigh, R. G. Winkler, and G. Gompper. Synchronization and bundling of anchored bacterial flagella. *Soft Matter* **8**, 4363–4372 (2012).

[390] R. Vogel and H. Stark. Rotation-induced polymorphic transitions in bacterial flagella. *Phys. Rev. Lett.* **110**, 158104 (2013).

[391] W. Gao, A. Pei, X. Feng, C. Hennessy, and J. Wang. Organized self-assembly of Janus micromotors with hydrophobic hemispheres. *J. Am. Chem. Soc.* **135**, 998–1001 (2013).

[392] A. Boymelgreen, G. Yossifon, S. Park, and T. Miloh. Spinning Janus doublets driven in uniform ac electric fields. *Phys. Rev. E* **89**, 011003(R) (2014).

[393] R. Soto and R. Golestanian. Self-assembly of catalytically active colloidal molecules: tailoring activity through surface chemistry. *Phys. Rev. Lett.* **112**, 068301 (2014).

[394] S. Badaire, C. Cottin-Bizonne, J. W. Woody, A. Yang, and A. D. Stroock. Shape selectivity in the assembly of lithographically designed colloidal particles. *J. Am. Chem. Soc.* **129**, 40–41 (2007).

[395] C. A. Grattoni, R. A. Dawe, C. Y. Seah, and J. D. Gray. Lower critical solution coexistence curve and physical properties (density, viscosity, surface tension, and interfacial tension) of 2,6-lutidine + water. *J. Chem. Eng. Data* **38**, 516–519 (1993).

[396] J. C. Clunie and J. K. Baird. Interdiffusion coefficient and dynamic viscosity for the mixture 2,6-lutidine + water near the lower consolute point. *Phys. Chem. Liq.* **37**, 357–371 (1999).

[397] H. S. Jennings. On the significance of the spiral swimming of organisms. *Am. Nat.* **35**, 369–378 (1901).

[398] J. Elgeti, U. B. Kaupp, and G. Gompper. Hydrodynamics of sperm cells near surfaces. *Biophys. J.* **99**, 1018–1026 (2010).

[399] R. Di Leonardo, D. Dell'Arciprete, L. Angelani, and V. Iebba. Swimming with an image. *Phys. Rev. Lett.* **106**, 038101 (2011).

[400] W. R. DiLuzio, L. Turner, M. Mayer, P. Garstecki, D. B. Weibel, H. C. Berg, and G. M. Whitesides. *Escherichia coli* swim on the right-hand side. *Nature* **435**, 1271–1274 (2005).

[401] L. Lemelle, J.-F. Palierne, E. Chatre, and C. Place. Counterclockwise circular motion of bacteria swimming at the air-liquid interface. *J. Bacteriol.* **192**, 6307–6308 (2010).

[402] S. van Teeffelen, U. Zimmermann, and H. Löwen. Clockwise-directional circle swimmer moves counter-clockwise in Petri dish- and ring-like confinements. *Soft Matter* **5**, 4510–4519 (2009).

[403] C. Weber, P. K. Radtke, L. Schimansky-Geier, and P. Hänggi. Active motion assisted by correlated stochastic torques. *Phys. Rev. E* **84**, 011132 (2011).

[404] P. K. Radtke and L. Schimansky-Geier. Directed transport of confined Brownian particles with torque. *Phys. Rev. E* **85**, 051110 (2012).

[405] A. Kaiser and H. Löwen. Vortex arrays as emergent collective phenomena for circle swimmers. *Phys. Rev. E* **87**, 032712 (2013).

[406] Y. Yang, F. Qiu, and G. Gompper. Self-organized vortices of circling self-propelled particles and curved active flagella. *Phys. Rev. E* **89**, 012720 (2014).

[407] E. Lauga and T. R. Powers. The hydrodynamics of swimming microorganisms. *Rep. Prog. Phys.* **72**, 096601 (2009).

[408] J. R. Blake. A note on the image system for a stokeslet in a no-slip boundary. *Math. Proc. Cambridge Phil. Soc.* **70**, 303–310 (1971).

[409] A. P. Berke, L. Turner, H. C. Berg, and E. Lauga. Hydrodynamic attraction of swimming microorganisms by surfaces. *Phys. Rev. Lett.* **101**, 038102 (2008).

[410] B. U. Felderhof. Comment on "Circular motion of asymmetric self-propelling particles." *Phys. Rev. Lett.* **113**, 029801 (2014).

[411] J. M. Yeomans, D. O. Pushkin, and H. Shum. An introduction to the hydrodynamics of swimming microorganisms. *Eur. Phys. J. Spec. Top.* **223**, 1771–1785 (2014).

[412] M. Yang, A. Wysocki, and M. Ripoll. Hydrodynamic simulations of self-phoretic microswimmers. *Soft Matter* **10**, 6208–6218 (2014).

[413] B. U. Felderhof. Dynamics of pressure propulsion of a sphere in a viscous compressible fluid. *J. Chem. Phys.* **133**, 064903 (2010).

[414] A. Najafi and R. Golestanian. Simple swimmer at low Reynolds number: three linked spheres. *Phys. Rev. E* **69**, 062901 (2004).

[415] R. Ledesma-Aguilar, H. Löwen, and J. M. Yeomans. A circle swimmer at low Reynolds number. *Eur. Phys. J. E* **35**, 70 (2012).

[416] P. K. Ghosh, P. Hänggi, F. Marchesoni, S. Martens, F. Nori, L. Schimansky-Geier, and G. Schmid. Driven Brownian transport through arrays of symmetric obstacles. *Phys. Rev. E* **85**, 011101 (2012).

[417] J. Alvarez-Ramirez, L. Dagdug, and F. J. Valdes-Parada. Effective diffusivity through arrays of obstacles under zero-mean periodic driving forces. *J. Chem. Phys.* **137**, 154109 (2012).

[418] S. B. Chen. Driven transport of particles in 3D ordered porous media. *J. Chem. Phys.* **139**, 074904 (2013).

[419] Y. Hatwalne, S. Ramaswamy, M. Rao, and R. A. Simha. Rheology of active-particle suspensions. *Phys. Rev. Lett.* **92**, 118101 (2004).

[420] J. P. Hernandez-Ortiz, C. G. Stoltz, and M. D. Graham. Transport and collective dynamics in suspensions of confined swimming particles. *Phys. Rev. Lett.* **95**, 204501 (2005).

[421] D. Saintillan and M. J. Shelley. Instabilities and pattern formation in active particle suspensions: kinetic theory and continuum simulations. *Phys. Rev. Lett.* **100**, 178103 (2008).

[422] A. Baskaran and M. C. Marchetti. Statistical mechanics and hydrodynamics of bacterial suspensions. *Proc. Natl. Acad. Sci. USA* **106**, 15567–15572 (2009).

[423] D. Saintillan. Extensional rheology of active suspensions. *Phys. Rev. E* **81**, 056307 (2010).

[424] L. Zhu, E. Lauga, and L. Brandt. Low-Reynolds-number swimming in a capillary tube. *J. Fluid Mech.* **726**, 285–311 (2013).

[425] S. Chattopadhyay, R. Moldovan, C. Yeung, and X. L. Wu. Swimming efficiency of bacterium *Escherichia coli*. *Proc. Natl. Acad. Sci. USA* **103**, 13712–13717 (2006).

[426] N. C. Darnton, L. Turner, S. Rojevsky, and H. C. Berg. On torque and tumbling in swimming *Escherichia coli*. *J. Bacteriol.* **189**, 1756–1764 (2007).

[427] R. D. L. Hanes, M. C. Jenkins, and S. U. Egelhaaf. Combined holographic-mechanical optical tweezers: construction, optimization, and calibration. *Rev. Sci. Instrum.* **80**, 083703 (2009).

[428] S. C. Takatori, W. Yan, and J. F. Brady. Swim pressure: stress generation in active matter. *Phys. Rev. Lett.* **113**, 028103 (2014).

[429] M. Iima and A. S. Mikhailov. Propulsion hydrodynamics of a butterfly micro-swimmer. *Europhys. Lett.* **85**, 44001 (2009).

[430] T. Sakaue, R. Kapral, and A. S. Mikhailov. Nanoscale swimmers: hydrodynamic interactions and propulsion of molecular machines. *Eur. Phys. J. B* **75**, 381–387 (2010).

[431] E. M. Purcell. The efficiency of propulsion by a rotating flagellum. *Proc. Natl. Acad. Sci. USA* **94**, 11307–11311 (1997).

[432] J. Lighthill. Flagellar hydrodynamics. *SIAM Rev.* **18**, 161–230 (1976).

[433] B. Liu, T. R. Powers, and K. S. Breuer. Force-free swimming of a model helical flagellum in viscoelastic fluids. *Proc. Natl. Acad. Sci. USA* **108**, 19516–19520 (2011).

[434] Y. Hong, D. Velegol, N. Chaturvedi, and A. Sen. Biomimetic behavior of synthetic particles: from microscopic randomness to macroscopic control. *Phys. Chem. Chem. Phys.* **12**, 1423–1435 (2010).

[435] P. T. Underhill, J. P. Hernandez-Ortiz, and M. D. Graham. Diffusion and spatial correlations in suspensions of swimming particles. *Phys. Rev. Lett.* **100**, 248101 (2008).

[436] K. C. Leptos, J. S. Guasto, J. P. Gollub, A. I. Pesci, and R. E. Goldstein. Dynamics of enhanced tracer diffusion in suspensions of swimming eukaryotic microorganisms. *Phys. Rev. Lett.* **103**, 198103 (2009).

[437] G. Miño, T. E. Mallouk, T. Darnige, M. Hoyos, J. Dauchet, J. Dunstan, R. Soto, Y. Wang, A. Rousselet, and E. Clement. Enhanced diffusion due to active swimmers at a solid surface. *Phys. Rev. Lett.* **106**, 048102 (2011).

[438] H. Kurtuldu, J. S. Guasto, K. A. Johnson, and J. P. Gollub. Enhancement of biomixing by swimming algal cells in two-dimensional films. *Proc. Natl. Acad. Sci. USA* **108**, 10391–10395 (2011).

[439] D. O. Pushkin and J. M. Yeomans. Fluid mixing by curved trajectories of microswimmers. *Phys. Rev. Lett.* **111**, 188101 (2013).

[440] M. Schienbein and H. Gruler. Langevin equation, Fokker-Planck equation and cell migration. *Bull. Math. Biol.* **55**, 585–608 (1993).

[441] M. Zhao. Electrical fields in wound healing–An overriding signal that directs cell migration. *Semin. Cell Dev. Biol.* **20**, 674–682 (2009).

[442] J. Li and F. Lin. Microfluidic devices for studying chemotaxis and electrotaxis. *Trends Cell Biol.* **21**, 489–497 (2011).

[443] J. P. Steimel, J. L. Aragones, and A. Alexander-Katz. Artificial tribotactic microscopic walkers: walking based on friction gradients. *Phys. Rev. Lett.* **113**, 178101 (2014).

[444] U. Erdmann, W. Ebeling, L. Schimansky-Geier, and F. Schweitzer. Brownian particles far from equilibrium. *Eur. Phys. J. B* **15**, 105–113 (2000).

[445] T. J. Pedley and J. O. Kessler. A new continuum model for suspensions of gyrotactic micro-organisms. *J. Fluid Mech.* **212**, 155–182 (1990).

[446] M. S. Jones, L. Le Baron, and T. J. Pedley. Biflagellate gyrotaxis in a shear flow. *J. Fluid Mech.* **281**, 137–158 (1994).

[447] F. Kümmel, B. ten Hagen, R. Wittkowski, D. Takagi, I. Buttinoni, R. Eichhorn, G. Volpe, H. Löwen, and C. Bechinger. Reply to Comment on "Circular motion of asymmetric self-propelling particles." *Phys. Rev. Lett.* **113**, 029802 (2014).

[448] F. Evers, R. D. L. Hanes, C. Zunke, R. F. Capellmann, J. Bewerunge, C. Dalle-Ferrier, M. C. Jenkins, I. Ladadwa, A. Heuer, R. Castañeda-Priego, and S. U. Egelhaaf. Colloids in light fields: particle dynamics in random and periodic energy landscapes. *Eur. Phys. J. Spec. Top.* **222**, 2995–3009 (2013).

[449] R. D. L. Hanes, C. Dalle-Ferrier, M. Schmiedeberg, M. C. Jenkins, and S. U. Egelhaaf. Colloids in one dimensional random energy landscapes. *Soft Matter* **8**, 2714–2723 (2012).

[450] M. P. Magiera and L. Brendel. Trapping of propelled colloidal particles. arXiv:1407.0983 (2014).

[451] J. Tailleur and M. E. Cates. Sedimentation, trapping, and rectification of dilute bacteria. *Europhys. Lett.* **86**, 60002 (2009).

[452] M. Paoluzzi, R. Di Leonardo, and L. Angelani. Run-and-tumble particles in speckle fields. *J. Phys.: Condens. Matter* **26**, 375101 (2014).

[453] N. Koumakis, C. Maggi, and R. Di Leonardo. Directed transport of active particles over asymmetric energy barriers. *Soft Matter* **10**, 5695–5701 (2014).

[454] H. H. Wensink, V. Kantsler, R. E. Goldstein, and J. Dunkel. Controlling active self-assembly through broken particle-shape symmetry. *Phys. Rev. E* **89**, 010302(R) (2014).

[455] N. H. P. Nguyen, D. Klotsa, M. Engel, and S. C. Glotzer. Emergent collective phenomena in a mixture of hard shapes through active rotation. *Phys. Rev. Lett.* **112**, 075701 (2014).

[456] A. M. Menzel. Collective motion of binary self-propelled particle mixtures. *Phys. Rev. E* **85**, 021912 (2012).

[457] F. Q. Potiguar, G. A. Farias, and W. P. Ferreira. Self-propelled particle transport in regular arrays of rigid asymmetric obstacles. *Phys. Rev. E* **90**, 012307 (2014).

[458] X. Ao, P. K. Ghosh, Y. Li, G. Schmid, P. Hänggi, and F. Marchesoni. Active Brownian motion in a narrow channel. *Eur. Phys. J. Spec. Top.* **223**, 3227–3242 (2014).

[459] J. C. Wu, Q. Chen, R. Wang, and B. Q. Ai. Rectification of self-propelled particles in entropic barriers: finite size effects. *J. Phys. A: Math. Theor.* **47**, 325001 (2014).

[460] M. Mijalkov and G. Volpe. Sorting of chiral microswimmers. *Soft Matter* **9**, 6376–6381 (2013).

[461] G. Volpe, S. Gigan, and G. Volpe. Simulation of the active Brownian motion of a microswimmer. *Am. J. Phys.* **82**, 659–664 (2014).

[462] W. Yang, V. R. Misko, K. Nelissen, M. Kong, and F. M. Peeters. Using self-driven microswimmers for particle separation. *Soft Matter* **8**, 5175–5179 (2012).

[463] K. K. Dey, S. Das, M. F. Poyton, S. Sengupta, P. J. Butler, P. S. Cremer, and A. Sen. Chemotactic separation of enzymes. *ACS Nano* **8**, 11941–11949 (2014).

[464] P. K. Ghosh, V. R. Misko, F. Marchesoni, and F. Nori. Self-propelled Janus particles in a ratchet: numerical simulations. *Phys. Rev. Lett.* **110**, 268301 (2013).

[465] B.-Q. Ai and J.-C. Wu. Transport of active ellipsoidal particles in ratchet potentials. *J. Chem. Phys.* **140**, 094103 (2014).

[466] G. Gompper, T. Ihle, D. M. Kroll, and R. G. Winkler. Multi-particle collision dynamics: a particle-based mesoscale simulation approach to the hydrodynamics of complex fluids. *Adv. Polym. Sci.* **221**, 1–87 (2009).

[467] J. Burdick, R. Laocharoensuk, P. M. Wheat, J. D. Posner, and J. Wang. Synthetic nanomotors in microchannel networks: directional microchip motion and controlled manipulation of cargo. *J. Am. Chem. Soc.* **130**, 8164–8165 (2008).

[468] S. Sundararajan, P. E. Lammert, A. W. Zudans, V. H. Crespi, and A. Sen. Catalytic motors for transport of colloidal cargo. *Nano Lett.* **8**, 1271–1276 (2008).

[469] S. Mitragotri and J. Lahann. Materials for drug delivery: innovative solutions to address complex biological hurdles. *Adv. Mater.* **24**, 3717–3723 (2012).

[470] L. Soler, V. Magdanz, V. M. Fomin, S. Sanchez, and O. G. Schmidt. Self-propelled micromotors for cleaning polluted water. *ACS Nano* **7**, 9611–9620 (2013).

[471] L. Soler and S. Sanchez. Catalytic nanomotors for environmental monitoring and water remediation. *Nanoscale* **6**, 7175–7182 (2014).

[472] G. A. Ozin, I. Manners, S. Fournier-Bidoz, and A. Arsenault. Dream nanomachines. *Adv. Mater.* **17**, 3011–3018 (2005).

[473] R. K. Soong, G. D. Bachand, H. P. Neves, A. G. Olkhovets, H. G. Craighead, and C. D. Montemagno. Powering an inorganic nanodevice with a biomolecular motor. *Science* **290**, 1555–1558 (2000).

[474] T. G. Leong, C. L. Randall, B. R. Benson, N. Bassik, G. M. Stern, and D. H. Gracias. Tetherless thermobiochemically actuated microgrippers. *Proc. Natl. Acad. Sci. USA* **106**, 703–708 (2009).

[475] S. Kjelstrup, J. M. Rubi, I. Pagonabarraga, and D. Bedeaux. Mesoscopic non-equilibrium thermodynamic analysis of molecular motors. *Phys. Chem. Chem. Phys.* **15**, 19405–19414 (2013).

[476] D. B. Weibel, P. Garstecki, D. Ryan, W. R. DiLuzio, M. Mayer, J. E. Seto, and G. M. Whitesides. Microoxen: microorganisms to move microscale loads. *Proc. Natl. Acad. Sci. USA* **102**, 11963–11967 (2005).

[477] N. Koumakis, A. Lepore, C. Maggi, and R. Di Leonardo. Targeted delivery of colloids by swimming bacteria. *Nat. Commun.* **4**, 2588 (2013).

[478] F. Jülicher, A. Ajdari, and J. Prost. Modeling molecular motors. *Rev. Mod. Phys.* **69**, 1269–1282 (1997).

[479] M. Schliwa and G. Woehlke. Molecular motors. *Nature* **422**, 759–765 (2003).

[480] P. Kraikivski, R. Lipowsky, and J. Kierfeld. Enhanced ordering of interacting filaments by molecular motors. *Phys. Rev. Lett.* **96**, 258103 (2006).

[481] J. L. Ross, M. Y. Ali, and D. M. Warshaw. Cargo transport: molecular motors navigate a complex cytoskeleton. *Curr. Opin. Cell Biol.* **20**, 41–47 (2008).

[482] X. Li, R. Lipowsky, and J. Kierfeld. Bifurcation of velocity distributions in cooperative transport of filaments by fast and slow motors. *Biophys. J.* **104**, 666–676 (2013).

[483] W. Xi, A. A. Solovev, A. N. Ananth, D. H. Gracias, S. Sanchez, and O. G. Schmidt. Rolled-up magnetic microdrillers: towards remotely controlled minimally invasive surgery. *Nanoscale* **5**, 1294–1297 (2013).

[484] B. Carrasco and J. García de la Torre. Improved hydrodynamic interaction in macromolecular bead models. *J. Chem. Phys.* **111**, 4817–4826 (1999).

[485] V. Bloomfield, W. O. Dalton, and K. E. van Holde. Frictional coefficients of multisubunit structures. I. Theory. *Biopolymers* **5**, 135–148 (1967).

[486] V. Bloomfield, K. E. van Holde, and W. O. Dalton. Frictional coefficients of multisubunit structures. II. Application to proteins and viruses. *Biopolymers* **5**, 149–159 (1967).

[487] B. Carrasco and J. García de la Torre. Hydrodynamic properties of rigid particles: comparison of different modeling and computational procedures. *Biophys. J.* **76**, 3044–3057 (1999).